CENTRIFUGAL PUMP AND ALIGNMENT PRACTICES

KAMESHWAR UPADHYAY
MSc ENGG (NIT–JAMSHEDPUR)

INDIA • SINGAPORE • MALAYSIA

Notion Press

Old No. 38, New No. 6
McNichols Road, Chetpet
Chennai - 600 031

First Published by Notion Press 2018
Copyright © Kameshwar Upadhyay 2018
All Rights Reserved.

ISBN 978-1-64324-618-5

THIS BOOK IS DEDICATED TO

My wife, parents and family members who sacrificed to support and provide congenial atmosphere.

All the personalities who helped directly or indirectly in materialistic or wisdom field, and trusted me during learning, working or nonworking span of my life.

CONTENTS

Chapter 8: Operation of Centrifugal Pumps 117

FOREWORD

A plant producing steel or power requires a lot of water as input for producing their end product from raw Materials. For producing one million tonnes of steel per annum, approximately 22 million M^3 of water is required as a process make up demand. Similarly, for producing one MW of power from fossil fuels, minimum 20 M^3/MW water circulations and 3–4 M^3/MW water make up is required. These requirements of water for industrial usage are transferred or circulated in system by mostly centrifugal pumps. Besides water pumping or transfer, centrifugal pumps attribute for transferring or pumping of any liquid like chemicals, by product or prime product liquids in industries.

Pump is one of the oldest machines invented by mankind first in the form of water wheels. However, pump is a machine which imparts energy into liquid to lift it to higher elevation or to transport that for desired useful circulation in pressurized form through piping system for the purpose designed for. Focus is made by narrowing on pumps requirements for steel plant and power plant usage in specific. Because of many and varied utilization of pump, it is obviously explained here that why so many verities of pumps are used in industrial applications.

In present industrialization era, all industries have a sizable portion of equipment like centrifugal pump. The plant engineers, engaged in such industries need the basic concepts of centrifugal pump's construction, installation and operation & maintenance. Though there are many authoritative text books, addressing on centrifugal pumps but those books are having more theoretical coverage. Writer has put his effort to cover all practical aspects of operation and maintenance fields of centrifugal pumps in a work book form to make the book more precised and easy to understand & apply in day to day work.

The pumps like boiler feed pump, utilized in power plants, need special attention in their installation. A precision alignment of driver and pump becomes essential hence brief alignment procedures are presented as a guide line to power engineers.

The purpose of this book will be fulfilled when it will be usefull for young and dynamic engineers in their practical work for achieving better availability and highest reliability of the process equipment.

Dr. Y. Chaudhary
Ex. Professor and HOD Mechanical
NIT (Jamshedpur), Jharkhand

Chapter 1

INTRODUCTION

1.1 Concept of Centrifugal Pumps

1.1.1 Centrifugal Force

A centrifugal pump is a device which generates pressure by accelerating fluid particles velocity and pressure energy. The velocity at which the appositin's foot ball player dashes at you and mass of the player determines the striking force. The combination of mass and velocity produces velocity energy. Newton's law explains, force (P) = mass (m) * acceleration (a). Fluid particles moving at high velocity contains velocity energy which can be felt by placing hand on the open end of a garden hose.

The fluid particles into the pump are continuously expelled from the tips of the impeller vanes at high velocity, and subsequently hit the inner casing of the pump where it gets decelerated lowering the velocity energy and raising the pressure energy, obeying the energy conservation rule. Simiar transformation of energy takes place with a solid lump tumling from top of a hill; lump speed gradually increases as it looses the elevation energy and transformed into velocity energy.

To understand the centrifugal force try an experiment as shown in the sketch, Take a plastic cup/container and poke a small pinhole at the bottom. Fill it with water and attach a string to its mouth and start spinning. The faster you spin, the more water will escape out of the pin hole. On spinning, the water is pressurized inside the cup by centrifugal force as happens in a centrifugal pump. In case of a pump, the rotational motion of the impeller projects fluid particles at high velocity into the casing space surrounding the impeller and velocity energy is converted into pressure energy. Thus, fluid particle's slow down after exit from Impeller tip throughout the system.

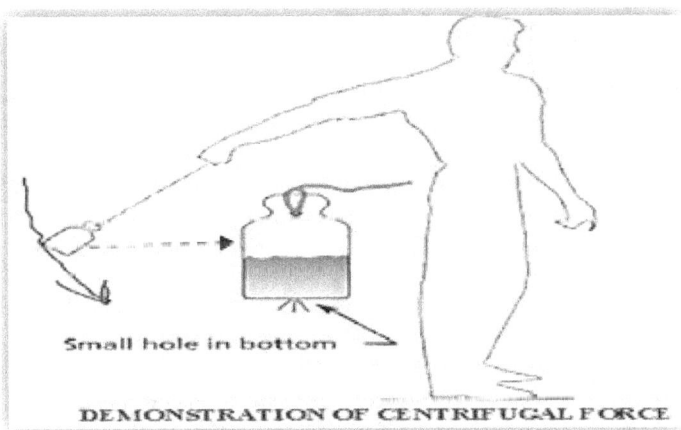

Small hole in bottom

DEMONSTRATION OF CENTRIFUGAL FORCE

Since, the speed of pump is normally constant, the pressure produced by impeller vanes will also be constant, corresponding to the particular conditions of the system (e.g. fluid viscosity, pipe size, elevation difference, etc.). Any deviation (viz. closing a discharge valve), in the system causes a flow reduction hence, there will be an increase in pressure at the pump discharge *as the impeller speed remains constant*. The pump produces excess velocity energy because it operates at constant speed, and velocity energy is transformed into pressure energy leading to pressure rise.

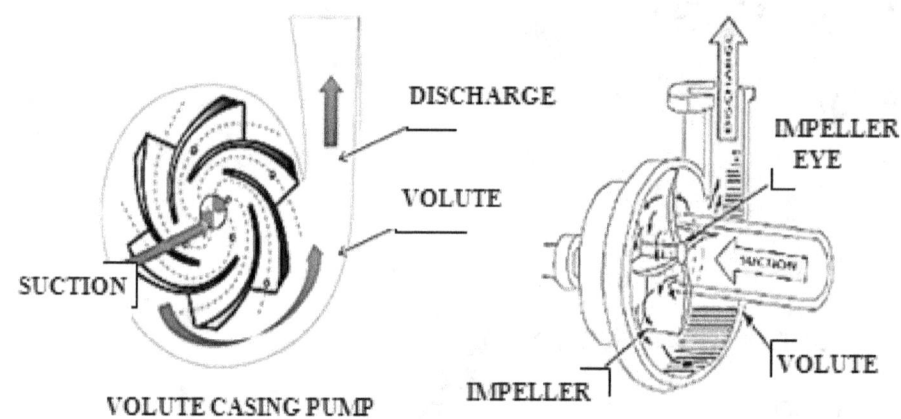

VOLUTE CASING PUMP

1.1.2 Energy In An Incompressible Fluid

The total energy in M or the pressure (kg/cm²) produced by a pump is measure of the energy added to the liquid. The energy difference between the Point of leaving the pump and the point of pump entrance can be expressed as amount of energy added to the liquid by the pumping system. The total energy at any point in a system is a relative term & measured above some arbitrarily selected datum plane.

An incompressible fluid can contains energy in the form of velocity, pressure, and or/potential heads. Bernolli's theorem for incompressible fluid states that in steady flow, without losses, the energy at any point is the sum of the velocity head, pressure head, and potential/datum head and that sum of these energies is constant along the flow stream. Therefore, the energy H in ft-lb, in gauge or absolute, at any point in the system relative to a selected datum plane is-

$$H = (V^2/2\ g) + (144p/\dot{\rho}) + Z$$
or in metric system $H = (V2/2g) + (P/\rho) + Z$

Where

- V = velocity, ft/s
- g = acceleration of gravity, approximately 32.17 ft/sec² (9.18 M/sec ²)
- p = pressure (+ or −) lb/in² gauge or kg/cm²
- $\dot{\rho}$ = specific weight of liquid, lb/ft³ or kg/M³
- Z = elevation above (+) or below (−) datum, ft or M

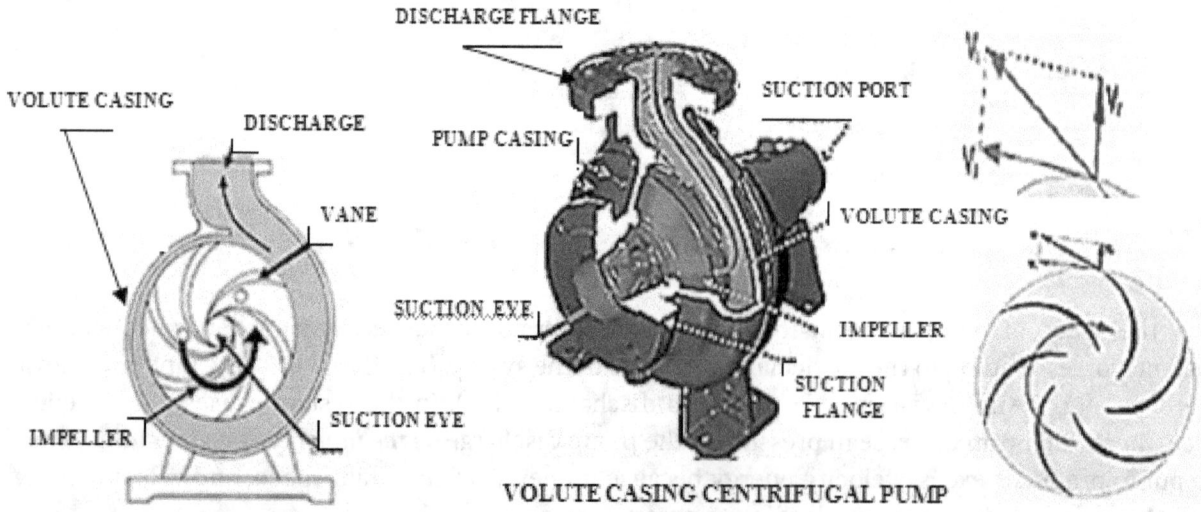

VOLUTE CASING CENTRIFUGAL PUMP

The centrifugal pump is a mechanical machine used to lift & move liquid from one place to other and pressurize the liquid for a numerous useful tasks. The pump is the mechanical mechanism for transfer or transport of liquid under various conditions. The rotating impellers work on principle of centrifugal force which convert mechanical energy into kinetic and potential energy. Transformation of mechanical energy into kinetic and potential energy is done by centrifugal pump. The force developed depends on peripheral speed of impeller and density of liquid. The amount of energy imparted per unit of liquid mass is independent of density of liquid being handled. Therefore, for given pump, operating at certain speed, handle a definite volume of liquid. *The energy required (Kg. M/Kg) is same for any liquid, regardless of liquid density but it do vary on change of viscocity of liquid.*

In any power plant and petrochemical plants, numbers of centrifugal pumps are in application e.g. for feeding water to boiler, condenser cooling system, condensate extraction, lube oil for bearings, chemicals and petroleum liquid handling etc. Therefore, reliability, availability and efficiency of a power station will broadly depend on the behavior and functioning of centrifugal pumps in process. In view of this, proper design, selection and implementation of systematic and predictive maintenance become essential. Hence, the skilled and trained maintenance personnel deployment is a must. In the following section of this book, attempt has been made to acquaint power engineers with different types of the pumps and systematic maintenance practices, trouble-shooting to enable the engineers to interact quickly to abnormal conditions/situation in centrifugal pumps.

1.1.3 Terminology

- **Barometric Pressure** is the level of atmospheric pressure above perfect vacuum. It is also the pressure at site being taken and changes with whether condition and altitude. Standard atmospheric pressure is 101.326 kilo Pascal (kpa) or 14.696 Lb/inch sq.

- **Gauge pressure** is measured above atmospheric pressure whereas *absolute pressure* is always referred below the absolute vacuum datum line or absolute atmospheric datum line.

- **Vacuum** usually expressed in inches of Hg column, is the depression of pressure below the atmospheric line.

- **Head**: The pressure at any point in liquid contained in vertical column of the liquid is caused by the weight of liquid column at that point. Height of liquid column is called static head, expressed in M or other linear units. **Static head = (pressure, psig * 2.31)/specific gravity of liquid.**

- **Suction Head**: Vertical difference in elevation (M or ft) from liquid supply level to the pump centre line is defined as suction lift.

- **Static Discharge Head**: Vertical difference in elevation (M or ft) from pump centre line to the point of free discharge of liquid is called as Static Discharge Head.

- **Total static Head**: It is sum of static suction head and static discharge head.

- **Frictional Head**: The head required to overcome the resistance to flow (frictional valves and fitting resistances) is called as frictional head.

1.1.4 Pump Grosseries

- **Velocity Head**: Energy of liquid transferred to its motion/velocity (M/sec), necessary to accelerate the liquid to reach the destination. Velocity Head (M) = $V^2/2\,g$, Where V = velocity of liquid (M/sec), g = Gravitational acceleration (M/sec^2)

  The kinetic energy per pound of liquid = ½ w v^2/w g = $v^2/2\,g$, measured in meters or feet head. This quantity is theoretically equal to the head of liquid required in a vessel above an opening or orifice, if the discharge have a average velocity 'v' (computed by dividing the flow by the cross sectional area of discharge).

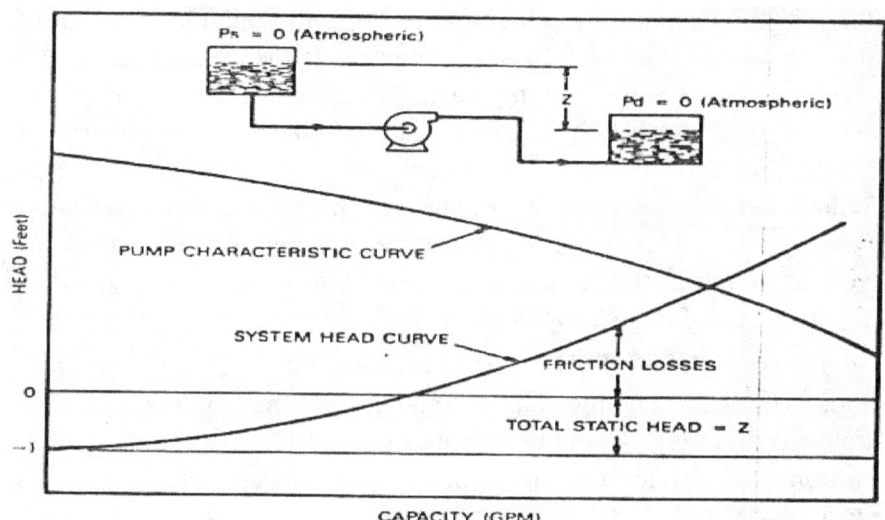

. Pumping system with fixed flow, pump speed, system static head, discharge below suction datum.

- **Pressure Head:** It is the discharge pressure (kg/cm²) converted in liquid column (head) known as pressure head, = 144 p/w, with units in FPS. Liquid having pressure is capable of doing work, e.g. on a piston having an area A and stroke L. The quantity of liquid required to complete one stroke = w * A * L. ***The work (force * stroke) per pound*** = 144 * p * A * L/w * A * L or (144 * p/w.)

Thus, pressure head also represents the work required per pound by a liquid under pressure. The work required by a pump to produce the pressure intensity in liquids having different specific weights varies inversely with the specific weight or specific gravity of the liquid. The liquid having low specific gravity i.e. lesser dense is raised to a higher column height to produce the same pressure at the same elevation compared to that of heavier liquid. The pressure at the bottom of each liquid column 'H' is the weight of the liquid above the point per unit cross sectional area A at pressure measurement point. Hence, A * H = w/A, in lb/in². It is simply H * w/144 in FPS system. If H is taken in M, weight of liquid in kg = w and cross sectional area A in cm², then p = w/A in kg/cm² and H ≈ 10 * p in M

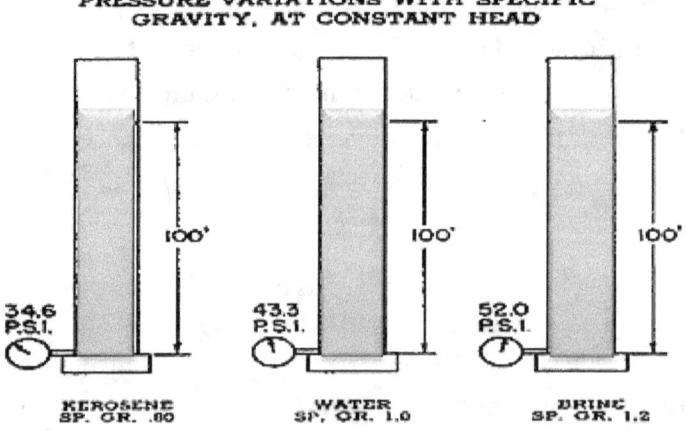

The pressure head in any liquid can also be expressed as height of the liquid column above the point of pressure measurement (density of the liquid throughout column being constant). Substituting

pressure intensity A * H * w/144 * A for p in equation 144 * p/w, can be observed as a pressure in terms of H is the liquid column height. Therefore, at the base of equal columns containing liquids with same density, the pressure (psi) and head (ft-column) will be same, but when liquid in different columns have different density will have different pressure intensity lb/in² for same column length. For this reason, it is necessary to identify the liquid while comparing pressure heads.

Fig indicates that discharge pressure of centrifugal pumps varies with specific weight of liquid pumped.

- **Elevation Head:**

  The elevation energy, or potential energy, of a liquid is the distance z ft measured vertically above or below an arbitrary selected horizontal datum plane. Liquid above the reference datum plane has positive potential energy since it can fall a distance z acquiring kinetic energy or vertical head equal to z meters. The w, kg of liquid requires w * z kg-M of work to lift above the datum plane by centrifugal pump. The work per kg is therefore w * z/w, or z meters. In a pumping system, the energy required to raise a liquid above a reference datum plane can be thought as being provided by a pump located at the datum elevation and producing a pressure is A * z * w/A or simply z * w kg/M². Since head is equal to pressure divided by specific weight, elevation head is z * w/w or z meters. Liquid below the reference datum plane has negative elevation head.

  In systems that are open ended and where there is a decrease in elevation from inlet to outlet, a portion of the system head curve will be negative. In this example the pump is used to increase gravity flow without a pump in the system, the negative resistance or static head curve is the driving head.

- **Total Dynamic Suction Head:** (Static suction head – Velocity head – suction frictional head) all expressed in M

- **Total Dynamic Discharge Head:** (Static discharge head + velocity head + discharge fictional head) in M head.

- **Total Dynamic Head:** (Total dynamic discharge head – total dynamic suction head)

- **Capacity;** It is volume liquid transfer/unit time, normally expressed as M³/hr. Since liquid is in compressible, relation of capacity and velocity of flow may be established as-

  Q = A * V

  where – A = area of cross section of pipe, V = Velocity of liquid

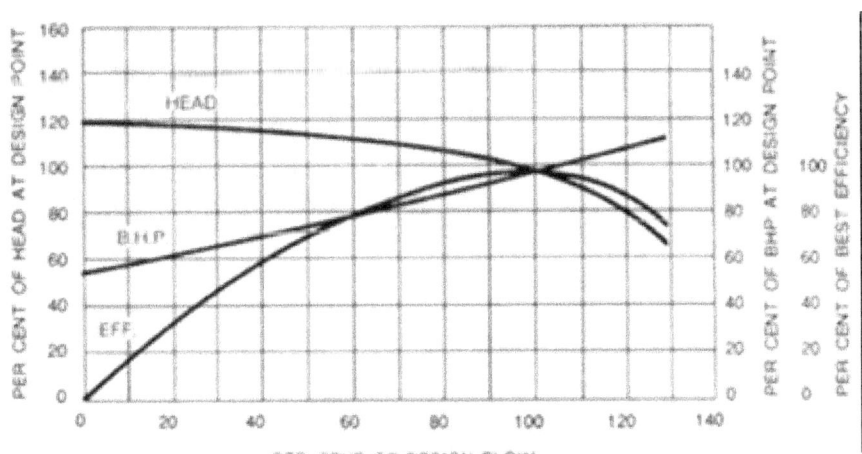

PER CENT OF DESIGN FLOW

Radial Flow Pump

> **Radial Flow Pump** - the characteristic curve of radial pumps is is relatively flat and head decrease or increase is geadual. BHP increases gradually over the flow range with max at point of max flow.

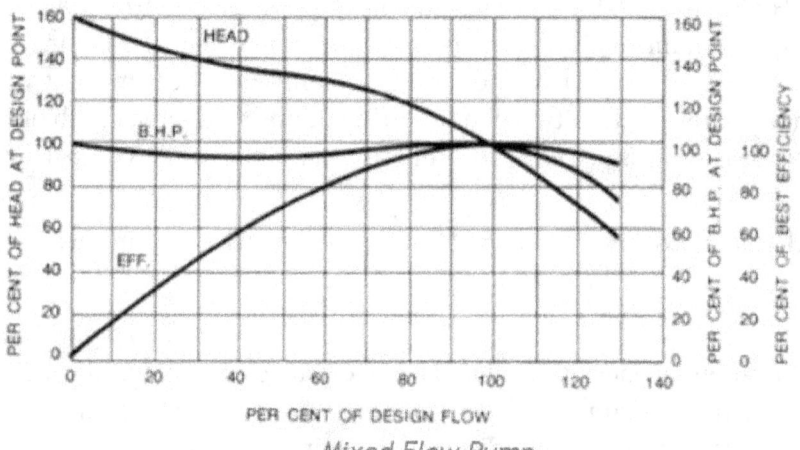

Mixed Flow Pump

Mixed flow pump – Head curve is steeper compared to that of radial pump The shut off heas is usually 150 – 200% of design normal operating head. BHP is fairly constant over large range of flow.

Axial flow pump – Head curve is steeper compared to that of radial or mixed flow pump The head and BHP both increase drastically near shut off operating zone. BHP is not constant over large range of flow.

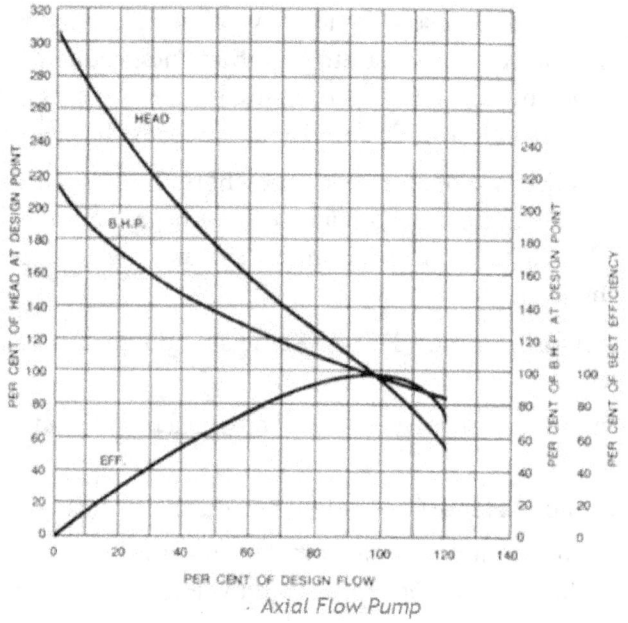

Axial Flow Pump

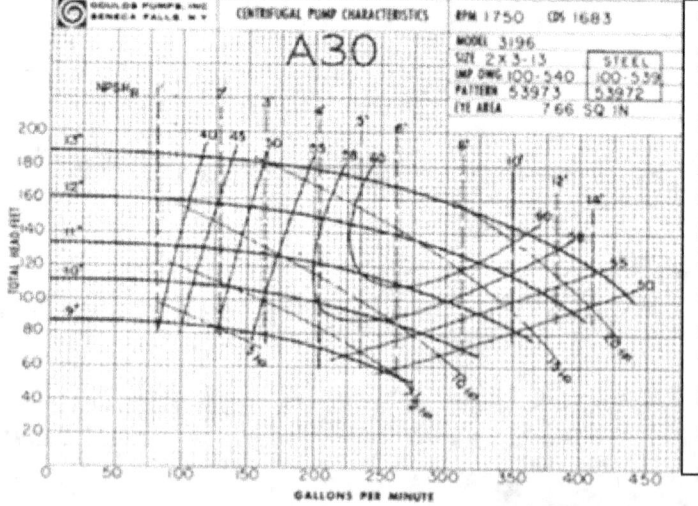

Typical composite performance pump curve, furnished by manufacturer indicates st s glance what pump will do with various impeller diameters from max to min. The constant HP, efficiency, and NPSH (r) lines are super imposed over various head curves. The curve is plotted from individual test results of different diameters.

Composite Performance Curve

Heads in Centrifugal Pumps

- **Static Suction Head (hS):** Head resulting from elevation of the liquid relative to the pump center line. If the liquid level is above pump centerline, **hS** is positive. Otherwise, when level is below pump centerline, **hS** is negative. Which is commonly denoted as a "suction lift" condition.

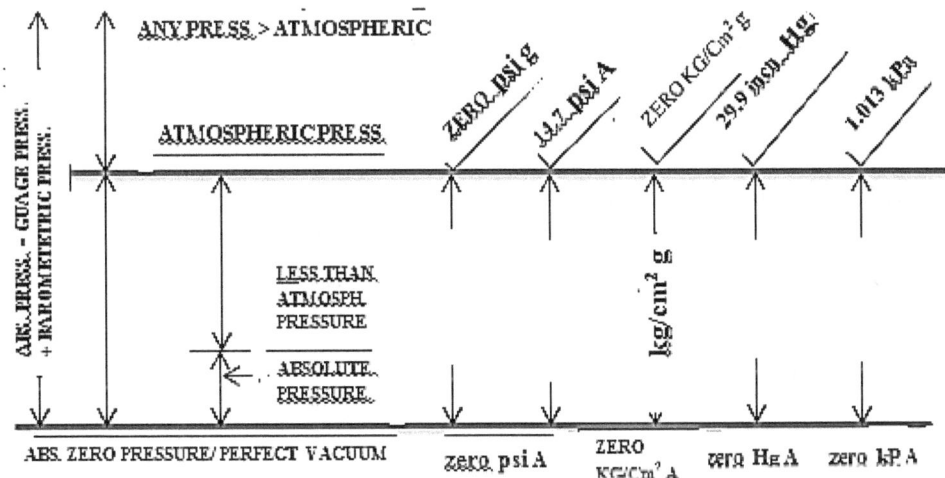

- **Static Discharge Head (hd):** It is the vertical distance in feet between the pump centerline and the point of free discharge or the surface of the liquid in the discharge tank.

- **Friction Head (hf):** The head required to overcome the resistance to flow in the pipe and fittings. It is dependent upon the size, condition and type of pipe, number and type of pipefittings, flow rate, and nature of the liquid.

- **Vapor Pressure Head (hvp):** Vapor pressure is the pressure at which a liquid and its vapor co-exist in equilibrium at a given temperature. The vapor pressure of liquid can be obtained from vapor pressure tables. When the vapor pressure is converted to head, it is referred to as vapor pressure head, **hvp**. The value of **hvp** of a liquid increases with the rising temperature and hvp, opposes the pressure on the liquid surface, (i.e. positive force which tends to cause liquid flow into the pump suction is reduced).

- **Pressure Head (hp):** Pressure Head must be considered when a pumping system either begins or terminates in a tank which is under some pressure other than atmospheric. The pressure in such a tank must first be converted to head of liquid. It is denoted as absolute pressure head **hp**, on the surface of the liquid reservoir supplying to pump suction, (converted to feet or M of head. If the system is open, hp equals atmospheric pressure head.

- **Velocity Head (hv):** Refers to the energy of a liquid as a result of its motion at some velocity 'v'. It is the equivalent head in feet or meter through which the water would acquire the same velocity, or in other words, the head necessary to accelerate the water. The velocity head is usually insignificant and can be ignored in most high head systems. However, it can be a large factor and must be considered in low head systems.

- **Total Suction Head (HS):** The suction reservoir pressure head (**hpS**) plus the static suction head (**hS**) plus the velocity head at the pump suction flange (**hVS**) minus the friction head in the suction line (**hfS**) is known as total suction head (HS).

$$HS = hpS + hS + hvS - hfS$$

The total suction head is the reading of the gauge on the suction flange, converted to feet of liquid.

- **Total Discharge Head (Hd):** The discharge reservoir pressure head (**hpd**) plus static discharge head (**hd**) plus the velocity head at the pump discharge flange (**hvd**) plus the total friction head in the discharge line (**hfd**) is known as total discharge head (Hd).

$$Hd = hpd + hd + hvd + hfd$$

The total discharge head is the reading of a gauge at the discharge flange, converted to feet of liquid.

- **Total Differential Head (HT):** It is the total discharge head minus the total suction head or

$$HT = Hd + Hs \qquad \text{(with a possitive suction lift)}$$

$$HT = Hd - Hs \qquad \text{(with a negative suction head)}$$

The $NPSH_{(R)}$, efficiency, and BHP vary with flow rate (Q). The **_Best Efficiency Point (BEP)_** is at the capacity of characteristic curve where pump has the highest efficiency. All points to the right or left of BEP have lower efficiency. Normally, designer opt the operating poit closed to the best efficiency point.

1.1.5 Type of Pumps & Application

Location	Name & Type of Pump in Use	Function of the Pump	Remarks
Intake channel make up	Vertical, Single Stage,	To maintain water	
Discharge Channel (C.T. Pump)	Vertical, Double Stage, Propeller type, Mix Flow	Pumping water through Cooling tower	When Cooling Towers are required to maintain water temperature
Control Structure	H.P.5 Stages, Vertical. L.P. Pumps 3-stages, Vertical, Volute type	To supply water to various systems like fire fighting, E.S.P. for flushing, bottom ash removal, ash pump sealing and cooling system	
Clarifloculator (Clarified water pump)	Veridical Centrifugal pump. Deliver to mill	To supply water to treatment plant for de mineralizing and for supplying clarified water to plant equipment cooling	
Intake channel inside Turbine House (Circulating water pump	Vertical, Mix flow, two stage	To Provide cooling water to condenser	
Hot well (Condensate Extraction Pump)	Vertical, Multi stage, Centrifugal	To pump water from hot well through ejector to L.P. Heater to De aerator	For recycling the condensate to Boiler
Meter Turbine (Drip Pump)	Single- Stage, centrifugal, Split casing, Horizontal	To maintain L.P. Heater level	
Ohmmeter, B.F.P. (Boiler Feed Pump)	Multi stage, Barrel type Centrifugal	Feeding water to boiler drum through feed heaters.	
Meter, S.O.P. (Starting oil pump)	Vertical/Horizontal	To Provide oil to bearings during start up.	Workings as taking oil pump

Location	Name & Type of Pump in Use	Function of the Pump	Remarks
Meter (Centrifuge/Oil Purifier pump)	Vertical, Bowl type, centrifugal	To purify oil for recycling in the system	To Provide purified Clean oil to turbine oil system for bearings and governing systems
Boiler Bay pumps			
i) Boiler Fill Pump	i) Centrifugal, Multi stage, Guide vane type, or Single stage, Volute type	i) For filling the boiler for testing & start ups.	
ii) DM make up pump	ii) Centrifugal, Single Stage, Volute type	ii) To provide make up water to the system	
Ash pump House (−4 M level) (Slurry pump)	Closed impeller, Centrifugal, or Single stage Positive head	To remove slurry from ash channel to lagoons	
13 Meter (Boiler) (Phosphate dosing pump.)	Reciprocating	To inject tri sodium phosphate to boiler drum water.	
Meter, Turbine (Booster pump)	Single Stage, Radial Impeller	For bearing cooling of condensate pump, boiler feed pump A.C. Lube oil pump, and to provide cooling water to Ball Mills.	

1.1.6 Guide Line for Preparing Specification

If rated head and flow are known, a preliminary, selection can be made for pumps that fall within normal ranges of operation, as shown in Figure. This preliminary selection exercise gives a good assessment of the pump which may meet the required process conditions. The specific pump selected will influence the standard specifications under preparation. Some industrial, national and international specifications for centrifugal pumps are presented in following standards–

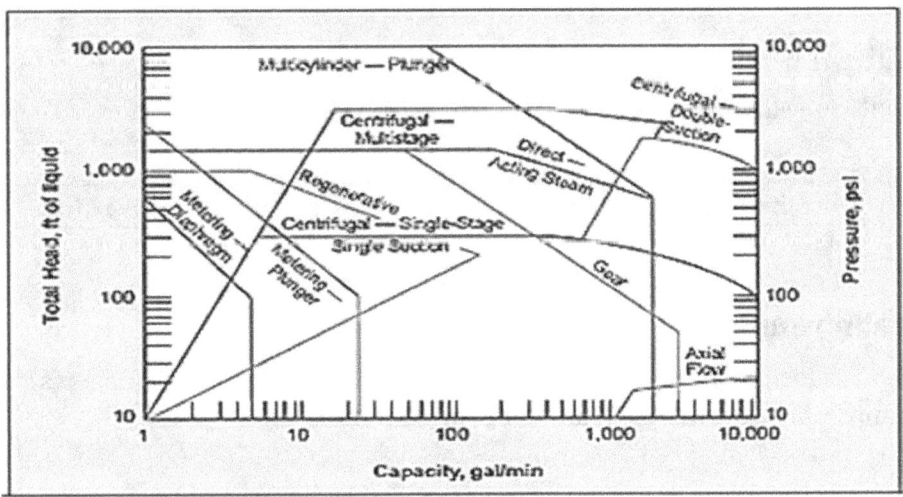

Selection Guide for Centrifugal Pumps

- ASME/ANSI B73.1, "Specification for Horizontal End Suction Centrifugal Pumps for Chemical Process," and ASME/ANSI B73.2 M, "Specification for Vertical In-Line Centrifugal Pumps for Chemical Process" (www.asme.org).

- API-610, "Centrifugal Pumps for Petroleum, Heavy Duty Chemical and Gas Industry Services."

- PIP RESP73H-97, "Application of ASME B73.1 M–1991 Specification for Horizontal End Suction Centrifugal Pumps for Chemical Process," and PIP RESP73V-97, "Application of ASME B73.2 M Specification for Vertical In-Line Centrifugal Pumps for Chemical Process" (www.pip.org). API-610 sets standards that create a more robust and expensive pump. Individual company policies may have additional centrifugal pump specifications that must be met, as well.

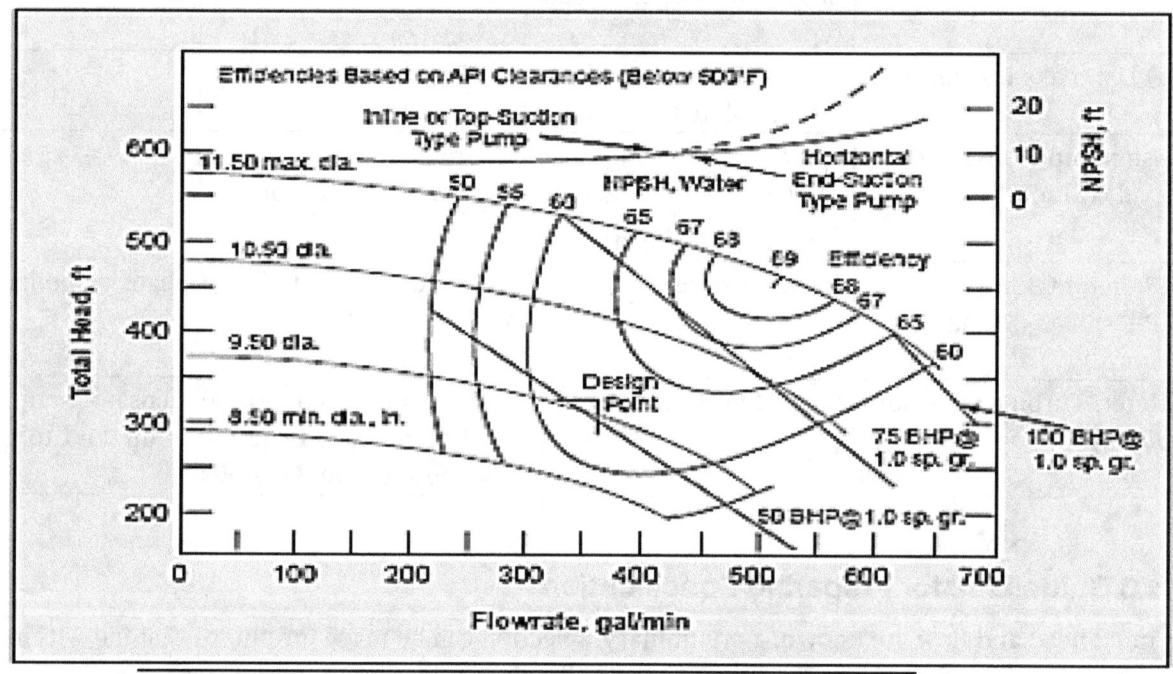

Typical Operating Curves for Centrifugal Pump

The correct selection of the impeller diameter, specification, to select a pump with a rated design point at the best efficiency point (BEP), is determined (refer graph above).

1.2 Classification of Pumps

Pumps are basically categorized in following types

i) Reciprocating	ii) Rotary,	iii) Centrifugal	iv) Misc. types

1.2.1 Reciprocating Pump

In this type energy is added to the liquid by to and fro movement of piston(s), plunger(s), diaphragms etc. Reciprocating pumps can be sub-divided into following varieties

i) Piston pumps	ii) Plunger pumps	iii) Ram pumps	iv) Diaphragm pumps

1.2.2 Rotary Pumps

Here the pumping action is caused by relative movement of rotating element against stationary elements of the pump. They can be sub-divided into following categories-

i) Gear pump	ii) Screw pump	iii) Vane pump	iv) Lobe pump

Rotary pumps are a positive displacement pump which develops the required pressure employing rotary motion e.g. gear pump and vane pumps. Hence, we can classify a gear- pump as a rotary & positive displacement pump. Most hydraulic systems employ rotary pumps (gear/vane/plunger types). The screw type rotary pump is also a positive displacement pump but it is constructed differently compared to other positive displacement pumps. However, that has similar operating characteristics. Usually, screw type pumps are popularly known as transfer pump because of their large capacities.

Classifications of Rotary Pumps

a. Gear Pumps

Gear pumps, sometimes called external gear pumps, are probably the most common type of rotary pumps used for industrial applications like hydraulic or lube oil systems. Many machine tools use gear pumps for bearing lubrication as well as for applying cutting fluid to the various points on the machine. In addition, the oil pump in the engines of most fork lift trucks is. a gear type pump.

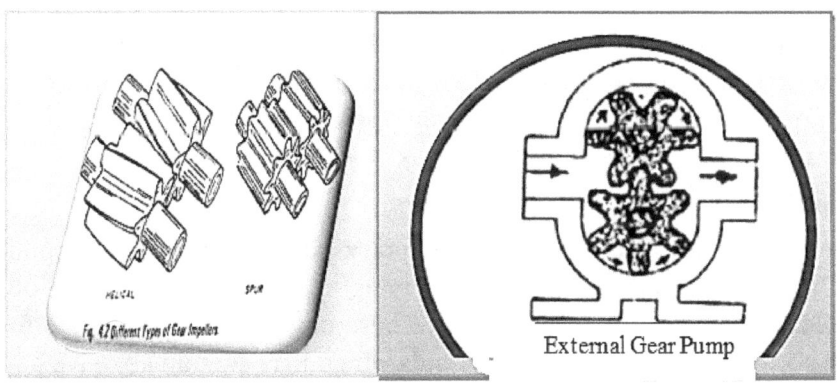

External Gear Pump

The operation of a gear pumps (External and internal), as shown in fig is simple to understand The rotary gear impellers are generally spur/helical or herringbone gears. Commonly, there is misunderstanding people have about flow in gear pumps about fluid flow. They think that the fluid being pumped is forced through the space between the rotors teeth in meshing and the pump casing and delivered out to the discharge port. But by this action (i.e. meshing of the gears & sealing with casing), fluid is prevented from flowing back to the suction side of the pump. In fact, fluid is trapped between spaces of gear teeth while in unmeshed position and forced to discharge port. If gear rotation is reversed, pump will start pumping in reverse i.e. discharge port to suction port subject to fluid is available at discharge port.

• Internal Gear Pumps

Another type of rotary gear pumps is the internal gear pump. This pump is entirely different in construction than external gear pump. The internal gear pump consists of two gears in mesh with each other. One gear of the set protruding out of pump chamber is the driving gear. The internal gear is the driven or idler gear of the pump. The crescent keeps the gears separated and reduces eddy currents, which increases the pump efficiency. In some models, it is movable to allow the pump to operate in either directions.

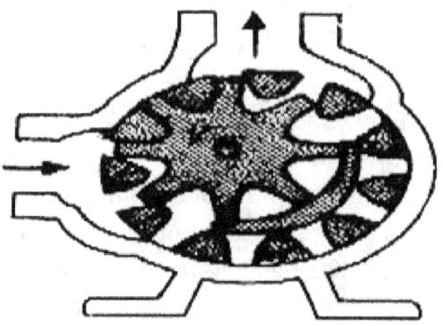

Internal gear pump.

In operation, the rotating internal gear opens the space between the teeth of both gears at the intake port area. Fluid is drawn in through the intake port and passes around the crescent area of the pump. As the gear teeth again come in contact with exit port, and fluid is discharged. Notice that the driven gear has fewer teeth than the driving gear. However, the gears mate smoothly at all times – without causing interference irrespective of equal or unequal pitch diameter of gears.

The gear pumps clearance between side face of gears with side plates and crown of teeth with casing internal diameter is maintained very closely (approx 0.012 mm, depending on viscosity of media) to prevent any leakage of liquid from these faces of stationary and rotary parts in working conditions.

b. Vane Type Pumps

The vane type pump is another type of rotary pump used in many applications in industrial plants. They are generally used for transferring hydraulic or lubricating oil, They are also used for chemical or solvent transfer purpose. These pumps are also suitable for transfer of viscous materials such as paint or other heavy fluids which may contain abrasive particles. The vanes are replaceable on wear therefore made of a softer material than the pump casing. On wear when pressure or discharge is reduced then vane plates are replaced at a relatively low cost.

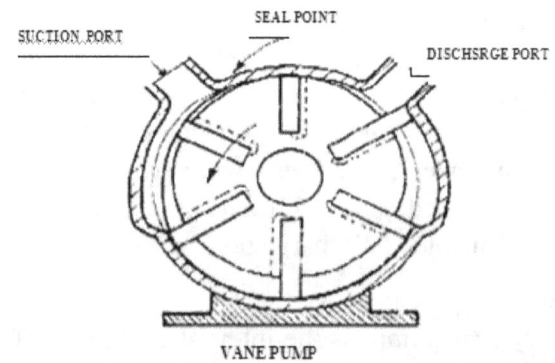

VANE PUMP

The, vane type pump shown in Fig is very simple in construction. The impeller rotates from its offset centre (above the pump centre line) and allows the vanes or blades to extend by gravitational force and draw in fluid on one side of open space between the impeller and the housing at the bottom of the pump, allowing movement of the fluid getting trapped into the two consecutive bottom vanes of rotor. As the impeller continues its rotation, the vanes are pushed back gradually in their slot as they approach the top of the housing. This constriction forces the fluid out to the discharge port.

The pump inlet and discharge ports are oval shaped and extend about three fourths across the casing width (less than full opening) to contain the vane within the pump casing. The oval shape ports smoothens the fluid flow with least resistance to flow (low head loss). The pump vanes slide into slots of rotor by gravity while in rotation, keeping slots clean & free from foreign materials is the prime requirement. These pumps are not recommend to handle dirty or crude liquid otherwise vanes may get jammed into their slots.

c. **Screw Pumps**

Screw 'pumps are a special design of rotary positive displacement pump where the flow through the pumping elements is truly axial. The liquid is carried between screw threads on one or more rotors and is displaced axially as the screws rotate in mesh. In other rotary, pumps, the liquid is forced to travel circumferentially, but the screw pumps with its unique axial flow pattern and low internal velocities will have advantages in many application areas where liquid agitation or churning is undesirable. Fig shows arrangement of screw pump.

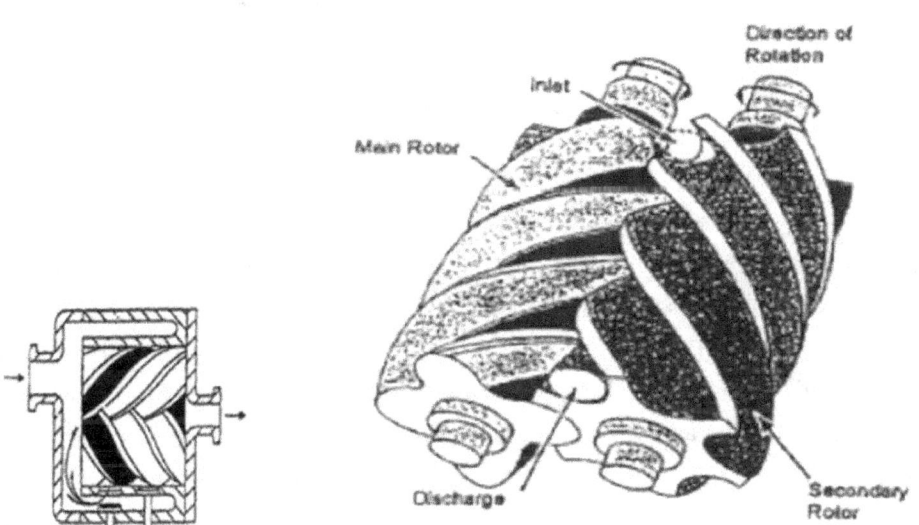

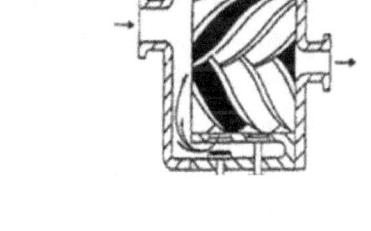

The screw pumps cover a wide range of applications viz. industrial oil, burners; lubricating-oil service, chemical process, petroleum and crude oil industries, power hydraulics for machine tools, utilities fuel-oil service, marine cargo, navy, marine and many other services. The screw pumps can handle liquid ranging from high to low viscosity e.g. molasses to gasoline, as well as synthetic liquids in a pressure range of 3.5 to 350 kg/cm² and flows up to 1300 L/min.

The screw pumps have relatively low inertia of their rotating parts and that are capable of operating at higher speeds than other rotary or reciprocating pumps of comparable displacement. Some turbine-attached screw pumps operate at 10,000 rpm. Like other rotary positive displacement pumps, screw pumps are also self-priming design and have a delivery flow characteristic independent of pressure.

1.2.3 Reciprocating Pumps

Reciprocating pumps have two sections- the fluid section and the drive section. The fluid section does the pumping work where drive section provides the driving force by air, electric power or steam energy, to operate the fluid section.

- Although the pumping and driving ends may vary in construction from pump to pump but their general operating characteristics and designs are similar. The terms used to describe the different components in the two sections (pumping and driving) are the same in most cases.

- Piston converts the steam or air pressure into mechanical energy. The piston in the pumping end uses mechanical energy for fluid movement and pressure development.

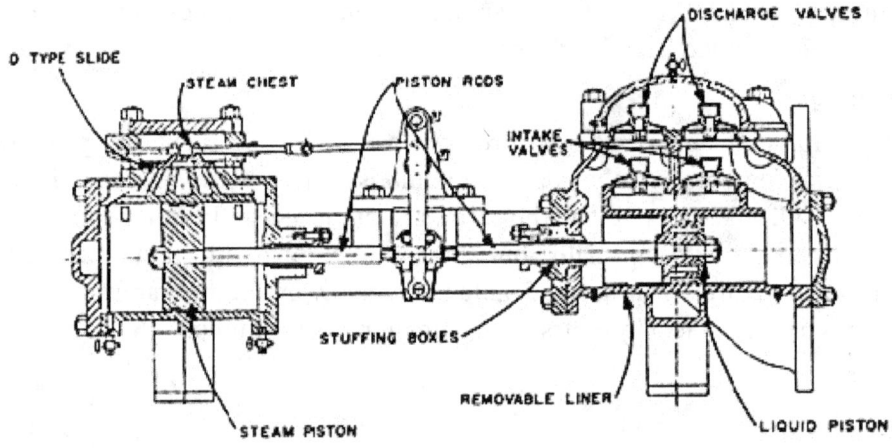

Fig.   Reciprocating-pump

- Piston Ring acts as seating element between the piston and the cylinder walls. Cylinder is the tubular chamber that contains piston which cause fluid to flow and convert mechanical/electrical/ thermal energy into pressure or kinetic energy.

- Gland packing or stuffing box is located where the connecting rod/piston rod pass through the cylinder head. The gland packing prevents leakage of the steam or liquid from the cylinder.

- Connecting Rod connects the piston on the pumping end to the drive section. If the driving force is a crankshaft, the liquid-section connecting rod is usually constructed in two pieces.

- Valves used are of two types. Either they control the power (steam or air) flow into the driving section or the liquid flow into the pumping section. Although both valves which control flow are quite different in construction and operation, the valves at the driving side of the pump are mechanically actuated, while that at the pumping side are materially actuated.

1.2.4 Diaphragm Pumps

Diaphragm pump is a positive displacement but it does not deliver liquid in continuation like rotary positive pumps. Flow of liquid is intermittent. It has similarity with single stroke reciprocating pumps.

- The pumping action of a diaphragm pump is similar to a single acting reciprocating pump. When diaphragm is drawn to the upper portion of the pump cavity, fluid is drawn through the check valve in the suction line and on down movement of diaphragm, the fluid is forced out of the cylinder through the discharge end check valve. On repeat of cycle, the discharge valve is closed by the pressure in the discharge line and the suction is created by the intake stroke.

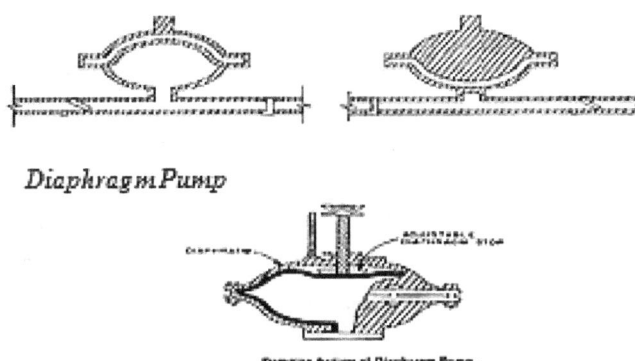

Diaphragm Pump

Pumping Action of Diaphragm Pump.

- Most diaphragm pumps are operated mechanically or by air. The compressed-air or Vacuum-air, depending on the pump can be opted to operate. These pumps are very popularly used for metering, transferring or dirty fluids, scurries, sewage handling.

- The travel or stroke of the diaphragm in the mechanically operated pump is controlled by an adjustable connecting rod or other mechanism but in case of air operated pumps that is controlled by an adjustable diaphragm stop which limits the upward travel of the diaphragm to control flow capacity and pressure.

- For handling chemicals or other corrosive fluids, pump is considered with acid resistance lining or manufactured from corrosion resistant metals. If should be remembered that the flow from a diaphragm pump is not steady, but pulsating.

1.3 Centrifugal Pumps

The definition is applied to all types of pumps with an impeller housed in a suitable shaped casing so that on rotation of the impeller, momentum is applied to liquid in the pump casing and liquid is transported from inlet to outlet of pump by changing velocity into pressure energy. The centrifugal pumps are subdivided in various categories. Most of the pumps used in power stations are centrifugal type. Therefore, the discussions will be centered around centrifugal pumps.

Majority of the pumps used in Power Stations are of this category. It is the machine that moves fluid by spinning its impeller into the casing which has a central inlet and a tangential outlet. The path of the fluid is an increasing spiral from inlet (at the centre) to the outlet tangential annulus. The pressure head develops against inside wall of the annulus; thus curved wall forces the fluid to move in a circular path by converting velocity head to pressure head.

Useful work comes from the pump when some of the spinning fluid flows from the casing tangential outlet into the piping system. Power from the motor accelerates the fluid speed fluid coming to impeller & fluid is delivered to the annulus space. Some of the power is lost in fluid friction in the casing and impeller. Head (pressure) of pump depend on impeller RPM, impeller diameter, and flow rate.

1.3.1 Classification of Centrifugal Pumps Based on Construction

a. **Volute Casing**

In the case of a volute pump a spiral casing is provided around the impeller. The water which leaves the vanes is directed to flow in the volute chamber circumferentially. The area of the volute chamber gradually increases in the direction of flow. Thereby the velocity is converted into pressure energy. As the water reaches the delivery pipe, a considerable part of kinetic energy is converted into pressure energy. However, the eddies are not completely avoided, therefore some loss of energy takes place due to the continually increasing quantity of water through the volute chamber. A radial flow impeller has small specific speeds which is suitable for discharging relatively small quantity of flow against high heads.

In case of a diffuser pump, the guide wheel containing a series of guide vanes or diffusers is an additional component. The diffuser blades are constituted of gradually enlarging passage surround the impeller periphery. They serve to augment the process of pressure built up that is normally achieved in the volute casing. Diffuser pumps are also called turbine pumps in view of their resemblance to a reaction turbine. Multistage pumps and vertical shaft deep-well pumps fall under this category. Centrifugal pumps can normally develop pressures up to 1000 kpa (100 m). If higher pressures are required there are three options–

i.   Increase of impeller diameter.

ii. Increase of RPM.

iii. Use of two or more impellers in series.

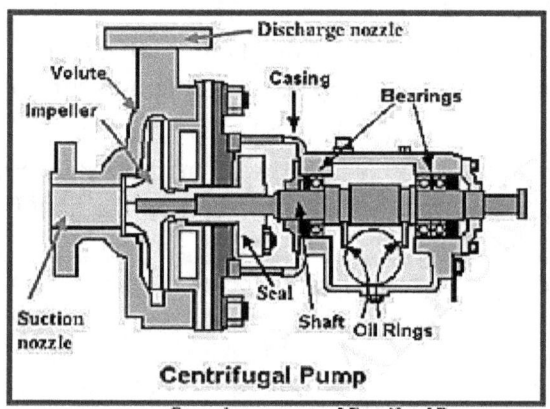

General components of Centrifugal Pump

General components of a Centrifugal Pump

In a diffuser pump the guide wheel containing a series of guide vanes or diffuser is an additional component. The diffuser blade design provide gradual enlarging passage surround the impeller in direction of rotation. They serve to augment the process of pressure buildup which is normally achieved in the volute casing. Diffuser pumps are also called turbine pumps in view of their resemblance to a reaction turbine.

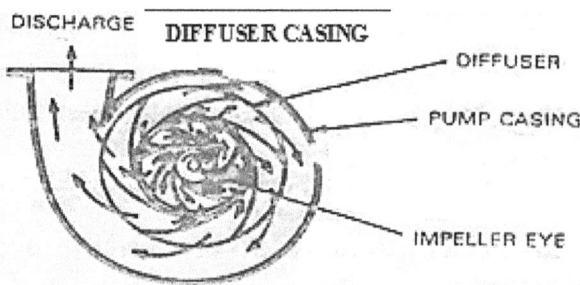

b. Split Casing: This design of pumps are easy to maintain and operate but costly.

The split casing pumps can be further sub classified on the basis of constructional features e.g.

•   Single stage or multistage or based on number of impellers mounted on the shaft.

•   Single suction or double suction depending on the number of liquid entry path;

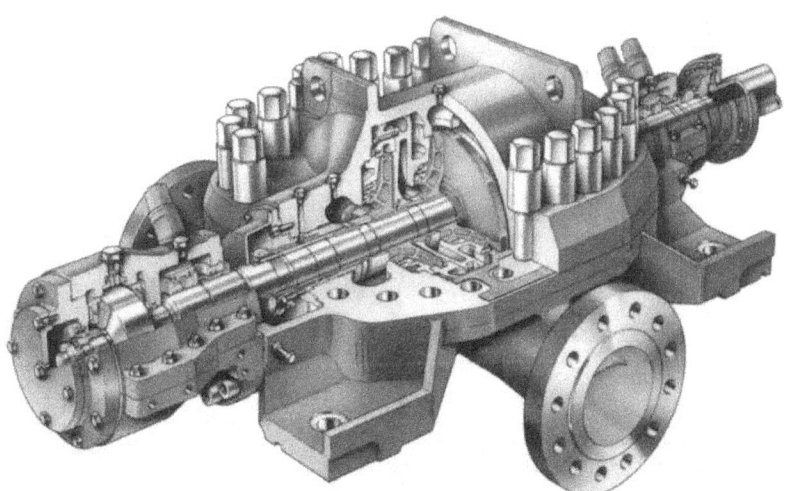

– Axially split or radically split construction of casing

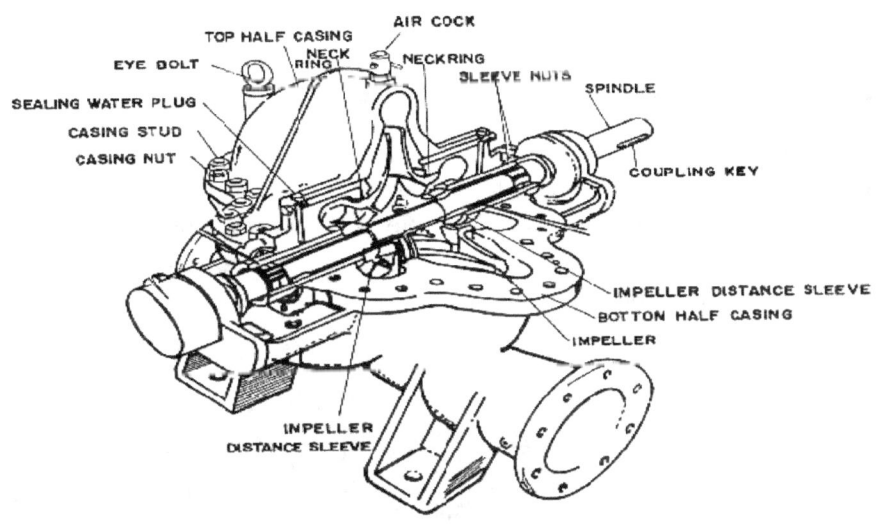

ₜFig. 3.2c CUT – AWAY SECTION OF SINGLE STAGE AXIALLY SPLIT CASING PUMP

c. **Multi Stage Pumps**

Multistage pumps and vertical shaft deep-well pumps fall under this category. Centrifugal pumps can normally develop pressures up to 1000 kpa (100 m). The unit is called a multistage pump when it discharges the same quantity of fluid as a single stage pump but the head developed is high. When a centrifugal pump consist of two or more impellers the pump is known as a multistage centrifugal pump.

There are centrifugal pumps up to 54 stages. However, generally not more than 10 stages are required. The important functions of a multistage centrifugal pump are:

(i) To produce high head ... (Pumps in series)

(ii) To deliver or discharge large quantities of a liquid ... (Pumps in parallel)

• **If higher pressures is required there are three options**

a. Increase of impeller diameter.

b. Increase of RPM.

c. Use of two or more impellers in series.

(impellers of the same size mounted on the same shaft).

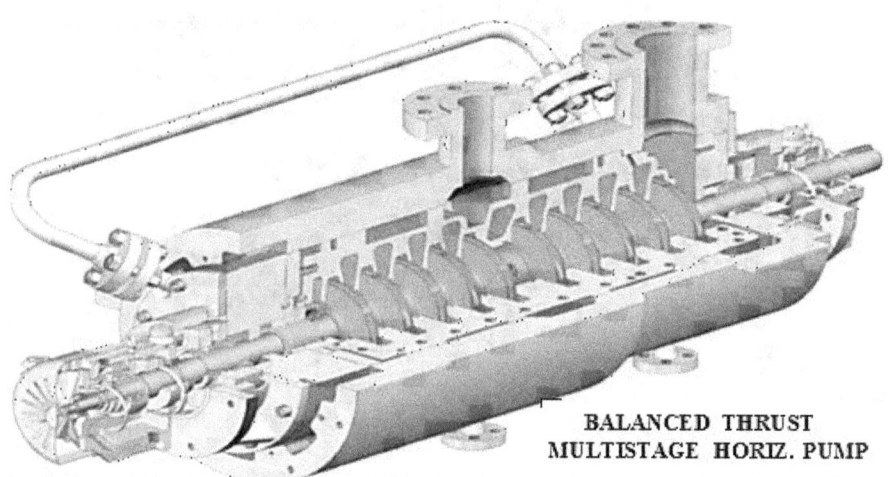

BALANCED THRUST
MULTISTAGE HORIZ. PUMP

1.3.2 Impeller Construction and Applications

- **Double suction impeller**- Two impellers are set back to back. The two suction eyes together reduce the intake. The two suction eyes also reduce the intake velocity and decline the risk of cavitation. Besides, mixed flow type double suction axial flow pumps are capable of developing higher heads. For convenience of operation and maintenance, horizontal shaft settings are the preferred setups for centrifugal pumps.

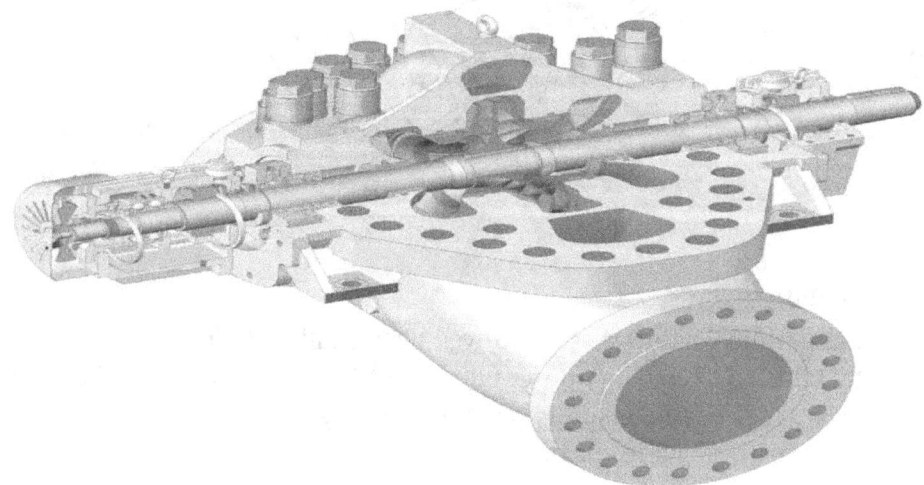

Exceptions to this are the deep-well turbine pumps and axial flow pumps which have vertical shaft mounting. Usually for restricted space mounting of pumps, a vertical shaft setting is preferred.

- **Shrouded impeller** is constituted of vanes fitted between the shroud plates. The crown plate forms suction eye and the base plate is mounted on a sleeve which is keyed to the shaft. Only clear liquids, can be safely pumped by a shrouded impeller pump. An impeller constructed without crown plate is called as non-clogging or semi open type impeller.

1.3.3 Type of Impellers

PARAMETERS	RADIAL VANE	Backward Vane	Forward Vane
Vane Angle	when $\beta_2 = 90°$, the radial curved vanes of the impeller.	when $\beta_2 < 90°$, the Backwards curved vanes of the impeller.	when $\beta_2 > 90°$, the Forwards curved vanes of the impeller
Vector diagram			
Impeller Shape			
where	where : V = absolute velocity of the water. U = Tangential velocity of impeller (peripheral velocity).	V_f = velocity flow. N = Speed of impeller in (rpm). V_r = relative velocity of water to the wheel.	β = vane angle. α = angle at which water leaves.

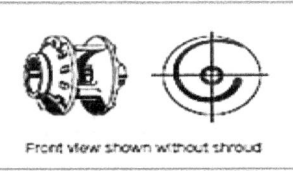

a: Closed single vane impeller for waste water containing solid or stringy substances

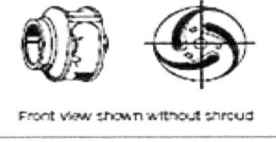

b: Closed non-clogging channel impeller for sludge or non-gassing liquids containing solids without stringy components

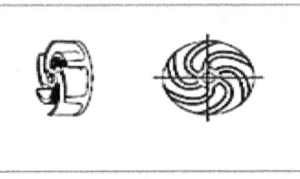

c: Free flow impeller for fluids with coarse or stringy solids and gas content

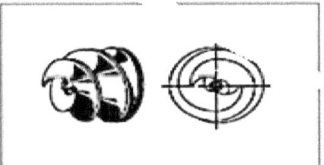

d: Worm type impeller for waste water containing coarse, solid or stringy substances or for sludge with up to 5 to 8% solids content

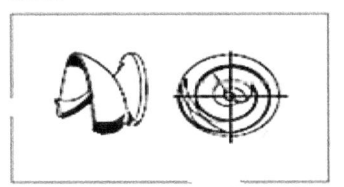

e: Diagonal impeller for waste water containing solid, stringy or coarse substances

TYPE OF IMPELLERS IN CENTRIFUGAL PUMPING SYSTEM

- **Semi-open impeller** is useful for pumping liquids containing suspended solids, such as sewage, molasses or paper pulp. Semi open impellers may be made of cast iron or cast steel. Open vane impellers are usually made of forged steel. If the liquid pumped are corrosive, brass, bronze or gun metal are the best materials for making the impellers.

- **Open vane type impeller** has no crown plate or base plate. That is employed for dredging operations in harbor and rivers. These impellers are very popular in sludge and sewerage handling. It is known as anti clogging impeller.

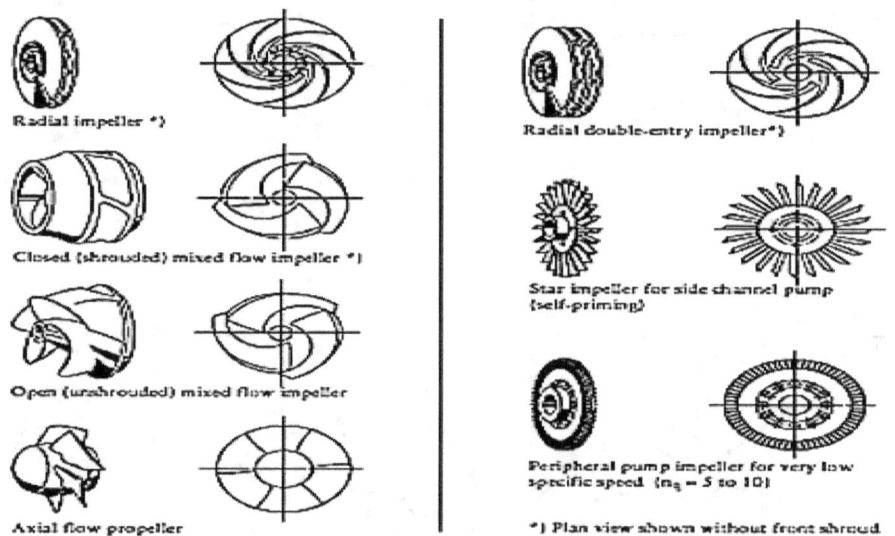

- If a larger head is required, more impellers are to be fitted in series (i.e. discharge from first impeller is guided into the inlet of the second impeller and it is repeated with the third impeller and so on until the required head is reached). Each impeller will increase the pressure by same amount. This type of pump is known as multi stage pump. A typical boiler feed pump may have sixteen stages. All the impellers are keyed to the same shaft and usually provided with diffusers in all stages of casings. The discharge from each diffuser is either circumferential or radial, and collected by vanes attached to the casing which direct, the liquid to the suction of the next impeller. The last diffuser will discharges to delivery pipe.

1.3.3 Advantages of Vertical Pumps

i. Space saving in case of vertical pumps. Very good option for limited space in project e.g. offshore platform.

ii. Lower $NPSH_{(R)}$ is needed in vertical pumps i.e. low $NPSH(R)$ impeller at 1st stage which increases $NPSH_{(A)}$ with lower pump datum but this is very specific case subject to pump design.

iii. Vertical pump creates trouble in motor dismantling and reassembly whereas horizontal pump are easy to maintain.

iv. Vertical pumps are subjected to higher vibration and dynamic stress.

1.3.4 Horizontal Vs. Vertical Pumps

Following are just a quick rundown of comparison:

1. The horizontal pump is a better accepted standard because of traditional acceptance, cost, & availability.

2. There is no perceptible difference in NPSH requirements between both versions. Both can handle equal NPSH situations – if properly selected and installed.

3. Vertical pump's orientation demands overhead space for maintenance – especially in the heavier horse power m/c. For larger motors we require an overhead rail and hoist. Horizontals pumps are easier to handle lift & shift around.

4. Once we commit for a vertical pump, for all practical **purposes there will be increase in investment and inventory in spare** parts. Vertical design is very specific as to type of bearings – primarily thrust – and that force to maintain a parallel inventory for vertical pumps. We don't get any the benefit of exchanging motors and couplings in vertical pumps we can standardize that in case of horizontal pumps.

5. Some vertical design pumps are used to reduce overall height but it poses maintenance problem subsequently.

6. Vertical pumps require special piping design due to pump weight.

   In such cases, we prefer independent supporting of vertical pump i.e. independent support/foundation to reduce piping stresses.

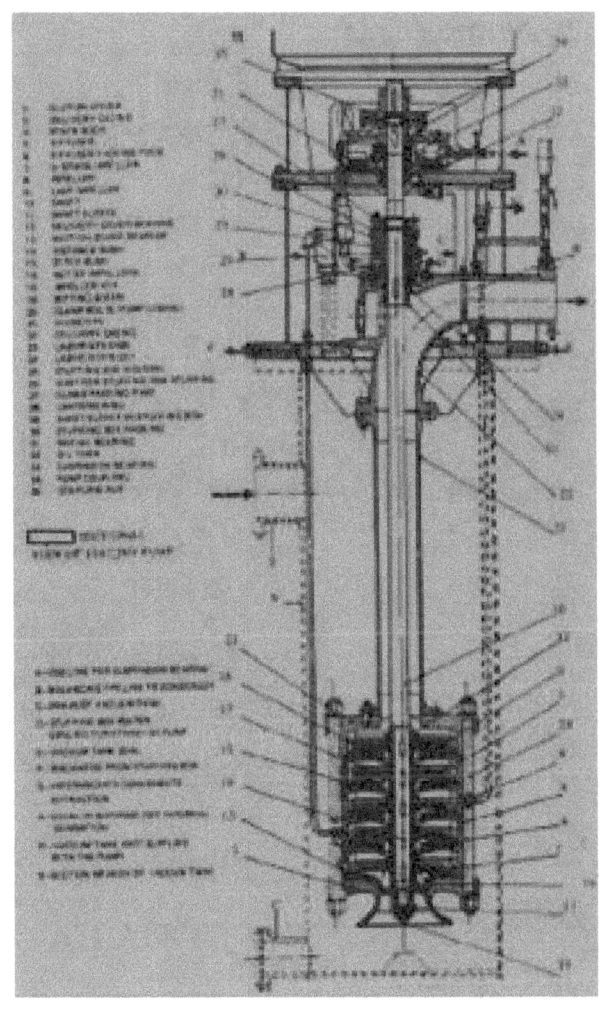

$$Q_z = f_Q * Q_w; \qquad Hz = f_H * Hw; \qquad \eta z = f\eta * \eta w$$

The factor **f** is designated with k; n = speed of rotation of pump;

Ns = specific speed of impeller.

Bernoulli equation, in terms of energy per unit mass

$$\underbrace{\frac{P_s}{Q} + Z_s \cdot g + \frac{C_s^2}{2}}_{\substack{\text{Pump suction} \\ \text{nozzle}}} + Y = \underbrace{\frac{P_d}{Q} + Z_d \cdot g + \frac{C_d^2}{2}}_{}$$

Pump suction nozzle   Pump

Energy input   Energy output (pump discharge nozzle)

Where:

C = velocity;   Q = Flow;   H = Head;   g = Gravitational constant

Z = Potential head;   Suffix S → suction;   Suffix d → Discharge

1.3.5 Pump Efficiency

The impeller flow Q_I generally comprises three components:

Q_E = the useful flow rate (at the pump discharge nozzle):

Q = the leakage flow rate (through the impeller sealing rings):

Q_L = the balancing flow rate (for balancing the axial thrust)

Taking into account the hydraulic losses, the power transferred to the fluid by the impeller is defined as:

$P_L = \rho \, (Q + Q_L + Q_E) \, Y/\eta_h$ where Y = Power transferred per unit mass flow

$\eta_v = Q/(Q = Q_L + Q_E)$

Power input required by the pump can be written as:

$$P = \rho \, (Q + Q_L + Q_E) \, Y/\eta h$$

Volumetric Efficiency of pump is given as

$$\eta_h = Q/(Q + Q_L + Q_E)$$

- **Power input required by pump**

$$P_{(LOSSES)} = (\rho \, Q \, Y/\eta_h \, \eta_v) + PRR + Pm + PER$$

where- **PER** = frictional losses in the balancing device,

PRR = disc friction losses (impeller side discs, seals);

P_m = mechanical losses (bearings, seals);

η_v = volumetric efficiency

η_h = hydraulic efficiency

- **Power required by pump**

$P = \rho \, Q \, g \, H/\eta$

where:

-   P = input power required (W)
-   ρ = fluid density (kg/m3)
-   g = acceleration of gravity (9.80665 m/s2)
-   H = Head added to the flow (m)
-   Q = flow rate (m3/s)
-   η = Efficiency of the pump plant as a decimal

Pump Efficiencies

Hydraulic Efficiency (ϵ_H)	$\dfrac{Pump's\ Total\ Head\,(H)}{Euler\ Head\,(H_e)}$	$\dfrac{gH}{V_{w2}U_2}$	60- 90%
Manometric Efficiency (ϵ_{Man})	$\dfrac{Pump's\ Manometric\ Head\,(H_m)}{Euler\ Head\,(H_e)}$	$\dfrac{gH_m}{V_{w2}U_2}$	---
Volumetric Efficiency (ϵ_V)	$\dfrac{ACTUAL\ VOLUME\ DELIVERED}{VOLUMETRIC\ CAPACITY\ OF\ PUMP}$	$\dfrac{Q}{Q+\Delta Q}$	97 – 98%
Mech. Efficiency (ϵ_{MECH})	$\dfrac{Power\ into\ the\ impeller}{Power\ at\ the\ shaft}$	$\dfrac{\rho(Q+\Delta Q)V_{w2}U_2}{Power\ Shaft}$	95 – 98%
Overall Efficiency (ϵ_0)	$\dfrac{P_{out}}{P_{in}} = \dfrac{\gamma QH}{T.\omega}$ $= (\epsilon_V)\,(\epsilon_{Man})\,\epsilon_H)$	$\dfrac{P_{out}}{P_t} \times \dfrac{P_t}{P_{in}} = \dfrac{P_t}{P_{in}} \times \dfrac{\gamma QH}{\gamma(Q+Q_l)h'}$	71 – 86%

Chapter 2
NET POSITIVE SUCTION HEAD

NPSH of a pump defines the corresponding cavitations criterion, referred to discharge head drop at 3%. For the majority of applications, the available NPSH ($NPSH_A$) for continuous operation should be greater than NPSH at 3% discharge head drop with safety margin (i.e. $NPSH_R$) to avoid loss of performance, noise, vibrations or cavitations, erosion.

NPSH(a) is a measure to prevent liquid vaporization. The manufacturer usually tests the pump with water at different capacities by throttling at suction side. When the first sign of vaporization is induced, the cavitation begins from that moment which can be noticed by suction pressure deviation. The head is published on the pump curve and is referred as the "net positive suction head [i.e. NPSH(a)] or sometimes in short spelled as the NPSH.

2.1 Following Points Correspond to NPSH Situations

a. Cavitations inception can be recognized when $NPSH_{(R)}$: at a given suction head allows first bubbles formation on the impeller vanes. At higher suction head no more cavitations occurs

b. As the suction head is reduced, the bubble formation spread-up over the length of the impeller vane.

c. On start of head drop, $NPSH_0 \rightarrow 0\%$ head drop.: If the suction head is reduced below the value where $NPSH_R < NPSH_0$, the head will start to fall off.

d. At 3% head drop, $NPSH_{(A)}$, is at widely spread cavitations criterion and this point of situation is much easier to measure than the normally gradual onset of fall-off in the head,

e. Full cavitations, $NPSH_{(FC)}$: at a certain suction head, the head falls off very steeply.

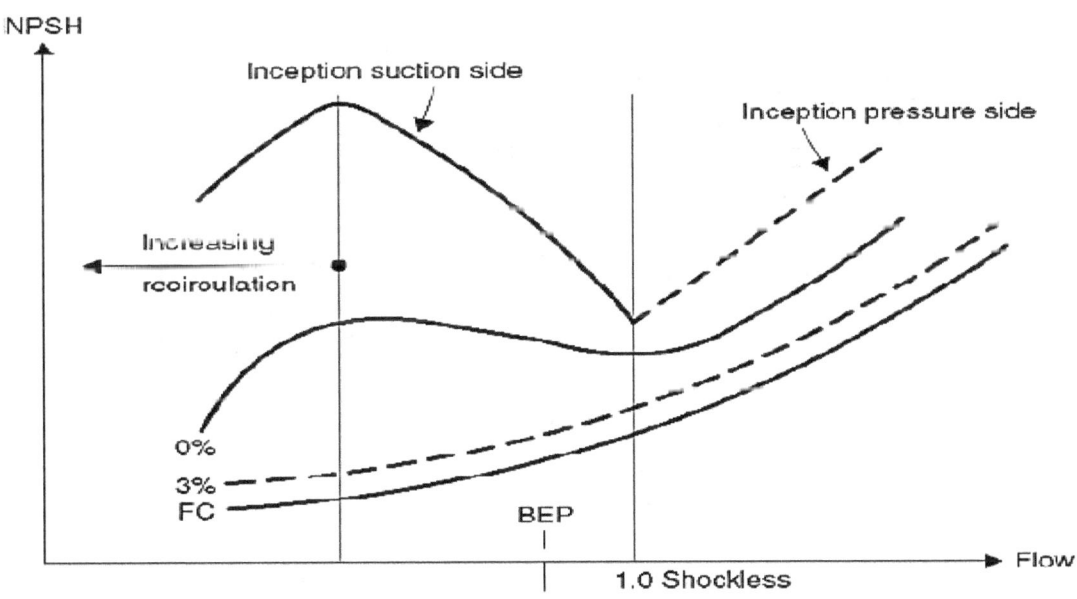

Typical NPSH curves

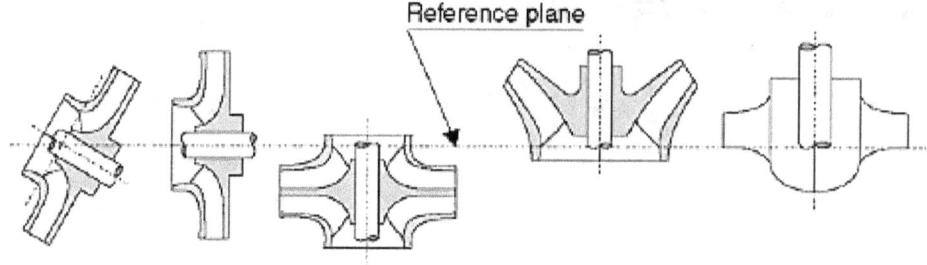

Reference plane

Reference plane for NPSH value according to ISO

$$NPSH_{(R)} = \{(Ps_{ABS} - P_D)/\acute{\rho} * g\} + v^2/2 * g = \{(P_b + P_S + P_D)/\acute{\rho} * g)\} + v^2/2 g$$

where:

PS_{ABS} = absolute pressure in the suction pipe, related to the pump centerline $PS_{ABS} = P_b + P_S$

P_S = gauge pressure (negative figures for pressures below atmosphere)

P_b = atmospheric pressure

P_D = absolute vapor pressure at the fluid temperature

v = flow velocity in the suction nozzle

2.1.1 NPSH Required

NPSH required, ($NPSH_R$**)** of the pump defines the total pressure at the suction nozzle necessary for the pump to operate under 1% head drop. Without specifying the cavitation criterion, the $NPSH_R$ of a pump in principle, is meaningless. The $NPSH_R$ value for a particular pump depends on the flow rate for a given speed and fluid. Typical $NPSH_0$ value for 3% discharge head drop can be determined from graphs given below:

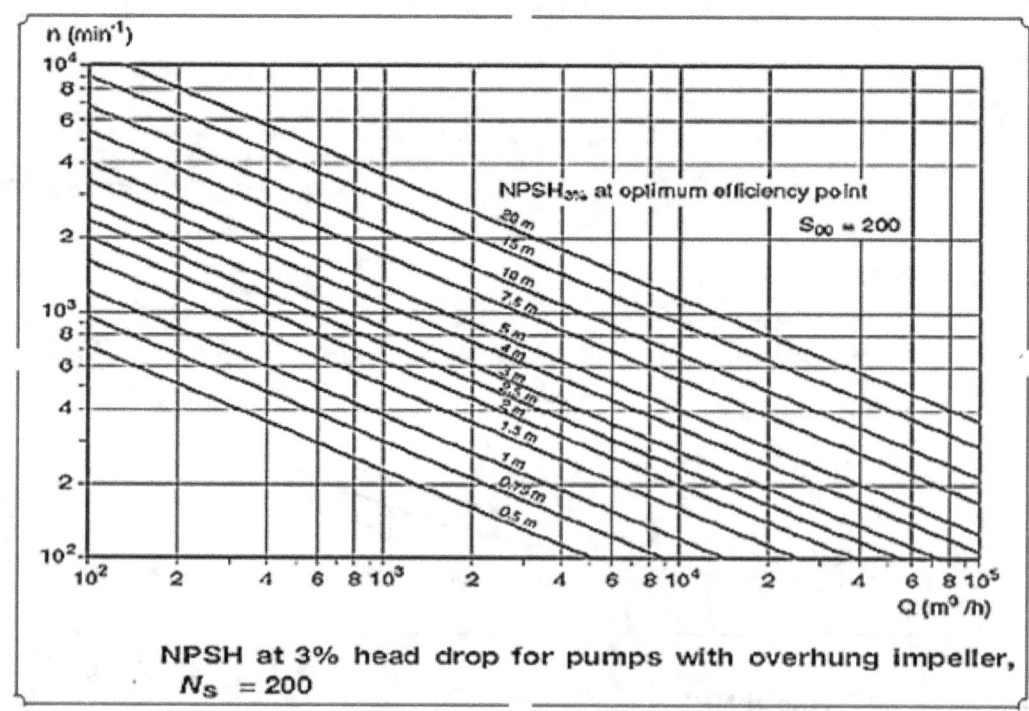

NPSH at 3% head drop for pumps with overhung impeller, $N_S = 200$

2.1.2 NPSH$_{(R)}$ is a Function of Pump Design

The Required Net Positive Suction Head NPSH$_R$ *is the total head at the suction flange of the pump less the vapor pressure (at given liquid temperature) in fluid column height minus frictional and other head losses in suction pipe up to the pump eye.*

NPSH$_{(R)}$ is a function of the pump design and it is determined while performance test of pump is carried out at manufacturer's works. As the liquid passes from the pump suction to the eye of the impeller, the velocity is increased and the pressure is decreased. Further, some pressure-loss takes place due to shock and turbulence caused by liquid striking the impeller which should be accounted. The centrifugal force applied by the impeller vanes enhances the fluid velocity furthermore, resulting decrease of the pressure of the liquid passed through impeller. The NPSH$_{(R)}$ is the positive suction head in meter absolute, required at the pump suction to overcome the pressure drop in the pump (margin of liquid head at suction eye maintained above the liquid vapor pressure = + 0.8 to 1.5 M). The NPSH$_{(R)}$ is always positive, expressed in terms of absolute fluid column height. The term "Net" refers to the actual pressure head at the pump suction flange and not the static suction head.

NPSH$_{(R)}$ is determined through closed loop or with valve suppressed suction while performance test conducted by pump manufacturer at test bench. The manufacturer usually tests the pump with water at different capacities by throttling at suction side. The test consist of lowering NPSH$_{(R)}$ to the point that cavitation of pump is initiated i.e. 3% discharge head reduction in single stage pump or 3% discharge head reduction in first stage impeller of multi stage pumps.

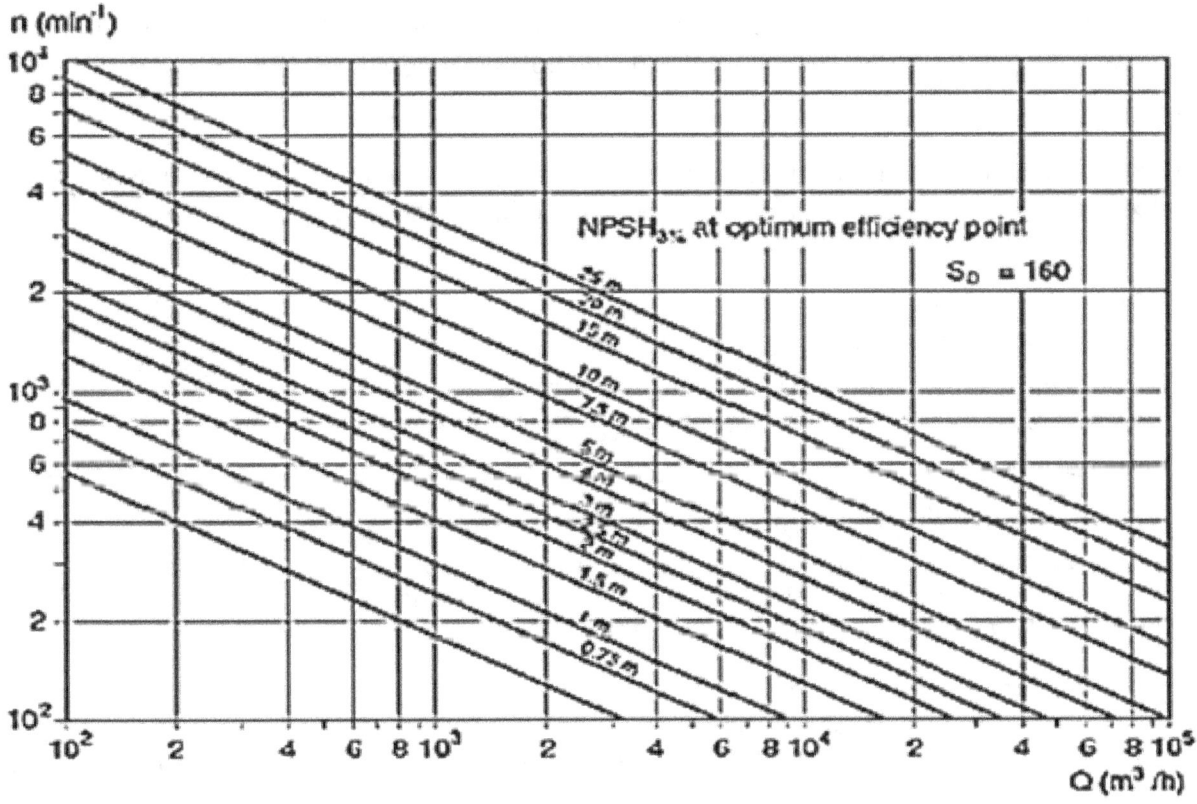

NPSH at 3% head drop for pump impeller between two bearings, Ns =200

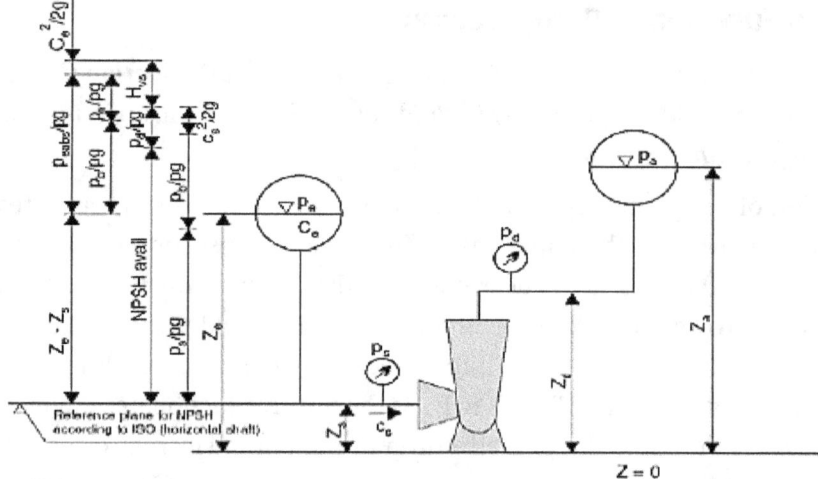

Pumping plant

$$NPSH_{(R)} = \{(Pe_{ABS} - P_D)/\rho * G\} + Ze - Zs + v_e^2/2 * g - Hvs$$

Where:

Pe_{ABS} = absolute pressure on the fluid surface at the intake

v_e = mean flow velocity in the intake vessel (in the suction tank, usually very small).

2.1.3 Physical Principles

- When the pump is operating on the suction lift (water level below the impeller inlet), Ze is negative. The term (Ze + Zs) is then defined as the geodetic suction head.
- If a boiling fluid is being pumped, the pressure above the water level is the vapor pressure (**pe abs** = p_D); the pump can then only be operated with positive suction head (i.e. Ze > Zs).

2.1.4 NPSH$_{(A)}$ Safety Margin Variation

The **NPSH$_{(A)}$** value of the plant defines the total head at the pump suction nozzle available for the given fluid characteristics at a certain flow rate. The **NPSH$_{(A)}$** value of the plant is independent of the pump design.

- **Effect on NPSH$_{(A)}$**
  - Increases with increasing peripheral speed at the impeller eye;
  - Reduced if cavitation-resistant materials are used;
  - Increases with increasingly corrosive media;
  - Depends on the operating conditions of the pump and the type and temperature of the medium being pumped.
  - Stroboscopic observations of cavitation can be employed at suction of impeller to observe bubble formation. To avoid cavitation erosion, the necessary positive suction head is usually provided by a booster pump.

- **NPSH$_{(A)}$ is a function of system at site**

  Net Positive Suction Head available, **NPSH$_{(A)}$** is a function of the system condition of pump at site. It is the excess pressure of the liquid head in M absolute over its vapor pressure plus suction losses at the pump suction eye, to ensure no cavitation of pump. It is calculated based on system or process conditions. **NPSH$_{(A)}$ > NPSH$_{(R)}$**

2.1.5 Selecting NPSH$_{(A)}$ Approximate Values for in the Plant

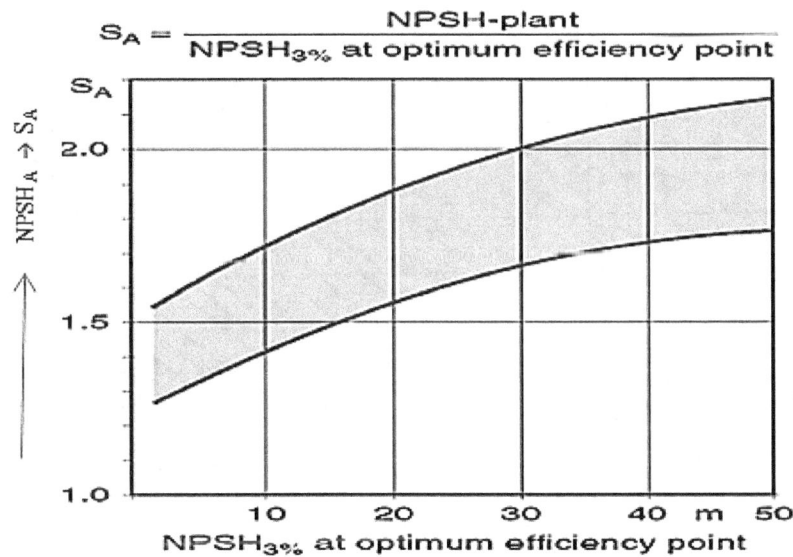

2.1.6 Improving Suction Performance by

– Using special suction impellers;

– Installing a double-entry to first stage impeller of multi stage pump;

– Installing an inducer;

– Using a booster pump.

$$NPSHa_S = hp_S + h_S - hvp_S - hf_S$$

• hp_S - Pressure Head i.e Barometric Pressure of the suction vessel converted to Head

• h_S - Static suction Head i.e.the vertical distance between the eye of the first stage impeller centerline and the suction liquid level.

• hvp_S - Vapor pressure Head i.e. vapor pressure of liquid at its max. pumping temperature converted to Head

• hf_S - Friction Head i.e. friction and entrance pressure losses on the suction side converted to Head

NPSH$_{(A)}$ = Suction Pressure Head – Vapor pressure head of liquid (at given temperature) – Friction and other (valves and fittings) head losses in the suction piping,. "All terms in M absolute."

2.1.7 The NPSH$_{(A)}$

The NPSH$_{(A)}$ can also be approximated by a gauge on the pump suction using the formula:

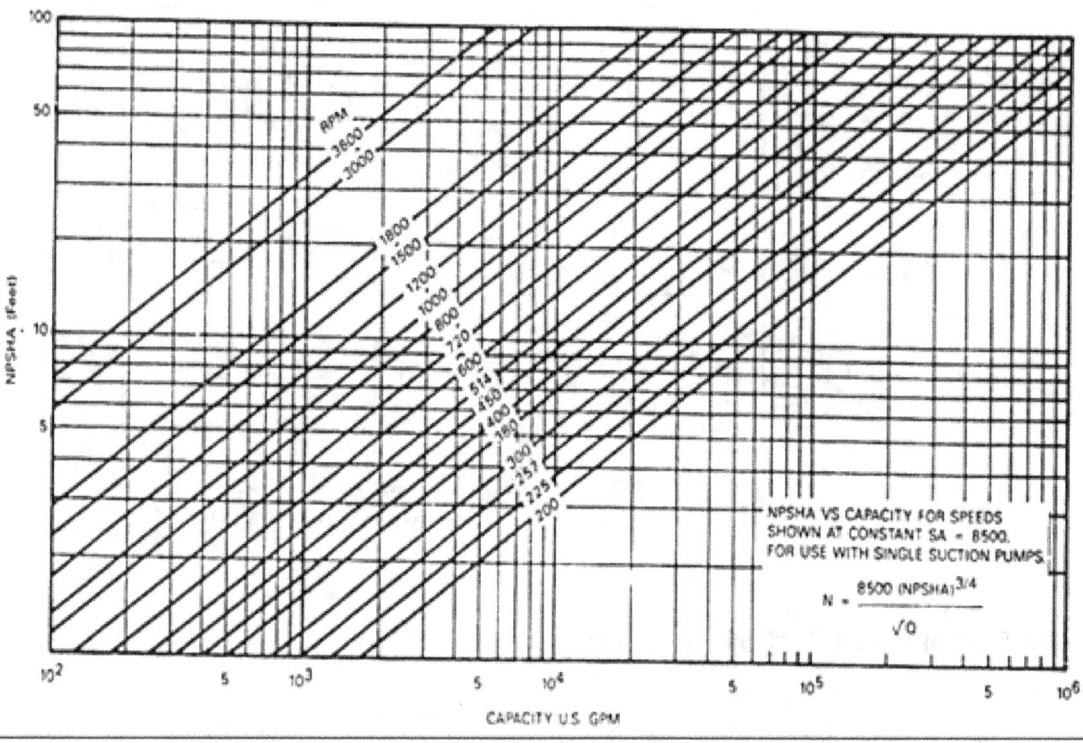

RECOMMENDED MAXIMUM OPERATING SPEED FOR SINGLE SUCTION PUMP

$$NPSH_{(A)} = hpS - hvpS \pm hgS + hvS$$

Where:

- **hvS** = Velocity head in the suction pipe at the gauge connection, expressed in feet.
- **hvpS** = Vapor pressure of the liquid at maximum pumping temperature, in feet absolute
- **hpS** = Barometric pressure in feet absolute
- **hvS** = Velocity head in the suction pipe at the gauge connection, expressed in feet.
- **hgS** = Gauge reading at the pump suction expressed in feet (plus if above atmospheric, minus if below atmospheric) corrected to the pump centerline.

*The NPSH **available** must always be greater than the NPSH **required** for cavitation free operation of the pump. It is normal practice to have at least 0.8–1.5 M of margin on calculated NPSH **available** at the suction flange to avoid cavitation problems due to deviation of duty point. NPSH$_{(A)}$ depends on actual design and configuration of piping system at site.*

- √ NPSH$_{(A)}$ = P − (Vp +Ls + hf)………. for negative suction/suction lift condition
- √ NPSH$_{(A)}$ = P + Lh − (Vp + hf)………for positive suction head condition

Where

- **p** = absolute pressure at surface of liquid (M), barometric reading in opened tank suction
- **Vp** = Absolute vapor pressure of liquid at max pumping temperature (M)
- **Ls** = Max static suction lift (M)
- **Lh** = Max static suction head (M)
- **hf** $_{(M)}$ = Frictional loss of head (M) in suction pipe at required capacity.

NPSHa for Suction Lift Operation

For suction lift operation.

the pump is installed above the suction-side water level. The value of NPSH₃ can be calculated from the conditions in the suction tank (index e) as follows

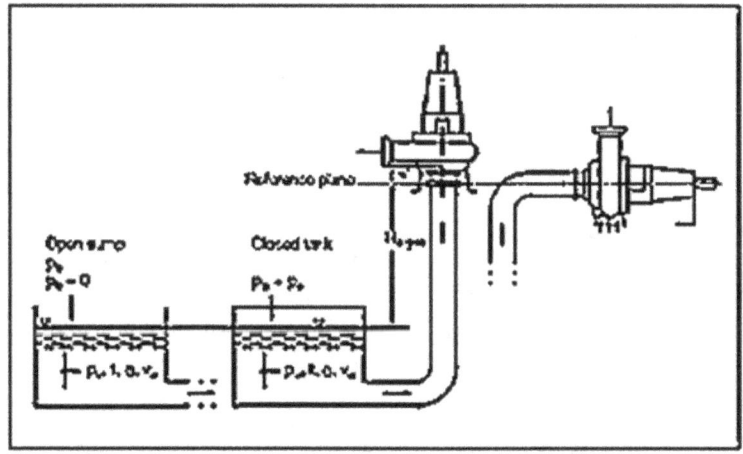

Calculation of the NPSH₃ for suction lift operation for horizontally or vertically installed pumps

$$NPSH_a = (p_e + p_b - p_v)/(\rho \cdot g) + v_e^2/2g - H_{L,s} - H_{s\,geo} = s'$$

Where-

- – Hs = geo Height difference between the fluid level in the suction tank or sump and the centre of the pump inlet in M
- – s' = Height difference between the centre of the pump inlet and the centre of the impeller inlet in m

  pe = Gauge pressure in suction tank in N/m²

  pb = Absolute atmospheric pressure in N/m² (consider effect of altitude!)

  pv = Vapour pressure in N/m² (as absolute pressure!)

  $\dot\rho$ = Density in kg/m³

  g = Gravitational constant, 9.81 m/s²

  ve = Flow velocity in the suction tank or sump in m/s

  HLs = Head loss in the suction piping in m

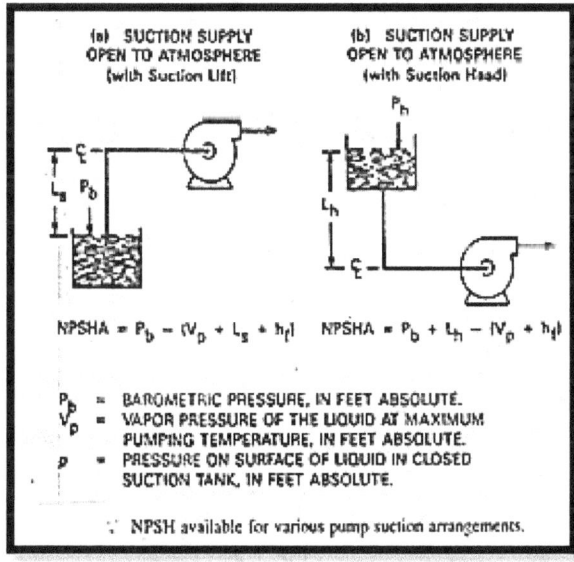

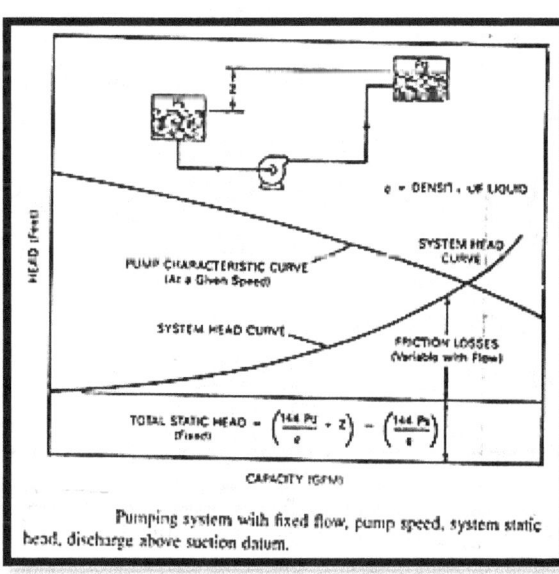

Pumping system with fixed flow, pump speed, system static head, discharge above suction datum.

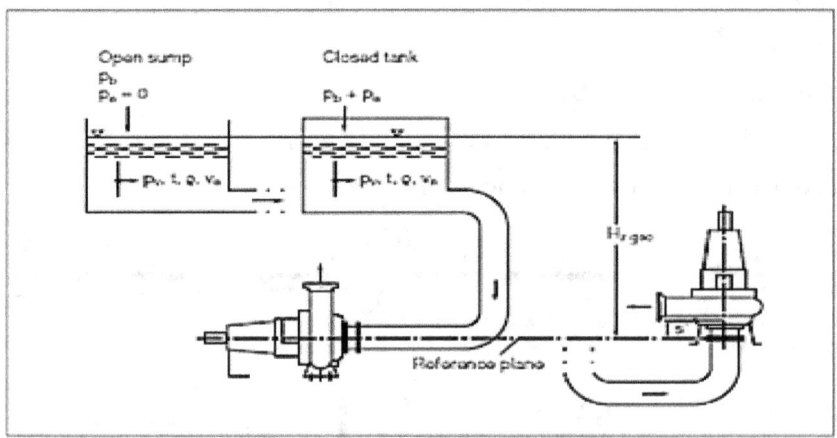

(a) SUCTION SUPPLY OPEN TO ATMOSPHERE (with Suction Lift)	(b) SUCTION SUPPLY OPEN TO ATMOSPHERE (with Suction Head)	(c) CLOSED SUCTION SUPPLY (with Suction Lift)	(d) CLOSED SUCTION SUPPLY (with Suction Head)
$NPSHA = P_b - (V_p + L_s + h_f)$	$NPSHA = P_b + L_h - (V_p + h_f)$	$NPSHA = p - (L_s + V_p + h_f)$	$NPSHA = p + L_h - (V_p + h_f)$

P_b = BAROMETRIC PRESSURE, IN FEET ABSOLUTE.
V_p = VAPOR PRESSURE OF THE LIQUID AT MAXIMUM PUMPING TEMPERATURE, IN FEET ABSOLUTE.
p = PRESSURE ON SURFACE OF LIQUID IN CLOSED SUCTION TANK, IN FEET ABSOLUTE.

L_s = MAXIMUM STATIC SUCTION LIFT IN FEET.
L_h = MINIMUM STATIC SUCTION HEAD IN FEET.
h_f = FRICTION LOSS IN FEET IN SUCTION PIPE AT REQUIRED CAPACITY.

2.2 Influence of Altitude

Influence of altitude above mean sea level on annual average, atmospheric pressure and corresponding boiling point (1 mbar = 100 Pa = 0.1 kPa or 1 bar = 0,1 MPa)

The cold water and open sump at sea level the equation can be simplified for practical purposes as:

$$NPSH_{(A)} = 10 * hLs - Hs\ geo \pm S'$$

2.2.1 Cold Water and Open Tanks

For cold water and open tanks of suction at sea level the equation can be simplified as

$$NPSH_{(A)} = 10 * hLs - Hz\ geo \pm S'$$

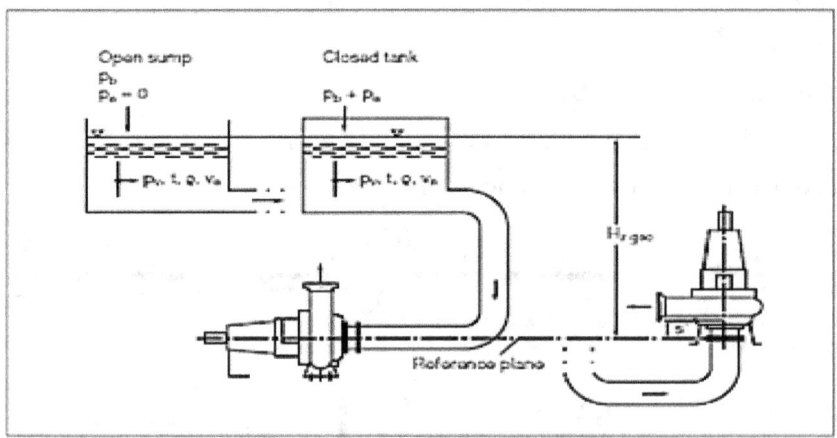

. Calculation of the NPSH_a for suction head operation for horizontally or vertically installed pumps

2.2.2 Pumps Can Pump Only Liquids, Not Vapors

When vaporization in the liquid being pumped does not occur at any process operating condition, the operation of a pump is recognized as satisfactory. The liquid plus vapor results the volume increase abruptly in multifold e.g. 1 M^3 of water at room temperature becomes 1700 M^3 of vapor at the same temperature. It is obvious that for pumping of liquid, media must be maintained in the liquid condition.

2.2.3 Rise in Temperature and Fall in Pressure Induces Vaporization

The vaporization in pumping system begins when the vapor pressure of the liquid at the operating liquid temperature equal to external system pressure in an open suction system i.e. always equal to atmospheric pressure or barometric pressure. Any decrease in external pressure or increase of operating liquid temperature can initiate vaporization and the pump may reduce/stop pumping under cavitation. Hence, the pump always requires a sufficient positive suction head to prevent the vaporization at the pump suction eye.

The NPSH **required** can vary with speed and the operating flow of the pump. On increase of flow capacity, there will be increase in $NPSH_{(R)}$ head as in such circumstances the velocity of the liquid will be increased and system becomes more closed to cavitation. Secondly, on reduction of the pressure (head) also, the velocity of a liquid is increased and render more NPSH **required**. The pump manufacturer's curves, normally furnishes all these information in operating manual. However, the NPSH **required** is independent of the fluid density.

Note: The net positive suction head required, $NPSH_{(R)}$ on the pump curves is given for the fresh water at 20°C and not for the fluid or combinations of fluids being pumped.

2.2.4 $NPSH_{(A)}$ for Suction Head Operation

For computation of $NPSH_{(A)}$ in case of operation with positive inlet pressure/suction head (i.e. the pump installed below the liquid suction datum level), (– hs geo)) is replaced with (+ hs geo) in given equation of $NPSH_{(A)}$

- $NPSH_{(A)} = 10 * hLs – hs\ geo ± S'$ for negative suction head. Thus, the equation will form as
  $NPSH_{(A)} = 10 * hLs + hs\ geo ± S'$ for positive suction head at suction.

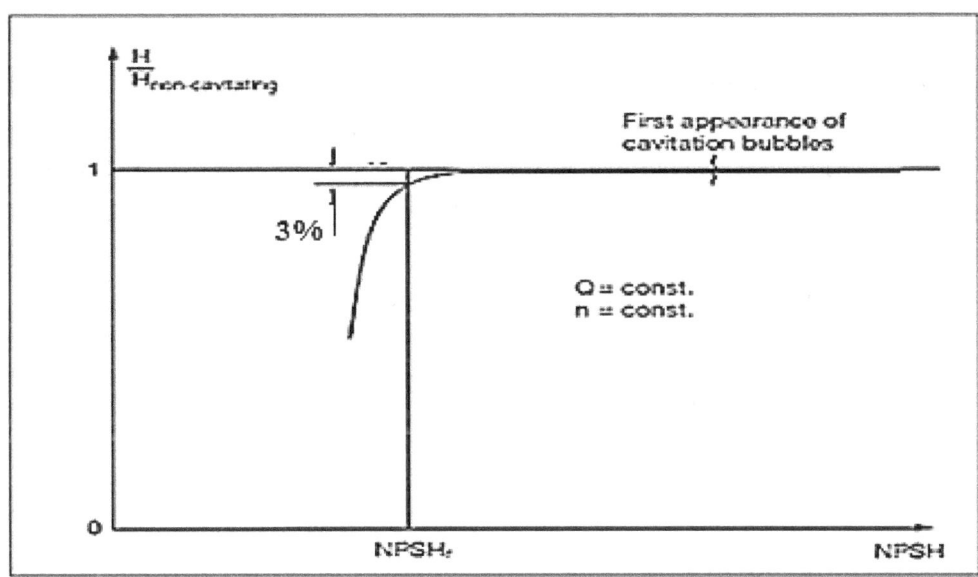

. *Experimental determination of the NPSH$_r$ for the criterion*
ΔH = 0.03 H$_{non-cavitating}$

When suction pressure drops, cavitation bubble starts developing much before the visible effect of hydraulic performance in the pump. The discharge head drop of 3% resulted by cavitation is considered tolerable (shown in fig above). The realization of cavitation can be observed from sound level, material erosion and reduction in the pump efficiency. Immediate course of action is needed when limit of cavitation reaches beyond the specified limit of 3% discharge head loss.

2.2.5 Corrective Measures

- The numerical values of $NPSH_{(A)}$ and $NPSH_{(R)}$ are based on the design geometry of the system and the pump respectively, which cannot be changed at the particular operating point. Alteration in installation of suction system of centrifugal pump will have a financial impact.

- OTHER CORRECTIVE OPTIONS FOR AVOIDING CAVITATION:

a. **Change of working point**

In sketch A_1 to A_2 points on H-Q curve indicate insufficient $NPSH_A$; Case -1 at point 1, have deficit of $NPSH_A$ shown as $NPSH_{A(1)}$. In case -2, $NPSH_{A(1)}$ is increased to $NPSH_{A(2)}$ and pumps useful operating range is increased from Q_1 to Q_2 and operating point B is now reached but in both the cases $NPSH_A$ is just equal to $NPSH_R$, Therefore, on deviation of Q out of Q_1–Q_2 range or other parameters like liquid temperature will lead to cavitations.

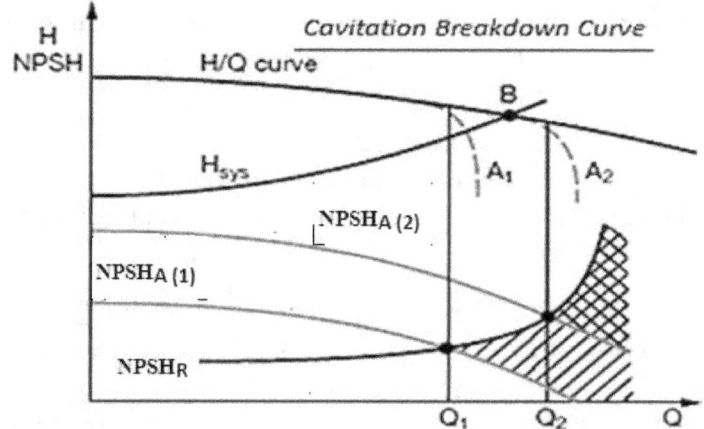

b. **Increase of system pressure**

In closed loop system (eg. heating loop), $NPSH_{(A)}$ can be improved by increasing system pressure (if permit). Alternatively, pump may be installed at lower level or suction system losses are reduced by reducing bends, reducers in suction pipe length to improve $NPSH_{(A)}$.

c. **Application of Inducer**

The $NPSH_R$ can be reduplication of "inducer" as shown here.

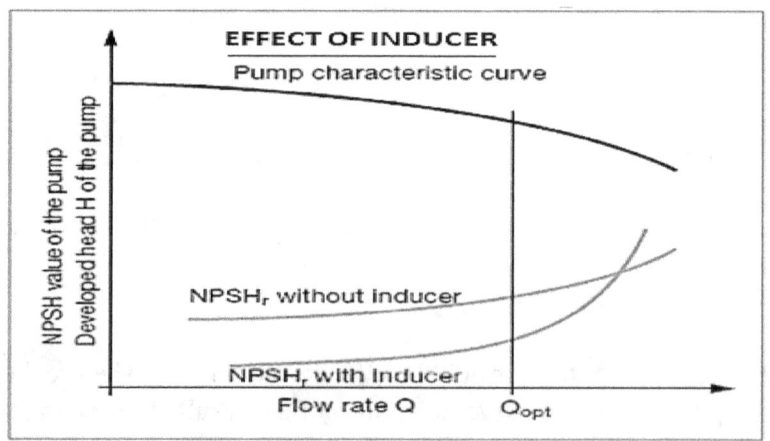

d. Increasing H$_z$ geo or reducing hs geo

- By mounting the tank at higher elevation have no impact on pump flow. In closed flow loops (e.g. heating system), by increasing the system pressure NPSH$_{(A)}$ can be improved, if design of pump can sustain higher pressure.

- Further, by elevating suction tank level or installing the pump at a lower point, minimizing the pressure losses in suction piping (hLs) or replacing the pump with higher NPSH$_{(R)}$ the cavitation/vaporization problem can be resolved.

- Instead of changing pump with higher NPSH$_{(R)}$, using a special low-NPSH suction-stage impeller or installing an inducer/propeller in front of the impeller can improve NPSH$_{(R)}$ within limits. It must be kept in mind that reduction of NPSH$_{(R)}$ by application of inducer is cheaper and convenient.

- The resistance to cavitation erosion can be increased by choosing suitable materials for the impeller, particularly for larger pump sizes.

2.2.6 Impeller Configurations & Spec. Speed Characterizes the Shape of Pump, Operating Head-Capacity Curve

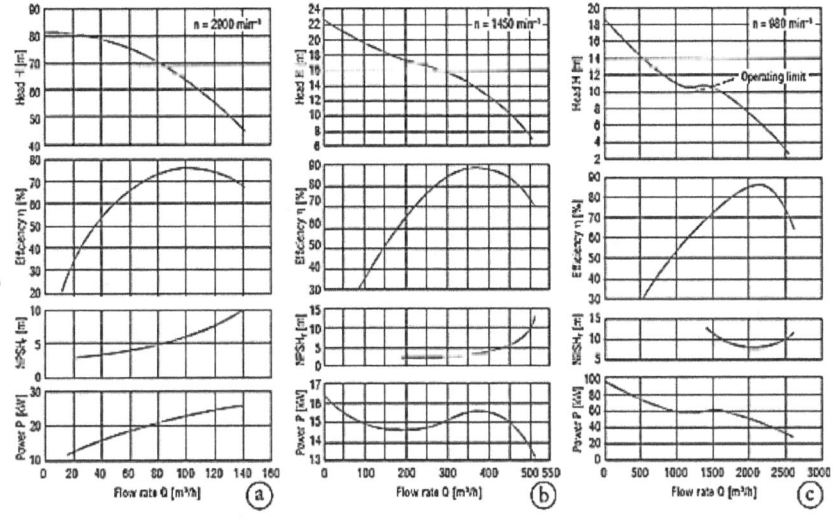

: Three examples of characteristic curves for pumps of differing specific speeds.
a: radial impeller| Ns ≈ 20; b: mixed flow impeller| Ns ≈ 80; c: axial flow impeller| Ns ≈ 200.

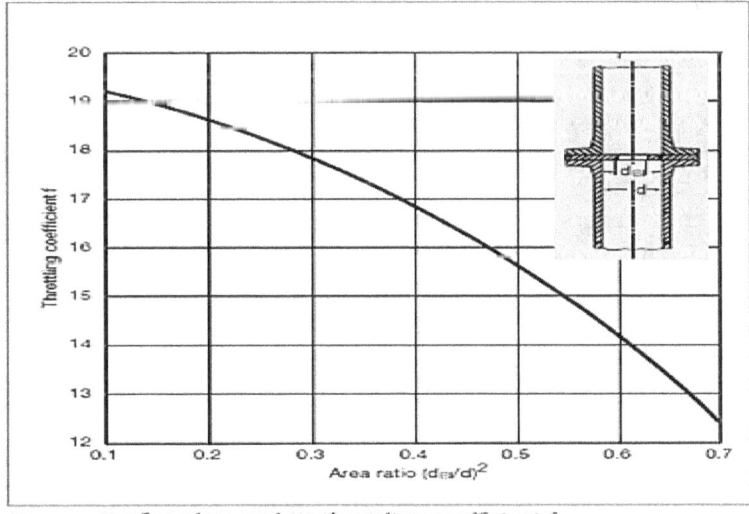

Fig. . Orifice plate and its throttling coefficient f

2.2.7 Cavitation Erosion

When cavitation bubbles implosion takes place on the surface of impeller or pump components, very high local pressure is generated (> than fatigue strength/yield strength or ultimate strength of the material). Thus, pitting of the material, known as cavitation erosion, is resulted. The erosion of material cab be affected by following points–

- Erosion is increased by increasing implosion energy. The intensity of erosion is approximately proportional to third power of (NPSH) or the 6th power of the speed;
- It decreases the tensile strength or hardness of the material with square of increasing implosion energy.
- Erosion increases with growing cavitations bubbles, assuming that the bubbles implode on the surface of the material and not in the liquid.
- Erosion decreases with increase of water temperature.
- Erosion decreases with increase in gas content in the fluid being pumped, as non-condensable gases reduce the partial implosion pressure.
- Increase of corrosive media is usually greater at part load compared to operating at optimum efficiency.

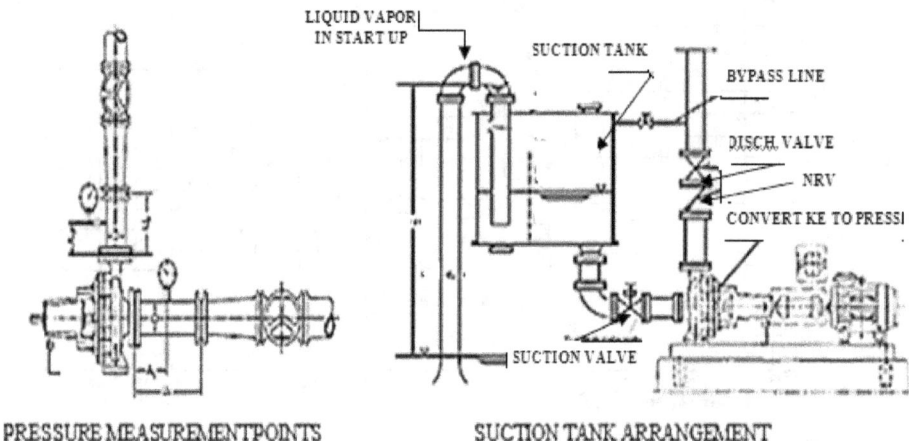

PRESSURE MEASUREMENTPOINTS   SUCTION TANK ARRANGEMENT

2.2.8 Radial Flow Impeller

The direction of flow at exit of the impeller is radial and it has comparatively smaller specific speeds (300 to 1000) which results relatively small amount of flow against high heads. The mixed flow type of impellers has a higher specific speed (2500 to 5000), constituted of large inlet diameter D and impeller width B to handle relatively large discharges against medium heads.

The axial flow or propeller impellers have the highest specific speed range (5000 to 10,000). They are capable of pumping large discharges against small heads.

Chapter 3
CONSTRUCTIONAL FEATURES

3.1 Construction of Centrifugal Pump

The construction of Centrifugal pump is simple, mainly consisting of stationary & rotating parts. The stationary parts are constituted of casing, bearings and stuffing box/mechanical seals whereas rotating parts are mainly impeller, shaft, shaft- sleeve and fitting parts eg. key, lock nuts, bearing housings, fittings etc.

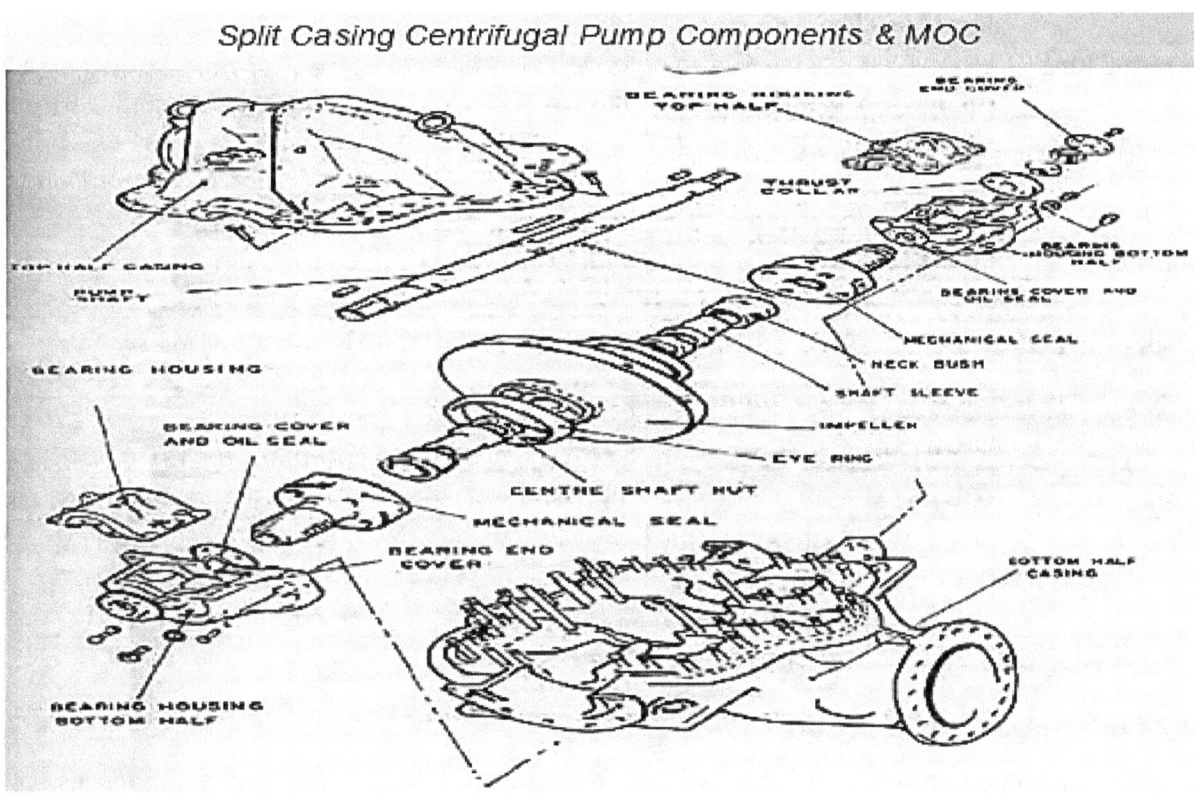

3.1.1 MOC of Pump Components for Water Applications

Pump Components	MOC for different Water Quality		
	Fresh Water	Blackish Water	Sea Water
Impeller	Bz, ASTM B584, alloyC92200 or SSASTMA 276–316	SS-ASTMA 351 Gr CF 3 M	SS-ASTMA 351 Gr CF 3 M
Shaft	SS-ASTM. A. 276 type 410	SS-ASTM. A. 276 type 316	Nitronic-50 AST m A 351-Gr CG 6 MMN

(Contd.)

Pump Components	MOC for different Water Quality		
	Fresh Water	Blackish Water	Sea Water
Shaft Sleeve	SS-ASTM. A. 276 type 410	SS-ASTM. A. 276 type 316	Nitronic-50 ASTm A 351-Gr CG 6 MMN
Shaft Enclosing Tube	SS-ASTM. A. 276 type 316	SS-ASTM. A. 276 type 316 L	SS-ASTM. A. 276 type 316 L
Radial Shaft Bearing	Bz backed cut less rubber	Bz backed cut less rubber	Bz backed cut less rubber
Suction Bell	CI, ASTM A 48 or Carbon Steel ASTM-A 283	SS-ASTM. A. 276 type 316 L	SS-ASTM. A. 276 type 316 L or ASTM A439-D 2 Ni resist
Column and Discharge. head	Carbon Steel ASTM-A 283 or A516	SS-ASTM. A. 276 type 316 L	SS-ASTM. A. 276type 316L or
Impeller Bowl	CI, ASTM A 48 or Carbon Steel ASTM-A 283	SS-ASTM. A. 276type 316 L	SS-ASTM. A. 276 type 316 L
Wearing Ring	Bz, ASTM B584, alloyC92200 or SSASTMA 276–316	SS-ASTM. A. 276 type 316 L	Nitronic 50-ASTM A 351 CGMMN and Nitronic60-ASTM A 351 CGMMN: use alternatively on stationary and rotary parts
Impeller	Bz, ASTM B584, alloyC92200	SS-ASTMA 351 Gr CF 3 M	SS-ASTMA 351 Gr CF 3 M
Shaft	CarbonSteel AISI 1045	SS-ASTM. A. 276 type 316 L	SS-ASTM. A. 276 type 316 L
Shaft Sleeve	SS-ASTM. B 582 type 416	SS-ASTM. A.743 type CF 8 M	SS-ASTM. A. 276 type 316 L
Casing	CI, ASTM A 48	ASTM A439-D 2 Ni resist	ASTM A439-D 2 Ni resist
Wearing Ring	Bz, ASTM B584, alloy C92200 on impeller and CI, ASTM A 48 on casing	SS 17–4 pH	Nitronic 60-ASTM A 351 CGMMN

3.1.2 Material of Construction

Strength, corrosion-resistance, abrasive & wear resistance, casting & machining properties, cost of manufacture are main considerations in the selection of material for centrifugal pump casings:

- Mostly cast iron is the preferred material for pump casings when evaluated against these parameters and initial cost factor. Cast iron produces sufficient strength for the pressures developed by general purpose pumps. For corrosion resistance and handling the volatile products, it may be necessary to use cast steel with internal epoxy coating or cast stainless steels.

- Cast iron castings for multistage pumps are limited to approximately 10 kg/cm² discharge pressure and temperature 175 °C. For temperatures above 175 °C and pressures up to 150 kg/cm², a cast steel is usually preferred material in cast or forged form for split, multi-stage casing pumps. For evaluation of cast iron versus steel casings, erosion during operation is considered a prime factor. Erosion can occur from either abrasive particles 'in- the fluid or from wire drawing across the flange of split-casing-pump. The initial cost of a steel casing is higher than that of a cast iron. However steel casings can often be salvaged by welding and machining of breakages or eroded portions. Salvaging a cast iron casing by welding is not practical and the casing usually must be replaced.

- The ductile irons casings are useful for higher pressure and temperature ratings compared to cast iron and cast steels. The modulus of elasticity for the ductile irons is essentially the same as that for cast iron but tensile strength is approximately double. The ductile iron has substituted the steels for the intermediate pressure and temperature. The ductile iron casing can be repaired by welding and machining. However, it is costly but quite effective in the field.

3.1.3 MOC of Pumps for Different Applications in Power Plant

a. **Impeller**

The, impeller can be described as a wheel having equally spaced *Blades or Vanes*, arranged around the shaft The inlet eye could be one side/either side axially. The blade tips of impeller run in a circular path at a specified tip velocity to impart kinetic energy to the media particles leaving the impeller vane known as centrifugal force.

b. **Criteria for Selection of the Material of the Impeller**

- Corrosion resistance
- Abrasive-wear resistance
- Cavitations resistance
- Strength to resist pressure developed by pump

c. **Selection of Pump MOC based on pH of media**

Components of Centrifugal Pumps

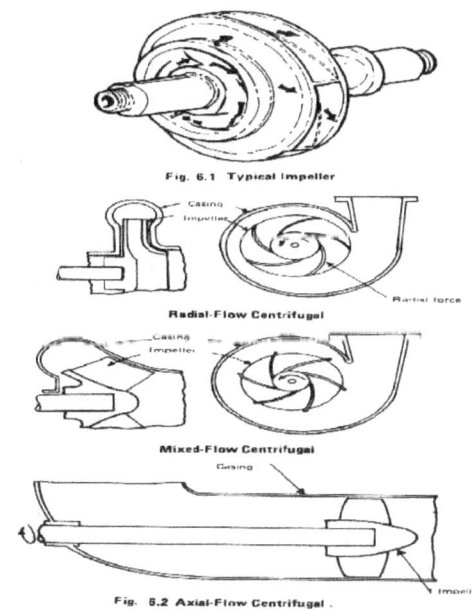

Fig. 6.1 Typical Impeller

Radial-Flow Centrifugal

Mixed-Flow Centrifugal

Fig. 6.2 Axial-Flow Centrifugal

pH range	Less than 3.5	3.5–6.0	6.1–8.0	More than 8.0
Material	Corrosion resistant steel	All grades of bronze	Bz, iron or combinations	iron & steels

3.1.4 Casing

Casing or housing casting of pump restrains water in an approximately circular or spiral path and collect water which is delivered from the periphery of the impeller. It is provided with the inlet/'Suction port,' the outlet/'Discharge port' and stuffing box arrangement to permit the leak proof projection of the spindle or impeller shaft.

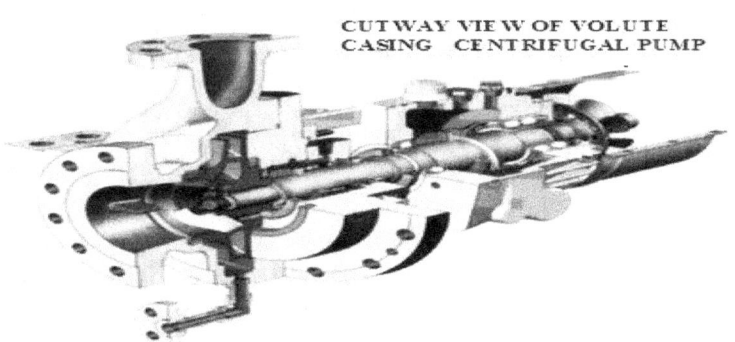

CUTWAY VIEW OF VOLUTE CASING CENTRIFUGAL PUMP

3.1.5 Casting and Machining Properties

Mostly for normal water and other noncorrosive services bronze satisfies the criteria of evaluations. As a result grade 13 bronze is widely used material of impellers for normal services but that is restricted for pumping liquid at temperatures more than 120 °C because the differential expansion between bronze and steel will produce an unacceptable clearance between the impeller and shaft.

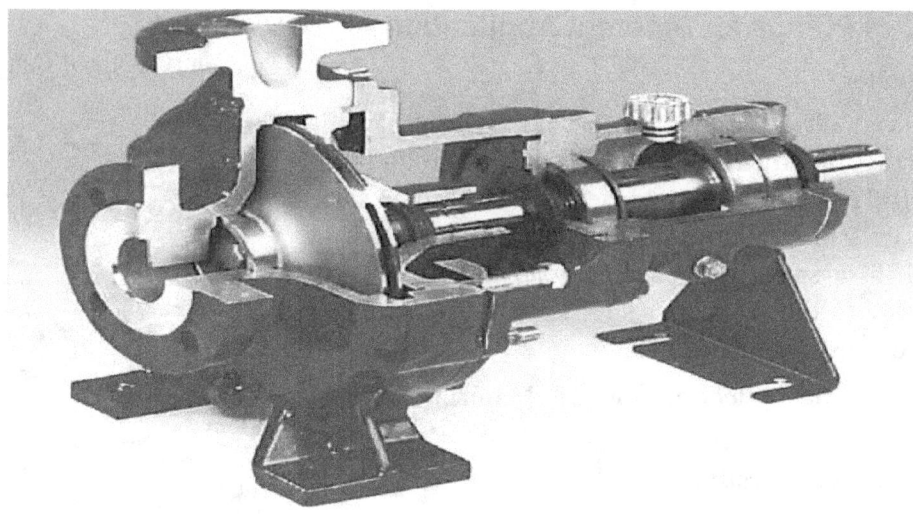

- Cast iron impellers are used to a limited extent in small low-cost pumps. The, cast iron is inferior to bronze in corrosion, erosion and cavitation resistance but low initial cost. The cast iron impeller can be chosen on justification and evaluated merits only.

- Stainless steel impellers are widely used where bronze could not satisfy the requirements of corrosion, erosion, or cavitation-resistance. The stainless steels are not used for sea water, or HCl transfer pumps since, the pitting- corrosion will limit the performance and life of impeller. The stainless steel impellers are used where the pumping temperature exceeds 120 deg C and differential expansion does not pose any problem at such elevated temperature of media.

- The austenitic stainless steel materials are the next option for corrosion and cavitations resistance but costly. Initial cost vs. increased performance life cost can be evaluated for breakeven point and based on that it may be selected.

3.1.6 Shaft

The basic function of the shaft is to transmit the torque and supporting the impeller and other rotating parts. The impeller is keyed to the shaft and shaft is supported at either ends by bearings. In overhung impellers are supported at one end of shaft with two bearings mounting.

- Criteria in selection of the material for **a** centrifugal pump shaft:

   Tensile strength & Endurance limit

   √   Corrosion resistance

   √   Notch sensitivity

- The endurance limits is the stress value below which the shaft will withstand an infinite number of stress reversals without failure. Since, shaft is subjected to fatigue stress, its stress level should always be within endurance limit otherwise, it may fail after application of some stress cycles.

Since, one stress reversal occurs in each revolution of the shaft this means that, ideally at least, the shaft will never fail if the actual maximum bending stress in the shaft is less than the endurance limit of the shaft materials. In actual practice. If corrosion and stress raisers such as threads, keyways and shoulders on the shaft are present or expected, the endurance limit is substantially reduced. In selection of the shaft material, the corrosion resistance of shaft material for the fluid being pumped as well as the stress concentrations and notches sensitivity should be considered in design stage.

3.1.7 Shaft Sleeves

Pump shaft usually protruded from casing is provided with, shaft sleeve& stuffing boxes between shaft and casing to prevent leakage from stuffing box area.

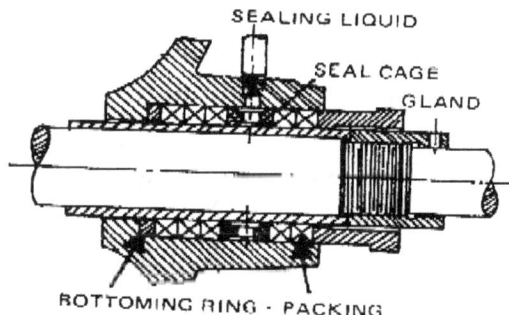

Shaft sleeves also serve to protect inter stage leakages. At some places it may function as distance sleeve also between the two stages in multi stage pump to form an inter stage leak proof joint.

3.1.8 Bearings

The function of bearings in centrifugal pumps is:

1. Reducing frictional force.
2. To keep the shaft in correct alignment with the stationary parts in radial and transverse loads.

In horizontal pumps, the bearings are usually designated as inboard and outboard. Inboard bearings are located at coupling end of pump housing. Due to heat generation by the bearing itself or heat transfer from the pumped liquid (more than 120 °C), the bearing temperature increases. To maintain bearing temperature within limit (**40 deg** C to **60 deg** C) either forced feed lubrication, oil ring lubrication or self lubrication is adopted. The cooler or Jacket cooling can also be useful for bigger pump bearings. Depending on the service conditions, selection of bearings are made.

3.2 Pump Sealings

Sealing's are provided to prevent any leakage at the point where the pump shaft protrudes out through the casing. The pump seals prevent air ingress into the pump if pump pressure is less than atmospheric and also function to prevent the liquid leaking out of the pump when pressure is above atmospheric.

Seals serve to provide leak proof segregation or segregation with minimum leakage between spaces at differential pressure.

3.2.1 Static Seals

Static seals provide surface contact between two sealing faces which are motionless or in limited axial/ radial motion in relation to each other. Such movements may be resulted by differential thermal expansion etc. Pressure-retaining parts are typically sealed with different type of seals depending on service condition and design factors:

   a. O-rings

   b. Flat gaskets, often with wire mesh inlay

   c. Spiral graphite seals

   d. Metallic seals/spiral metallic seal (graphite impregnated).

Seals (b), (c) and (d) can seal only in axial- direction and require considerable preload to perform their function. This affects the sizing of the fasteners.

(a) O-rings are capable of sealing axially and radially with small preload, but like U-packings they too can be employed within limited service temperatures. When absolute sealing is required, seal welds are provided. In certain cases, a barrier seal is sufficient. For such cases, two single seals are fitted in series, and a sealing fluid is forced into the space between seals at a pressure slightly higher.

3.2.2 Shaft Seals

Shaft seals provide sealing between rotating and stationary faces and acts to seal from inside of the pump to the atmosphere. Shaft seals may be subdivided into non-contact and contact types.

   a. **Non-contact shaft seals:** the simplest kinds of such seals are throttle- bush- seal which form a small clearance between the stationary bush and the rotating sleeve. The clearance is kept as narrow as possible to minimize leakage.

     In floating ring seals, reduction of the clearance reduces leak rate. Casing and shaft deformation are taken care by radial displacement of the individual floating rings while starting, allowing the clearance to be matched to the maximum shaft vibration amplitudes.

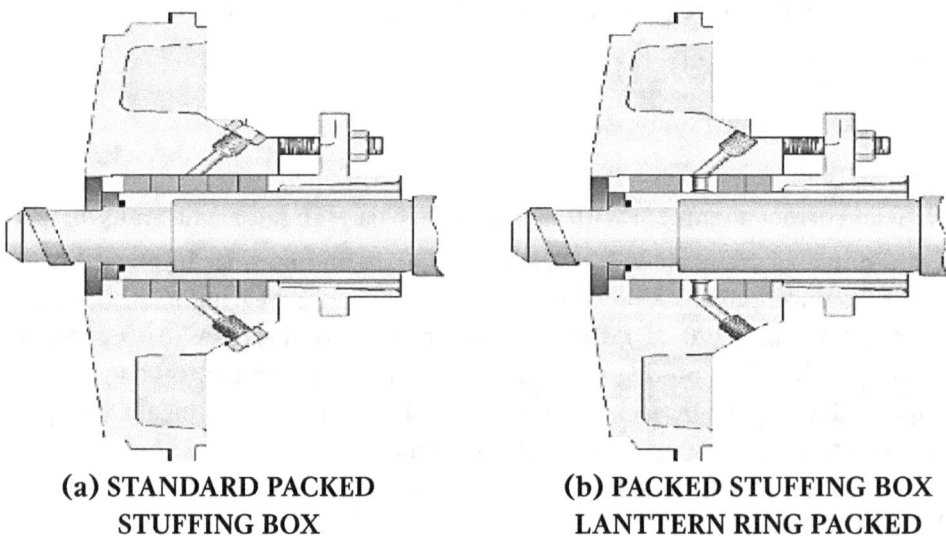

(a) STANDARD PACKED STUFFING BOX   **(b) PACKED STUFFING BOX LANTTERN RING PACKED**

   b. **Contacting shaft seals**

     This is simplest type seal, used popularly in centrifugal pump. It is known as packed stuffing box. Where stuffing box seals are no longer adequate on account of excessive pressure, circumferential speed or some other reasons (like high leakage rate) then rotating mechanical seals are employed. Both single- or double acting seals are used in centrifugal pumps.

Mechanical seal requires cooling and lubrication by clean liquid in order to achieve long life of seal. Small leakage rates are expected from mechanical seals. If the pumped liquid is and compatible with environment, single mechanical seals are normally used. These seals are flushed with liquid drawn from pump discharge. Coolers can be installed in the flushing piping to cool hot pumped liquid and cyclone separator to clean it. Jacket cooler is located inboard of the mechanical seal if the mechanical seal has to be protected from hot liquid in standstill position.

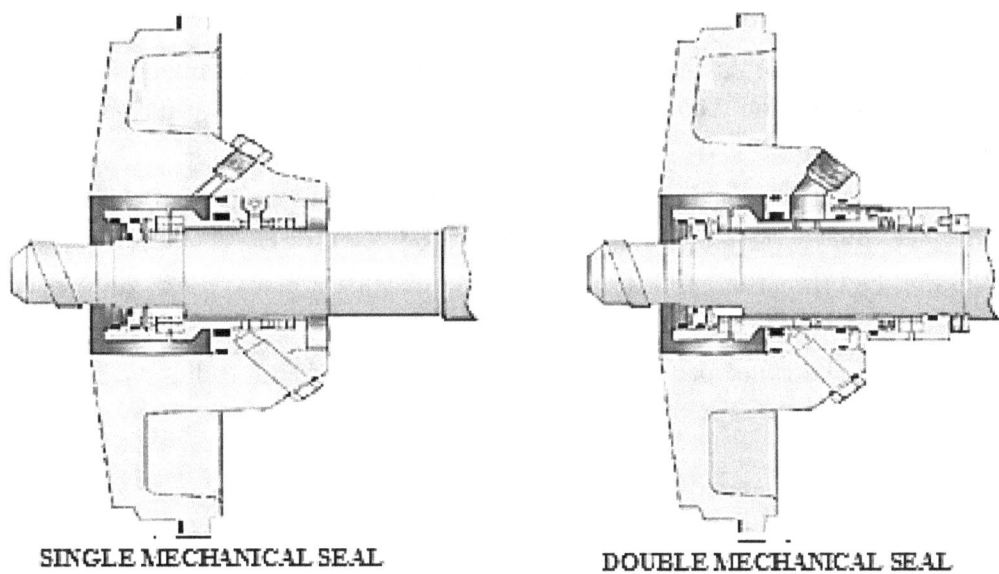

SINGLE MECHANICAL SEAL   DOUBLE MECHANICAL SEAL

Double mechanical seals are normally used if the condition for the single mechanical seals as explained above is not acceptable. Double mechanical seals need to be flushed with clean and cool sealing liquid which should be compatible with the pumped liquid and the environment. The sealing liquid is provided from an external source at a little higher pressure than the pressure to be sealed so that contamination of sealing liquid from pumping liquid is avoided.

3.3 Stuffing Box

A stuffing box consists of a number of rings of packing around the pump shaft, housed inside a cylindrical recess between the pump casing and the pump shaft. By adjustment of gland nuts in the axial direction, the packing are compressed to give the desired fit on the shaft. For sealing and cooling the gland packs, a lantern ring is provided at middle portion of pack length. Thus, lantern ring separates the packing rings in two equal sections. The gland cooling liquid is injected at centre location of lantern ring which passes through the holes of lantern ring to shaft and creep out through gland packings rendering a cooling effect to packings (heat generation by friction between shaft & glands). The best gland packing adjustment allows trickling leakage of water droplets.

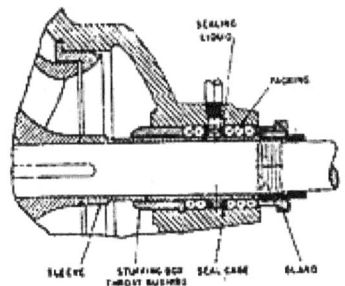

Conventional stuffing box with throat bushing.

Lantern Ring

- PACKINGS
  - √ Gland packings are normally supplied in spiral spool/coil, or die formed rings are made to specified dimensions. When it is available in continuous length, it is necessary to cut off in length to make the required length of rings with 45° end joint.
  - √ Place the packing around the shaft, or on a mandrel of the shaft diameter. The bore of specified packing spirals should conform to shaft diameter and OD to stuffing box bore for correct section of packing.
  - √ To assist in cutting packing rings, two guide lines parallel to the shaft axis and separated by a distance equal to the packing section may be drawn on the spiral coil of packing.
  - √ Cut the rings from the spiral at an angle of 45° diagonally across the guide lines to have no gap between the ends of cut ring.
  - √ Metallic and extruded packing rings are, spirally opened and readily available for fitting.

3.3.1 Fitting of Packings

1. Check the shaft and ensure that it turns freely on its bearings.
2. Fit each packing ring individually one after another for length and joints matching
3. Ensure that two adjacent packing joints are staggered by 120° or180°. After fixing each packing ring, turn shaft about one round for checking freeness.
4. Ensure that shaft turns freely after fixing each packing ring in position.
5. If lantern ring is provided, it must be fitted in correct position matching with centre line of gland cooling inlet pipe hole, allowing 2–3 mm compression of the inner packing rings. Bring the gland follower squarely against the last packing ring and tighten the nuts equally in sequential order. Keep on turning shaft by hand to see that it does not jam while tightening the follower nuts. Final micro adjustment of nuts is done in running condition of pump with gland cooling liquid pressure on. Allow trickling droplet leakage from gland.

3.3.2 Summary of Packing Types and Applications

SN	Type	Material	Construction	Temp. Max °C	Press. Max bar	pH	Suitable Media
1	Natural fibre	Natural fibre Cotton	Woven/braided	90		6–9	Water
		Hemp	Woven/braided	80		5–9	Water
		Flex	Woven/braided	70–120			Water
2	Synthetic fibre	Nylon, rayon etc.	Plaited or braided impregnated with PTEE	Application temp. < 120	10–15		Not used for valve packing

SN	Type	Material	Construction	Temp. Max °C	Press. Max bar	pH	Suitable Media
3	Asbestos	White asbestos	Planted or braided with graphite, mica or, oil lubricant	−30 to +300	10−15	5−12	Steam, liquors, etc
	Asbestos (reinforced)	White asbestos, inconel wire reinforcement	Plaited or braided with graphite, mica or oil lubricant	−50 to +750	250 to 650	4−11	Water, steam solvents, Hydrocarbons incorporating acids and alcohols
	Asbestos graphite	Asbestos fibers mixed with graphite	Wet- spun, dust free type preferred as smoother than dry mix.	−40 to +300	10−15	5−12	High temp. and steam services
4	PTFE	PTFE yarns or tape	Planted or braided	−200 to +250		0−14	Water, steam, hydro-carbon, soft packings
	PTFE yarns with lubricants						
5(a)	PTFE	PTFE yarns, treated with PTFE dispersion	Plaited or braided	−200 to +300	100	0−14	All media, but little used for valves.
5(b)	PTFE/ graphite	PTFE/graphite mix	Solid extrusion	+250	100	0−14	Water, oils hydrocarbons alkalis, acids, alcohols etc.
5(c)	PTFE/ aramid	Aramid fibers treated with PTFE dispersion	coated fibres braided or plaited and impregnated with lubricant	220 to +300	200 to +1000	1−14	Not particularly suitable for valve stem seals.
6(a)	Expanded graphite	Pure Expanded graphite	Tape form	−200 to +600	300	0−14	High Temperature services sealing gases
6(b)		Pure expanded 'graphite	Flexible plait	−200 to +600	300	0−14	Low Viscosity fluids: all services requiring superior leak tightness

(Contd.)

SN	Type	Material	Construction	Temp. Max °C	Press. Max bar	pH	Suitable Media
7	Carbon fibre	Amorphous carbon yarns treated with graphite powder	Twisted or plaited	-200 to +600	–	0–14	High Temp. services but little used for valves
8	Glass fibre	Glass fibre yarns	Braided with added lubricant			–	Corrosive media (except strong alcohols)
9	Alumina silicate	Alumina silicate	Plated with or without inconel wire reinforcement	1260	–	0–9	Extremely high temperature services.

3.3.3 Gland Packing Procedure

1. Preparing the equipment – Shut off the stuffing cooling water and drain water. Remove the gland follower nuts and clear the old packings.

2. Carefully withdraw the old packing, using paired extractor tools of the correct size, placed on opposite sides of the shaft. Remove all waste of the old packings and wipe off the stuffing box by air pressure.

3. Check the shaft concentricity with the stuffing box bore.

4. Check the shaft run out to be within ± 0.0125 mm (± ½ mil)

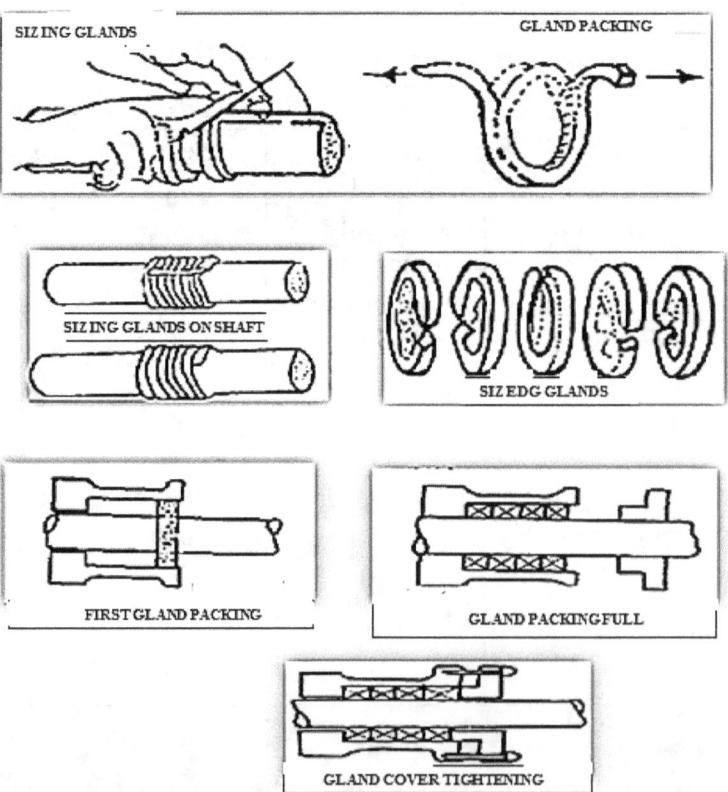

5. The shaft surface in gland area must be free from scores, pitting, grooves, dents, scratches or ridges.

6. Examine the gland follower for general condition and fitting. The inner radial clearance should be 0.25 mm to 0.4 mm and the outer radial clearance should be within 0.25 mm to prevent cocking or rubbing the shaft.

7. Check the clearance between the neck bush and shaft to be within 0.25 mm radial. It may be advantageous to employ a tin, a close clearance spacer ring in the bottom of the stuffing box, to prevent risk of packing extrusion.

8. Ensure that first ring fit properly into stuffing box, before cutting further rings.

3.4 Mechanical Seals

The stuffing box cannot eliminate leakage of fluid in comparatively high pressure pumps. The lubrication and cooling of the shaft and the packing may not be effective. The packing require periodic tightening and adjustments. The frictional losses between shaft and gland packing produces enough heat as energy loss. **Mechanical seals** have replaced gland packing to overcome the loss of friction. In mechanical seals, sealing surfaces are located in a plane perpendicular to the shaft.

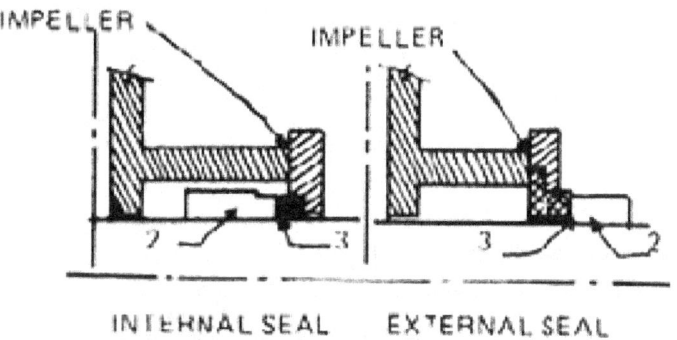

Three Sealing Points in Mechanical Seals

• This seal essentially consists of two, highly polished surfaces running adjacent, one surface moving (placed on the shaft) and the other one stationary (fitted into the pump casing). The lapped surfaces are made of dissimilar metals, held parallel and have regular contact by spring(s) force to form liquid seal between the rotating and stationary members. The frictional losses are negligible and liquid in cotact with scaling surfaces serve as lubricant as well as provide a cooling effect to remove small frictional heat.

3.4.1 Cooling of Mechanical Seals

• There are a number of reasons for cooling mechanical seals:
   √ To prevent the destruction of the liquid film between sealing faces due to temperature rise
   √ To prevent vaporization of the liquid at the seal faces
   √ To protect the seal faces by venting overheating and provide continuous flushing of surfaces.

• Basically there are two types of seal arrangements;
   √ Internal assembly is the one of which the rotating element is located inside the box and is in contact with the liquid being pumped
   √ External assembly the rotating element is located outside the box.

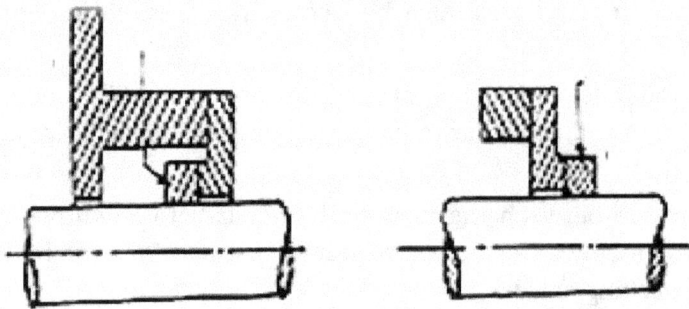

3.4.2 For Internal and External Mechanical Seals Some Important Guide Lines are given as

- The casing joint is sealed by conventional gaskets or some synthetic 'O' ring.

- Between the rotating element and shaft sleeve surfaces are sealed by 'O' rings,/some form of flexible wedges.

- Leakage between the mating surfaces of rotating and stationary seal element cannot be stopped 100% but that is reduced significantly by maintaining a very close contact and dynamic liquid film sealing.

- Whenever the temperature of liquid being pumped is above rated temperature of mechanical seal, the thin film of liquid entrapped between the seal faces flashes as steam on rise of temperature of seal faces when frictional heat is not being carried out fully. In such situation, the seal faces wear out rapidly.

  Therefore, it must be ensured that temperature of liquid around the seal area is limited to 70 deg C. However, while pumping liquid above 70 deg C, arrangement is made to cool the seal faces either by regular flushing with external cold water or by provision of a close circuit external cooler in sealing system.

3.5 Wear – Rings/Impeller-Neck-Ring

It is used between the pump casing-and impeller neck to provide a leak tight dynamic joint. Neck ring is a renewable part when clearance between impeller neck and ring increases beyond specified limit. The neck ring is normally fixed to the casing but in some cases it is fixed with impeller. In some rare cases, double rings are provided on impeller and casing both.

- There are various types of wearing ring design depending on the liquid to be handled, the differential pressure across the leakage joint, the running speed and the pump design. The various designs are adopted accordingly. Wear/neck ring and impeller materials combination should be non- seizing & non Galvanic series e.g. bz. with C I, steel with bz or SS with CI.

- These designs are based on the variation of the leakage points to provide resistance to leakage& flow from high pressure (discharge side) to low pressure (suction side) etc.

- The clearance between the wearing ring mounted on the casing or casing ring and the impeller is a function of the leakage joint diameter as shown in the figure. Following graph present guidance for tolerance of manufacture-

  i. For impeller neck + 0.00" and minus value as per graph and for neck ring + value as per graph and negative value 0.00".

  ii. The diametrical clearance is calculated as per table below &clearance developed due to tolerance of manufacturing.

Leak joint diameter (inch)	2 to 4	4 to 6	6 to 12	12 to 20	20 to 30	30 to 48	48 to 80
Diametrical clearance (inch)	0.012	0.014	0.016	0.018	0.022	0.026	0.030
Impeller neck tolerance (inch)	+0.00, −0.002			+0.00, -0.003		+0.00−0.004	+0.00−0.005
Neck ring tolerance (inch)	+0.002, −0.00			+0.003, −0.00		+0.004−0.00	+0.005−0.00

3.5.1 Neck Ring Maintenance

Most of the neck rings are now pressed on the impeller. Since, distortion may occur during the mounting process, it is advisable to check the shaft and impeller assembly on centers to check that new ring surfaces are true otherwise turn the wearing surface to the proper size after mounting.

- **Clearance**

    If the metals gall are used (like the chrome steels), provide the galling allowance = 0.002". In multistage pumps, the basic diameter clearance should be added with 0.003 in. for larger rings. The tolerance indicated will be 'plus' (+) for the casing ring and 'minus' (−) for the impeller hub or impeller rings in a single-stage pump with a joint of non galling components. The correct machining tolerance for a casing ring diameter can be specified in design. Further, the tolerance on impeller hub or ring is specified. Thus, max clearance between neck and casing wear ring will be between 0.018 and 0.024 inch.

- **Allowable Weared Clearance**

    It is difficult to generalize on the amount of wear allowable before a pump is dismantled and the wear ring clearance checked. It is advisable to replace neck rings based on guide lines given here because too many factors are involved in evaluation of pressure loss. Internal leakage through the rings, naturally means lower pump efficiency. Ring renewal could be solution while overhauling because cost involved will be lesser than that of power saving. Even though the clearance is not excessive and the pump can be reassembled without renewing the wearing ring joint, always check the empty neck diameter and the inside diameter of the casing for evaluating wearing ring eccentricity due to wear.

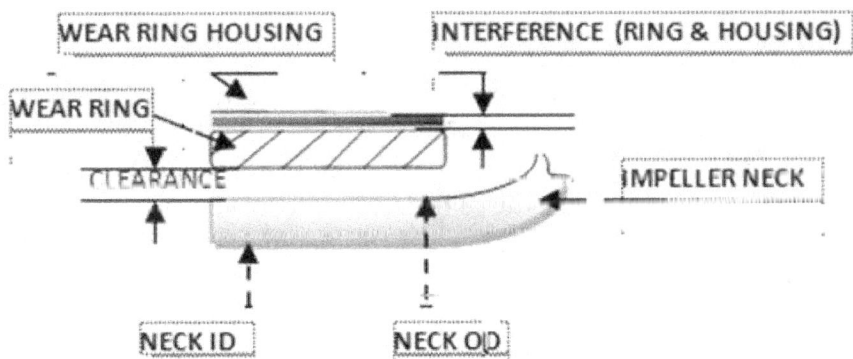

3.5.2 Measurement of Clearance

Wearing ring clearances may sometimes be measured by inserting a feeler gauge leaf. The wearing ring of L-type prevents insertion of feeler gauge. The clearance may be approximately checked without dismantling the rotor, in the following manner:

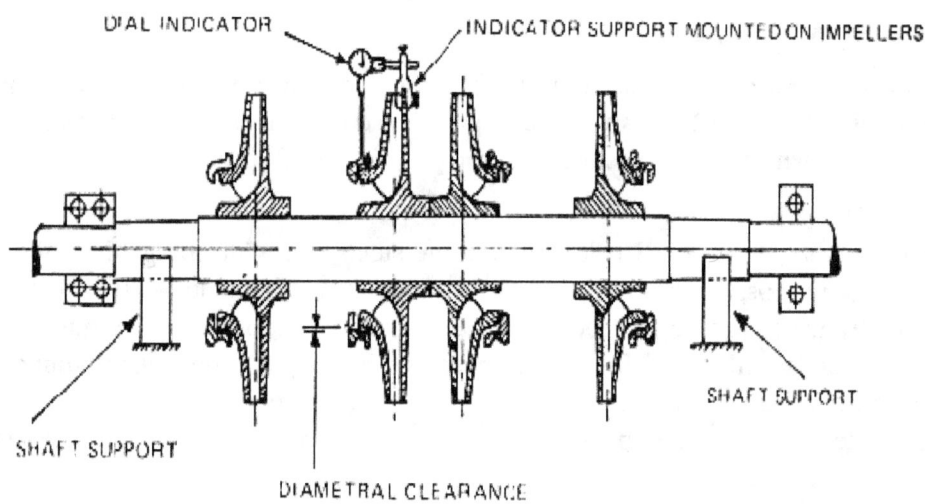

i.   Mount a dial indicator base on the impeller and stylus resting on stationary/casing of impeller neck, set the dial reading to zero.

ii. Without moving the impeller or dial indicator, push up the stationary ring from bottom side and record the maximum dial reading. This corresponds to the diametric clearance.

iii. Repeat this operation for every clearance joint and make a record of all readings. However, removal of rotor from the pump casing, is the best method for multi stage pumps because the stationary rings may be freely removed and the clearance determined by measuring the two diameters and calculating the difference.

3.5.3 Warning Notes

• This short-cut method gives no clue to the condition of clearance surfaces. In other words, burrs, grooves, or indentations, foreign matter passing through the clearances will go undetected. If the pump has been dismantled, normal procedure of measuring the ID of the wearing ring and the OD of the impeller neck can be measured by inside and outside micrometers respectively. Several measurements will determine whether or not the wearing ring or impeller has become worn in oval shape. The clearance is considered to be maximum difference of the maximum ID and the minimum OD readings.

• Clearances can also be measured directly by placing the impeller within the wearing ring and moving impeller neck laterally against a dial indicator to determine the total diametrical clearance. To determine inequality of wear around the circumference, the impeller should be rotated and the dial indicator rolling on several points of the stationary part. If the pump has been dismantled, the diameter difference' method is more reliable. The impeller and wearing rings should be at the same temperature while measurements are made.

• Some high pressure and high-temperature pumps usage shrunk-impellers on shaft which require heating of impeller to at least 204 °C or possibly to max. 260 to 316 °C to remove from shaft. Before taking any measurement this should be allowed to cool down to about 49 °C. But if the wearing ring is at 27 °C (say), and impeller neck at 29 °C deg C (differential temperature) that is considered quite significant for error. This error will, of course, be magnified if the impeller diameter is measured when its temperature is even higher than the 49 °C. The possibility of error due to differential temperature is frequently over-looked, as some people assume that such a small difference in metal temperatures does not have any consequence.

3.5.4 Restoring Wear Ring Clearances

a. **When no rings are used:**

To restore the clearance between impeller and casing, when no ring is provided, consider one of the following steps–

i. Buy new parts.

ii. Build up worn surfaces by welding, metal spraying and machining.

iii. Install a wearing ring or rings if sufficient metal is available in the 'casing and impeller neck.

b. **Pumps with single rings:**

There are three ways to restore the clearance of a pump with single flat or L- type wearing ring.

i. Obtain a new casing ring bored undersize from the manufacturer. Then, true up the impeller wearing ring hub by turning on a lathe.

ii. Build up the worn surface of the wearing ring by welding or metal spraying so that undersized bore can be made. Then true up the wearing ring and impeller neck diameter.

iii. True up the wearing ring by boring oversize, buildup the impeller wearing ring hub, and machine to give correct clearance with the re bored ring.

The last two methods are difficult and are only practicable with larger pump provided if facilities allow that work could be done in the premises. Usually building up of impeller wearing ring hub by welding is also very difficult, and double ring construction can be preferred. The first method is generally the best.

c. **Pumps with double rings:**

If the pump has double flat or L-type wearing rings, the clearance may be renewed by one of the following methods:

a. Obtain a new oversize impeller ring and use the old casing ring bore machining to larger size.

b. Obtain a new casing ring bored undersize and use the old impeller ring turned down.

c. Renew both rings if necessary.

d. Build up either the casing or impeller ring by welding or metal spraying and machine the other part. By altering the ring buildup the original leakage joint diameter can be closely maintained

3.6 Axial Thrust & Hydraulic Balancing

Axial thrust is the vector summation of unbalanced forces acting on impeller in the axial direction. The ordinary single suction radial flow impeller, with the shaft passing through the impeller eye, is subjected to axial thrust. The unbalance force can be visualized from fig shown here. Since, a portion of the front wall is exposed to suction pressure, hence the net axial force will be acting towards the suction side which is expressed as $(P_d' - P_s) \times$ hub area, (where P_d' = discharge pressure and P_s = suction side pressure).

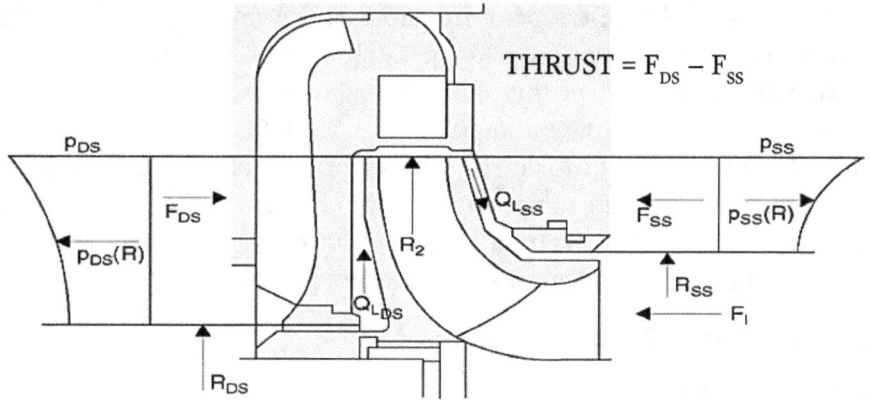

$$\text{THRUST} = F_{DS} - F_{SS}$$

Axial forces acting on the impeller

3.6.1 Axial Thrust

- **Axial Thrust in Double-Suction Pumps**

  Double-suction impeller is axially balanced i.e. the pressure on one side equal to that on the other side. Though this is theoretically true, but in practice, small axial unbalance still persists. To compensate the residual axial unbalance force, pumps are incorporated with thrust bearing(s) at NDE.

- **Axial thrust in multistage pumps**

  To eliminate the axial thrust of a single -suction impeller, pump is provided with both front and back wearing rings. Thrust is, equalized by providing the inner diameter rings, made of same pressure rating, (approximately equal to the suction pressure). It is maintained in a chamber located on the impeller side of the back wearing ring through drilled balancing holes through the impeller. Leakage past the back wearing rings is returned into the suction area through these holes. In case of large single-stage single-suction pumps, balancing holes are replaced by a piped connection to the pump suction.

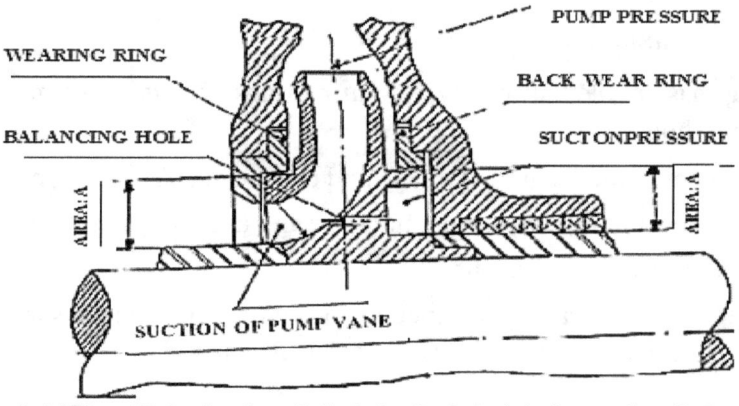

Axial Thrust Balancing through the balancing holes (single stage Impeller)

- From flow through the clearance between the impeller and the casing, in multistage, single-entry pump, the axial thrust increases with neck seal wearing in course of time. For such pumps, the axial position of impellers in relation to the diffuser is affected by manufacturing and assembly tolerances and ultimately variation of axial thrust.. Since, the forces acting on the impeller side plates are large in comparison with the residual thrust, even small deviation in tolerance may cause appreciable change in residual thrust.

3.6.2 The Axial Thrust Can Be Reduced or Partial Compensation By

a. Balancing holes and seals on the hub

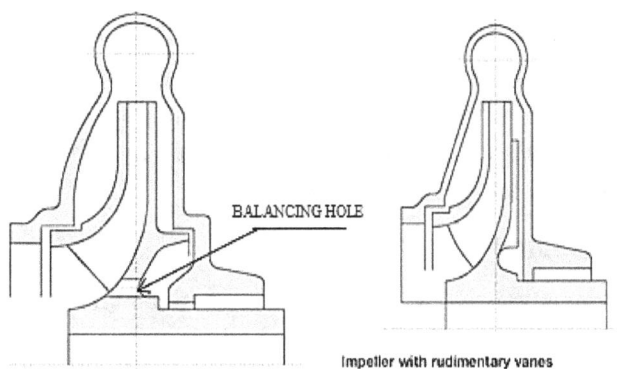

b. Balancing pistons or seals

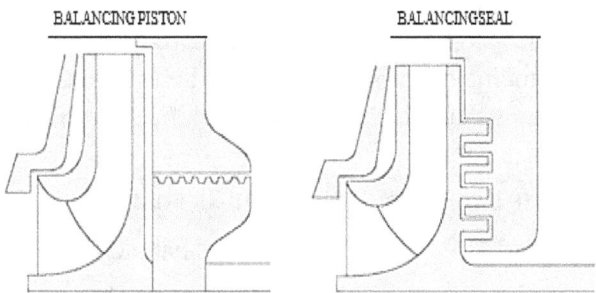

c. Balancing disc

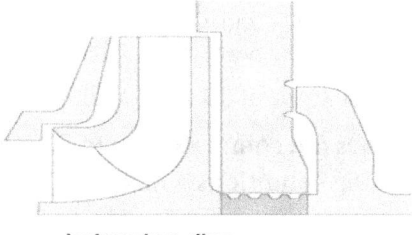

d. Opposed arrangement of the impellers in multistage pumps ("back-to back" design)

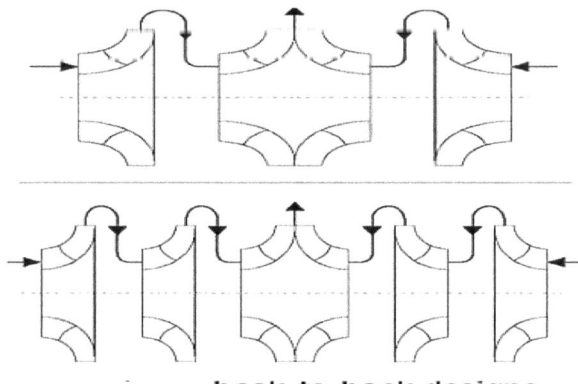

A multistage pump essentially consists of a number of single-stage impellers mounted on the same shaft.

3.6.3 Impellers Can Be Mounted In Two Ways

a. Several impellers mounted on one shaft, each facing its suction end in the same direction and stages following one after another in ascending order of pressure. In this case, the axial thrust is high which is balanced by hydraulic balancing device mounted on shaft.

b. An even number of single-suction impellers are mounted in one direction, and odd number impellers in opposite direction which develop two equal and opposite axial thrust (i.e. axial thrust created by odd impellers is balanced by the thrust generated by even impellers). See back to back design impellers arrangement shown in 3.6.2 (d).

3.6.4 Hydraulic Balancing Devices

Hydraulic balancing devices are mostly used in multistage pumps to balance the axial thrust and to reduce the pressure on the stuffing to the first-stage impeller. The hydraulic balancing device may be a balancing drum, a balancing disc or a combination of the two.

a. **Balancing Drum**

Balancing chamber is provided at the first stage impeller. The chamber is separated from the pump interior by a drum that is either keyed or screwed to the pump shaft and hence it rotates with the shaft. The drum is separated by a small radial clearance from the stationary portion of the balancing device. The balancing chamber is connected either to the pump suction or to the vessel from which the pump takes its suction. Thus, the back pressure in the chamber is slightly higher than the suction pressure

THE FORCES ACTING ON THE BALANCING DRUM ARE AS FOLLOWS:

i) Towards the discharge end: $F1 = Pd \times$ front balancing area

ii) Towards the suction end: $F2 = Pb \times$ back balancing area,

where: Pb = back pressure, Pd = discharge pressure.

The first force is greater than the second one, thereby counter balancing the axial thrust is essentially needed. Since, in practice, 100% balancing is unattainable, the balancing drum is often designed to balance thrust to the extent of 90 to 95% of total impeller thrust.

However, the balancing drum does not compensate automatically for any change in the axial thrust caused by varying pump operating conditions which is a disadvantage of drum balancing system.

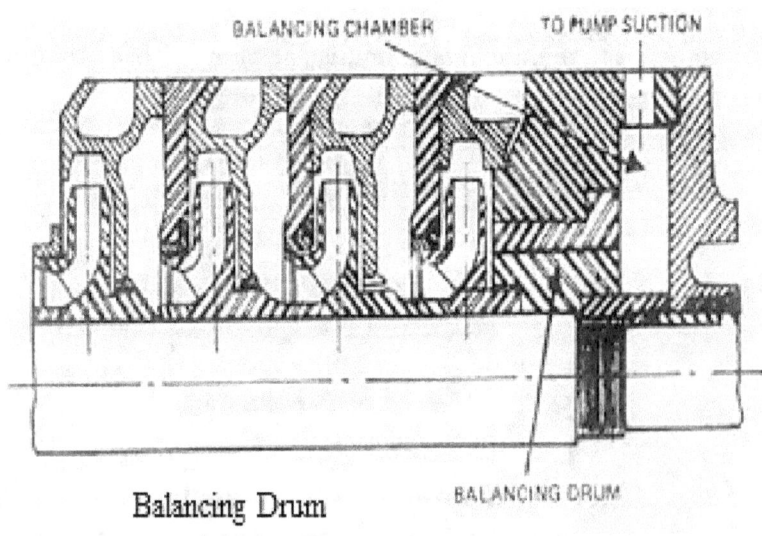

Balancing Drum

b. **Balancing Disc**

In this case balancing disc rotates with the pump shaft and it is separated from the balancing disc head (stationary part) by a smart axial clearance. The leakage through this clearance flows into the balancing chamber and from there flow to the pump suction or to a vessel from which the pump takes its suction. The back of the disc is subjected to the balancing chamber back pressure, whereas the disc face experiences a range of pressures (discharge pressure at its smallest diameter). The inner and outer diameters of the disc are so chosen that the total force acting on the disc face and that acting on its back wall will balance the impeller axial thrust. The balancing disc arrangement is shown in figure below–

BALANCING DISC

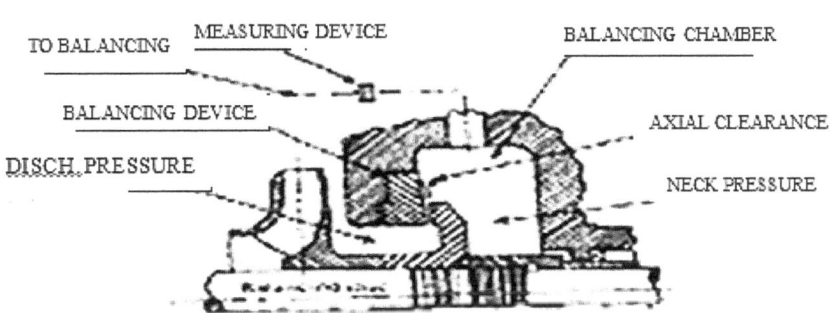

If the axial thrust increases during operation, the disc moves towards the disc head reducing the axial clearance. This will result in reduction in leakage and hence back flow the pressure in the balancing chamber. This automatically increases the pressure acting on the disc and it moves away from the disc head, increasing the clearance. Now, the pressure builds up in the balancing chamber and the disc is again forced to move towards disc-head until an equilibrium is reached. Thus, automatic compensation is ensured. In a balancing disc arrangement, the pressure on the stuffing box packing is variable and this is quite detrimental to the life of packing.

c. **Combination of Balancing Disc and Drum**

The combination of balancing disc and drum is developed to overcome the shortcomings of the disc while retaining the advantage of automatic compensation for any change in axial thrust. Combination of Disc and drum had proven satisfactory and effective in multistage boiler feed pumps used in power plant applications.

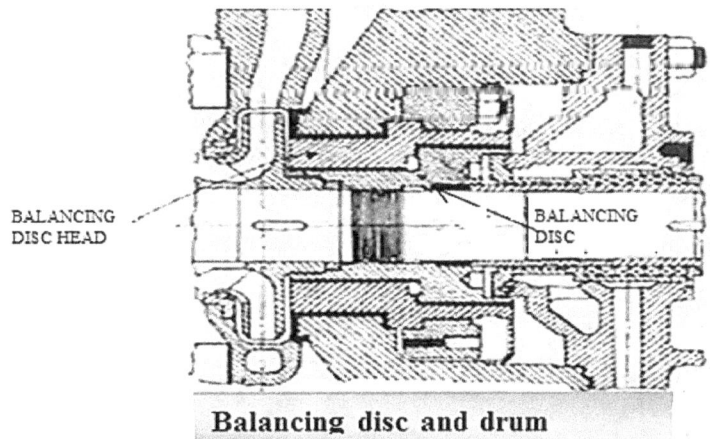

Balancing disc and drum

d. Balancing Device in Boiler Feed Pump

The axial thrust is balanced by the provision of balancing device. It consists of a rotating balancing disc with a small axial clearance of 0.08 to 0.10 mm against a static balancing disc. A part of the high pressure water from discharge of last impeller is allowed to pass through the throttling bush to act on the front surface of the balancing disc. The back surface of the disc is exposed to suction pressure. The difference in pressure on both sides of the disc exerts a thrust on the disc and tries to move the entire rotor towards right. Since this force is opposite and equal to axial thrust of the rotor in equilibrium position, maintaining the axial clearance of 0,08 to 0.10 mm depending on the type of the pump.

Chapter 4

DESIGN FEATURES IN CENTRIFUGAL PUMPS

4.1 Rotor Dynamics

The vibrational (lateral and torsional) behavior of rotor is termed as rotor dynamics. The critical speed is an important feature of lateral vibration. The speed at which the rotational frequency coincides with any natural frequencies of the rotor is known as critical speed which is restricted for operational range of the pump, especially when rotor damping is low. In order to ensure operational reliability of high-head-pumps, it is important that careful analysis of the vibrational behavior of the rotor is carried out to ensure that critical speed of rotor does not fall within operating range of speed ("N" ± 20%)

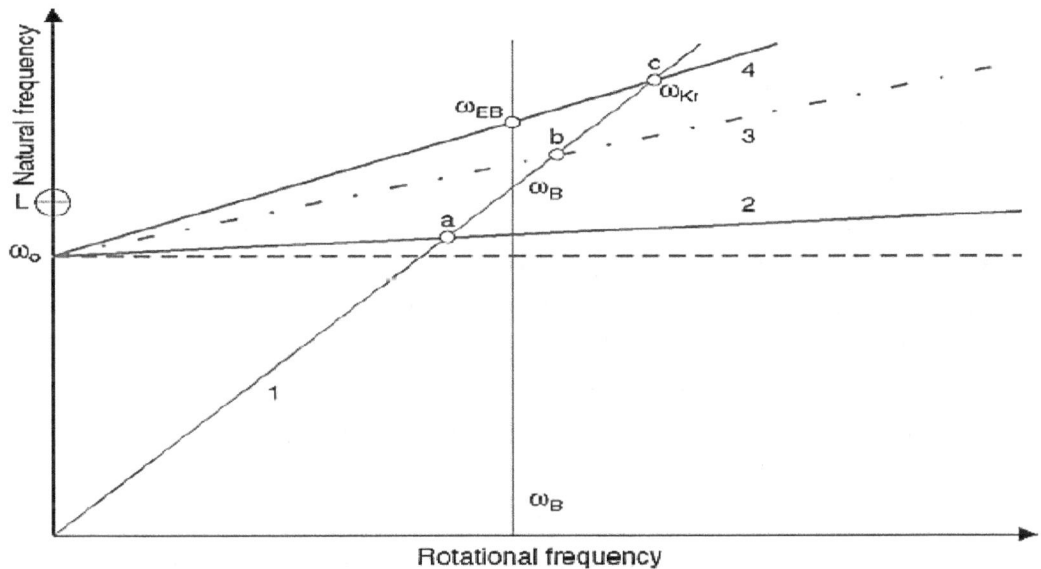

Natural frequencies of a multistage pump shaft as a function of operating speed in air and water and of variable seal clearances

Rotor dynamics, especially in the case of high performance critical pumps, the rotor design should be optimized. The natural frequency, vibration mode configuration and damping depend on factors as follows:

Rotor geometry: bearing span	Shaft diameter, mass distribution, overhung shaft sections
Radial bearings type	Relative clearance, loading, width to diameter ratio, oil viscosity (temperature)
Axial bearings	Mass, overhang, stiffness, damping, loading
Shaft coupling	Type, mass, overhang, alignment errors

(Contd.)

Annular seals, balance pistons	Seal clearance length diameter, clearance, surface geometry, differential pressure, Reynolds number, tangential inlet and outlet geometry of the gap, balance piston/sleeve, misalignment, pre-rotation. Rotor Dynamics
Hydraulic excitation forces due to hydraulic unbalance interaction	Forces due to hydraulic unbalance interaction (unequal blade loading) and vane passing restoring and lateral forces caused by impeller/diffuser, interaction dynamic axial thrusts flow separation, phenomena

4.1.1 Torsional Vibrations

Often given less attention to detect and monitor rotordynamics in pumps. However, torsional vibration may lead to a failure (e.g. coupling, or even a shaft rupture). Torsional excitation from the driver is important in some of the cases:

- **Startup of asynchronous electric motors** (switching to the grid):
  √ Excitation frequency
  √ Grid frequency

- **Short circuit at motor terminal at:**
  – Excitation frequencies:
  – Grid frequency
  – Twice grid frequency

- **Normal operation of variable speed electric motors**... Excitation frequencies depend on motor control. Examples:
  – Harmonics of rotational frequency
  – Slip-dependent frequencies

- **Normal operation of reciprocating engines (diesel)**
  √ Excitation frequency and harmonics of rotational frequency may originate torsional excitation while speed increasing/reducing through gears by pitch circle run out.
  √ The excitation magnitude depends on the quality of the gear.
  √ Excitation frequencies: The rotational frequencies & gear mesh frequencies have less importance.

- **Torsional excitation from pump impellers:**

  If impeller vane numbers and stator is selected properly in design stage, there will be remote chance of torsional excitation. It has importance for waste water pumps, having small & number of thick impeller vanes. While considering critical frequency (e.g. large size impeller with reliability in demand) a torsional natural frequency analysis is carried out for the entire shaft train including driver, couplings, gear and pump. Resonance with excitation frequencies should be avoided & torsional damping in shaft train is kept very low.

  If resonance situation can't be avoided, forced torsional vibration analysis need to be carried out. As a preventive measure, the fluctuating stress level is reduced & installation of damper couplings should be considered.

4.1.2 Problems in Rotor-Dynamics

a. **System-related problems**
  √ Unfavorable dynamic behavior of foundations, supporting structures or pipelines (e.g. resonance excitation by forces at rotational frequency);

√ Excitations from the coupling, especially due to misalignment;

√ Excitation from a component in the pipeline (valve, filter, etc.);

√ Excitations from the drive (motor, steam turbine, gearings);

√ High-pressure- pulsations due to hydraulic instability caused by entire system.

√ Unfavorable operational conditions like insufficient suction pressure (cavitation), swirling intake vortex, suction pipe with bends in more than one plane.

b. **Problems of the pump it-self:**

√ Mechanical unbalance of the rotating parts, careless assembly or operational influences (e.g. cavitation erosion, abrasion, deposits, corrosion in impellers, jammed ports);

√ Unfavorable dynamic behavior of the rotor due to excessive seal or bearing clearances;

√ Increased hydraulic forces when pump operated beyond the operating range (departing from the operating point on characteristic curve);

√ Defective bearings – Data analysis for design and installation of centrifugal pump in initial stage.

c. **Hydrodynamic Bearings on Larger Pumps**

It is a common practice to measure shaft vibrations, i.e. the relative movement of the shaft with reference to the bearing housing. The displacement is stated in micron (thousandths of a millimeter) or mils (thousandths of an inch). Such measurement, usually enable to judge the dynamic behavior of the rotor compared to the measurements on the bearing housing. However, shaft vibration measuremens can indicate false readings due to shaft ovality/eccentricity or nonhomogeneous material. For correct assessment of the measured results, the run out must be small. Often, each bearing is fitted with two transducers at right angles to each other. The orbit analysis of the shaft center can be plotted

a. **Major Noise Sources**

1. Drive motor, gears;

2. Structure-borne noise in adjacent light structures by excitation (e.g. chequered plates);

3. Radiation from the discharge pipe, and to some extent, from suction pipe. Acoustic insulation of pipes can reduce noise to large extent. For acceptable sound level of feed pumps refer VDI 3743.

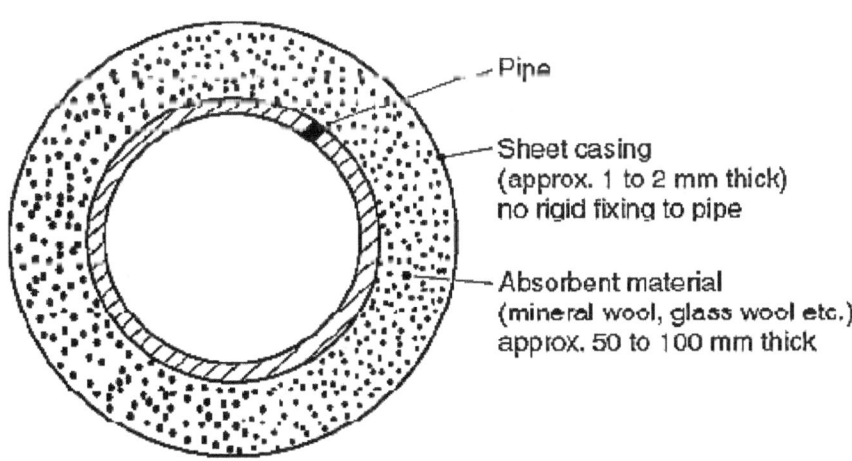

Acoustic insulation of a pipe

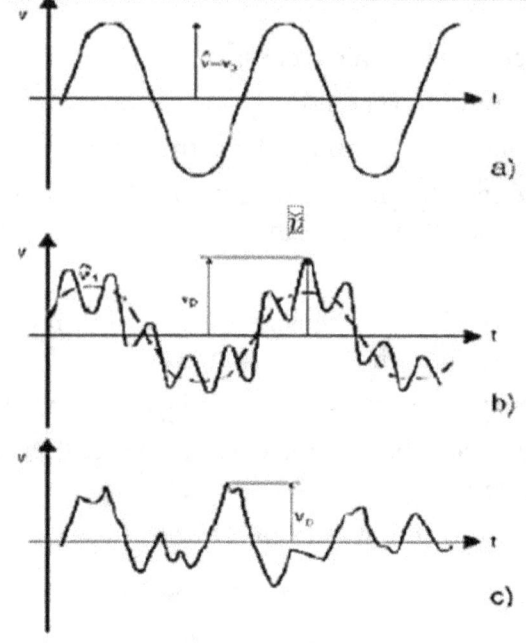

$$v_{RMS} = 0.707 \cdot \hat{v}$$

v_{RMS} = RMS value of vibration velocity (mm/s, in/s)

$\hat{v}$ = peak value of vibration velocity for harmonic signals (mm/s, in/s)

v_p = peak value (mm/s, in/s)

Where the vibration consists of several harmonic components (Fig. (b)), then:

$$v_{RMS} = \sqrt{\hat{v}_1^2 + \hat{v}_2^2 + ... \hat{v}_2^2}$$

For non-periodic or random signals, by definition:

$$v_{RMS} = \sqrt{\frac{1}{t_0} \int_0^{t_0} v^2(t)\,dt}$$

Vibration velocity as a function of time

4.1.3 Vibration Severity Charts, (ISO Vibration Guideline & Limits)

For further information refer standards. Please note that severity charts are a general guide line, not so accurate for vibration and alignment specification on m/c.

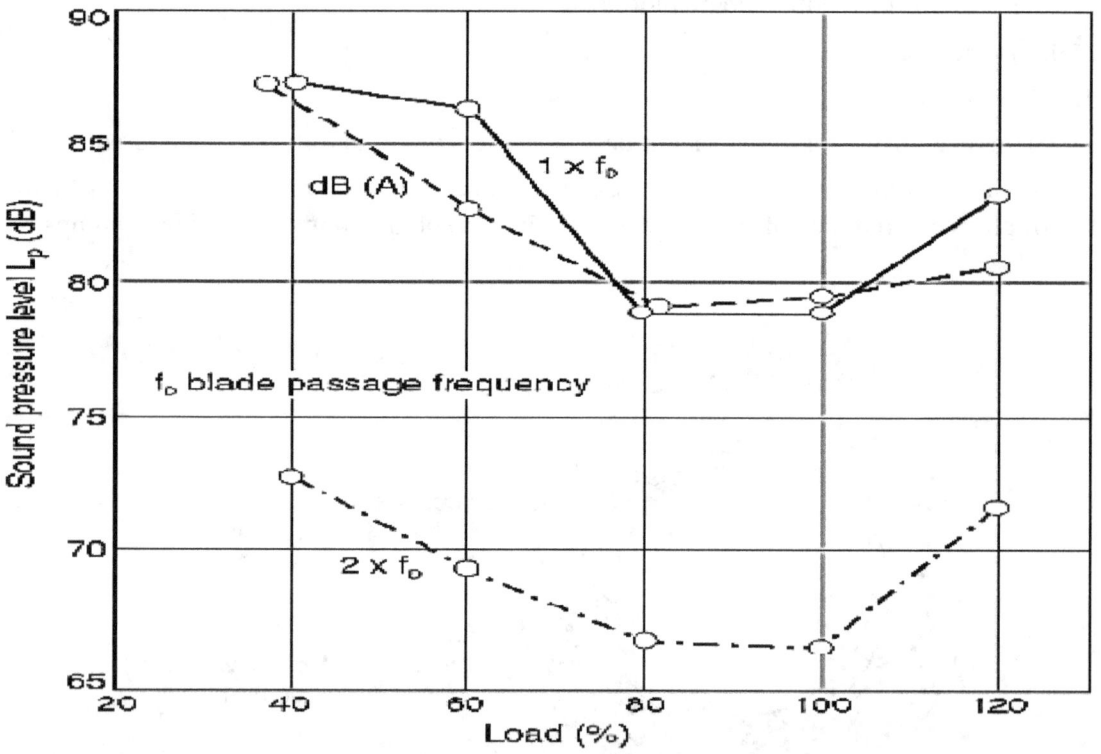

Sound pressure level as a function of load

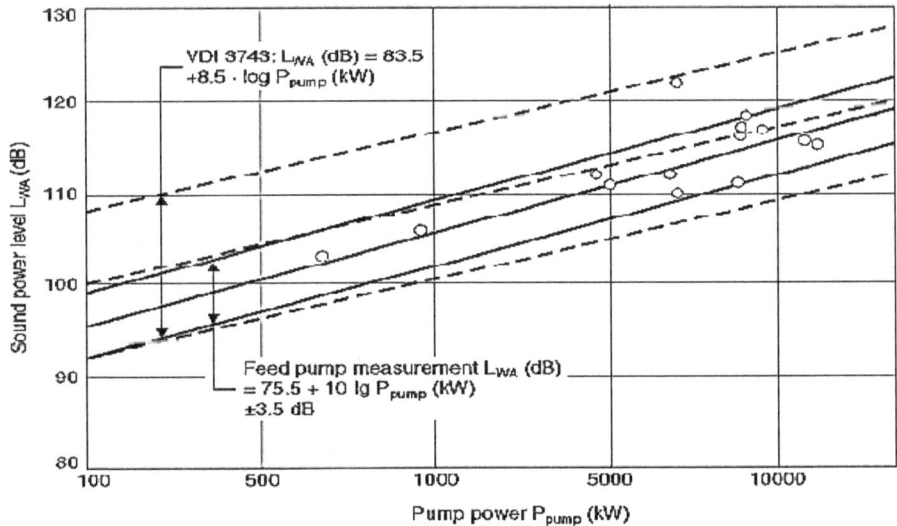

Sound power levels from feed pumps (compared with VDI 3743)

4.1.4 Acceptable Vibration Limits for Power Plant Equipment – (IPS ≡ Inch Per Sec)

ISO-13709 (API610) VIBRATION LIMITS		
Discrete frequencies	PUMP RPM ≥ 3600; POWER/ STAGE ≥ 300 KW	2.0 mils Peak to Peak
Allowable increase in vibration at operating point	30%	30%

SYSTEM DESCRIPTION	ACCEPTABLE LIMIT (inch/sec)	FIRST ALARM (inch/sec)
COOLING TOWER DRIVES		
Long shaft	0.15	0.5
Close coupled	0.09	0.3
Belt driven	0.12	0.4
COMPRESSORS		
Reciprocating type	0.135	0.39
Screw type	0.117	0.45
Centrifugal type	0.09	0.3
FANS AND BLOWERS		
Integral shaft	0.08	0.27
Direct driven	0.1	0.325
Belt driven	0.11	0.37

4.1.5 Vibration Limits for Vertically Suspended Machine

ISO 13709 (API 610) Vibration Limits for Vertically Suspended

Location of vibration measurement	Pump thrust bearing or motor mounting flange	Pump shaft adjacent to bearing
Pump bearing type		
	All	Hydrodynamic guide bearing adjacent to accessible region of shaft
Vibration at any flow rate within the pump's preferred operating range[a]		
Overall	$v_u < 5.0$ mm/s RMS	$A_u < (6.2 \times 10^6/n)^{0.5}$ μm peak to peak $[(10\,000/n)^{0.5}$ mils peak to peak] Not to exceed:
Discrete frequencies	$v_f < 0.67 v_u$	$A_u < 100$ μM peak to peak[a] (4.0 mils peak to peak) $A_f < n. A_f, 0.33 A_u$
Allowable increase in vibration at flows outside the preferred operating region but within the allowable operating region	30%	30%

[a]*Values calculated from the basic limits shall be rounded off to two significant digits*

where:
v_u is the unfiltered velocity, as measured
v_f is the filtered velocity
A_u is the amplitude of unfiltered displacement, as measured
A_f is the amplitude of filtered displacement
n is the rotational speed expressed in RPM

4.1.6 Zone and Limit of Vibration

ZONE AND LIMIT OF VIBRATION TAKEN AT BEARING HOUSING OF ROTODYNAMIC PUMP (> 1 KW AND NUMBER OF BLADES IN IMPELLER ≥ 3, RPM ≤ 1500)						Pump running speed ≤ 900. Spectrum μM peak to peak 0.5 N, N or 2 N value
Zone	Description	Vibration Velocity mm/sec RMS				
		Category		Category		
		> 200 kw	≤ 200 kw	≤ 200 kw	> 200 kw	Up to 400 kw
A	Newly commissioned m/c	2.5	3.5	3.2	4.2	60
B	Unrestricted long term operation	4.0	5.0	5.1	6.1	80
C	Limited operation	6.0	7.6	8.5	9.5	130
D	Risk of damage	6.6	8.6	9.6	10.6	> 130
Preferred operating range		2.5	3.5	3.2	4.2	50–60
Allowable operating range		3.5	4.4	4.2	5.2	75
Alarm range (1.25 * zone C)		5.0	6.9	6.4	7.6	80
Trip range		8.9	9.5	10.6	11.0	100

4.1.7 IRD Vibration Severity of Centrifugal Pumps

IRD Mechanalysis had developed a graph in general for all type of m/cs. Graph is plotted amplitude (displacement peak to peak in µM) against frequency (cpm) of vibration and also it provides severity of vibration (8-categories) against vibration velocity (mm/s PTP). One can check severity of vibration – Extremely smooth, Very smooth, Very good, Good, Fair, Slightly rough, Rough, & Very rough, knowing either amplitude & frequency of vibration or velocity of vibration. When vibration velocity measured on bearing housing of centrifugal pump reaches in very rough zone, m/c should be stopped.

- **Severity wise IRD recommendations:**

Severity Category	Recommendations
Extremely smooth to Fair	M/c is operating well. No action required
Slightly rough to Rough	By an opportunity of shut down, check for reason of vibration and repair
Very Rough	M/c to be stopped for investigation and repair
Note: 1.0 Peak to peak velocity(mm/s) = 0.71 mm/s RMS	

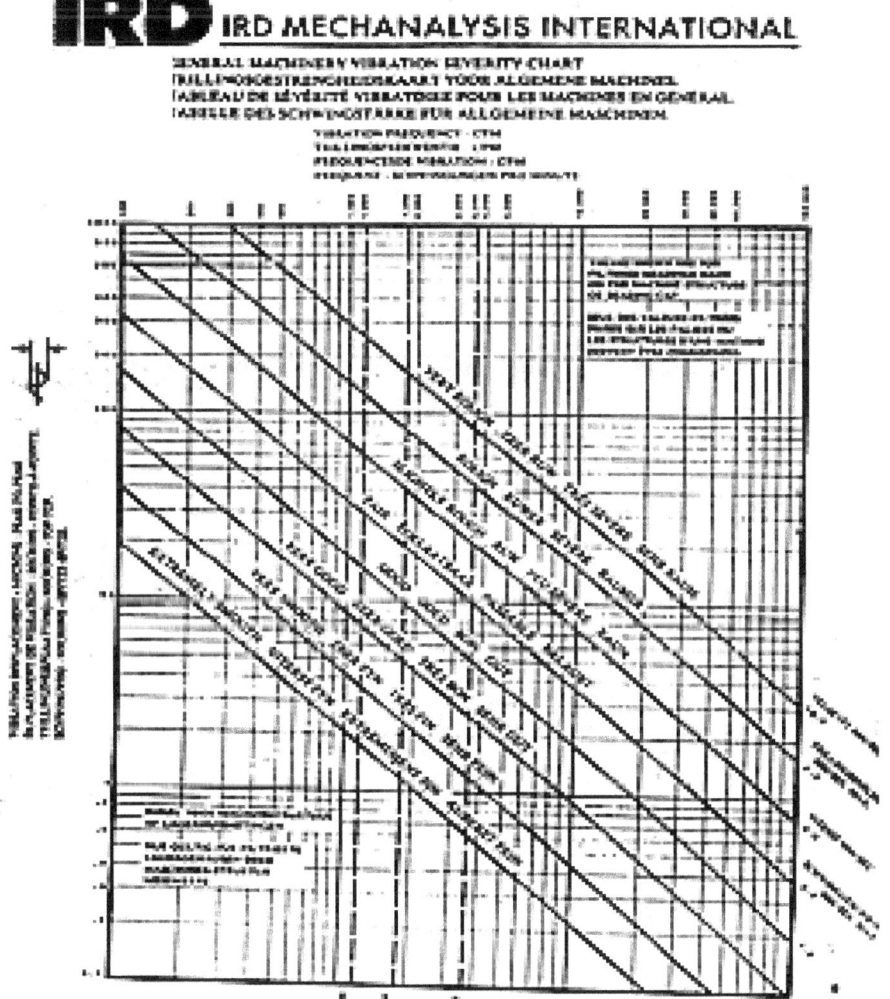

4.2 Suction Flow Constructions

The purpose of the intake design is to ensure liquid laminar approach to the pump impeller. The laminar approach of liquid to suction will result smooth operation and longivity of pump. If this requirement is not satisfied, it will render loss in capacity & efficiency, and cavitation may occur affecting damages to the pump components. The disturbances will demand increase in specific speed and the size of the pump.

4.2.1 Two Sources of Pre-Swirl Need Particular Attention

i. Asymmetric approach flow renders a pre-swirl and vortices which can be prevented by proper design of the intake structure.

ii. At part load operation of pump, impeller is subjected to recirculation of flow to inlet of impeller. The fluid recirculation from the impeller discharge induces a major pre-swirl which may render cavitating or noncavitating vortices in suction zone of pump. The cavitation and vibrations, can be suppressed to the great extent by swirl-breaking deices, upstream of the impeller (e.g. ribs or floor splitters). Hence, detrimental effects on pump performance can be observed.

The details of such designs are explained in following discussion points.

4.2.2 Anti Swirl Measures in Pumping System

a. **Flywheel**

Flywheel prolongs the rundown time of pumps. It is normally employed in pumps used for pumping systems containing up to 2 km piping length. Generally, the maximum flywheel size is dictated by the accelerating capacity of the motor and/or the load capacity of the bearings, mass moment of inertia of moving parts along with flywheel for enhancing the rundown time.

For separate intermediate flywheel bearings, only the acceleration has prime importance in design.

b. **A surge tank installation**

This takes over full pipeline flow surge with virtually no time lag. With a suitable surge tank, the minimum admissible pressure is maintained in the system. Low-pressure systems are normally exception to this.

– Surge tank installation and automatic water level monitoring is the simplest solution to the problem. If the level in the surge tank exceeds to maximum, the installation must be taken out of service till the level reaches to normal again.

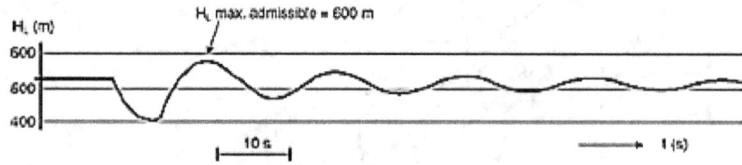

Pressure curve in the pumping station after power failure.
With surge tank.

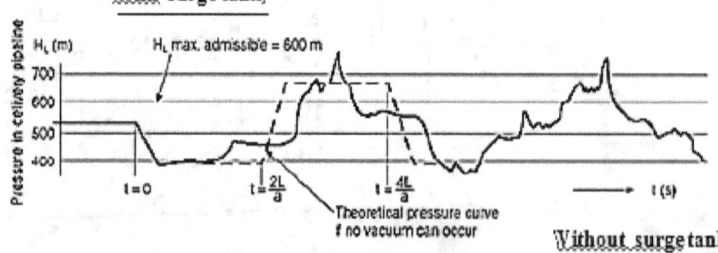

Pressure curve in the pumping station after power failure,

- Air valves – It is effective when atmospheric pressure reaches in the suction pipe at any point. Substantial volumes of air may be drawn into the system; consequently air valves are ruled out for water supply systems.

- Pressure relief valve- This must operate with minimum inertia and sufficient damping, otherwise pressure pulsations may be excited. When a specified pressure is developed, the valve open to atmosphere and allow water to escape from the system through valve. It function as one way surge tank.

- Check valve (design of shortest possible closing of the check valve) in main pipes. In the event of reverse flow due to long closing time of NRV, pressure pulsation may be developed.

4.2.3 Instrumentation for Centrifugal Pumps

The instrumentation level is decided based on need of controls, mode of operation and protection philosophy of planned installation:

- The basic rule "avoid any thing superfluous" should be stictly followed. Addition of number of devices do not enhance reliability. That should apply to critical processes and parameters, considered to be very crucial for process or safety of man and machine. On contrary multiple parallel instrumentation for any parameter in installation may not create high reliability rather that can create confusion to increase the possibility of trouble sources.

- Quantum of automation should be techno economicaly justified. Minimum manpower and maintenance costs should be aimed while designing instrumentation.

- Only a small selection of the possible measuring and monitoring instruments for protection of centrifugal pumps shoud be enroled.

4.2.4 External Conditions for Pump Protection

- The pump must be primed and must stay primed for instant operation.

- A minimum flow must be ensured to avoid excessive rise of the liquid temperature and risk of evaporation due to loss of internal energy in the pump. At zero flow situation during operation, depending on the pump type and specific speed, approx. 50% or even more of the driver's energy input at BEP is converted into heat.

- The operational protective devices, must be absolutely reliable & dependable. While choosing instruments model, brand, and product longivity to render service on-site must be verified.

- The clearly define the functional description of the entire pumping system (pump, motor, control system, protective devices and power supply) should be readily available for commissioning.

- The instrument constraints which affect equipment must be noted in particular.

- The pumps should be arranged with positive suction pressure where ever possible. If not feasible, provision should be made for auto priming to pump remaining primed even in idle condition. This simplifies the starting and renders higher reliability of pump.

4.3 Condensate Extraction Pump

4.3.1 Design Criteria

- Condensate temperature is considered for deciding flow and head of pump.

- Condensate flow required from hot well at MCR + 10% margin with full valves opening is considered in design stage.

- Condensate makcup flow is considered ±1–3% of rated condensate flow.

- Shoot blowing & Auxs steam flow and Secondary air heating flow if applicable should be accounted for capacity computation in design stage.

- Boiler feed pump seal water injection flow if applicable, should be considered for flow accounting.

- Flow margin of 10% on total sum flow, for future requirements without reduction in rated pump efficiency should be accounted in design stage.

- Considerations for stable parallel operation of CEPs, minimum rise of curve from design 20–25% and that of maximum rise to 40% is considered in design. Lower rise to shut off is acceptable if re circulation is provided.

- Specific speed of pump should be within 6000 to 12000 as per Hydraulic Institute, and peripheral vane tip velocity of first stage impeller should be limited to 19 M/sec.

- NPSH $_{(R)}$ of pump should be within 1.5 M to avoid cavitation problems.

- Suction head of pump from first stage impeller of vertical pumps or from centre line of suction eye in horizontal pump is considered in calculations to avoid cavitation.

- Deeper the suction, longer will be the cantilever of horizontal pump and probability of bearing and seals damage due to critical speed resonance in transient operation of pump is increased. Occurrence of cavitation may lead to deterioration or failure of pump components.

- Flow control can be considered by–
  i.   Throttle valve control in constant speed pump
  ii. Variable speed control device

- Head developed per stage should not exceed 100–110 M to avoid high vibration, cavitation or erosion problems.

- The pump shaft design should have stress resistance to 75000 psi for long life.

4.3.2 Selection of Condensate Extraction Pumps

- Condensate temperature should be considered for flow and head requirements.

- Margin of 10% on max condensate flow from condenser hot well or MCR with full valves opening should be considered in design.

- Condensate-makeup cycle flow is considered 1–3% of max condensate flow.

- Steam flow rate for shoot blowing is added to the capacity of CEP.

- Operating point capacity and head should be selected based on criteria's discussed above

- Selection of pump speed at design capacity and head is important parameter of CEP. It should operate normally at 1000–1500 RPM. Higher the speed of pump, NPSH $_{(R)}$ increases, and probability of cavitations is increased. NPSH $_{(R)}$ of CEP should be ≤ 1.5 M. The single pump operation should be reviewed at design stage.

- The best efficiency point should fall in operating zone of characteristic curve.

- Power required for single pump operation at MCR or working point and startup of pump with open discharge should be considered in selection of prime mover.

- Max Shut off head of pump is designed to take care of parallel operation of CEPs.
  The pump characteristic curves of all CEPs should be identical for Parallel operation.

- No critical speed band should fall within operating speed ±20%.

4.3.3 MOC of CEP Components

SN	Pump Component	MOC for Condensate water quality
		MOC of CEP Components
1.	Suction and discharge pipe	Carbon steel ASTM A 36, or ASTM A 283 Gr C for fabricated part and ASTM A 216 WCB for castings
2.	Bowls	Carbon steel ASTM A 216 WCB
3	Impeller	SS- 12 to 14% Cr steel ASTM A 743 or A 217 Gr CA-15 or CA-6 NM
4.	Shaft sleeve	SS 11.5–13.5% Cr, ASTM A 278 type 410 H
5	Shaft	SS 11.5–13.5% Cr, ASTM A 278 type 410 T
6	Wearing Ring	SS 12–14% Cr, ASTM A 582 type 416 H
7.	Bearings	Carbon or Graphite alloy
8.	Bowl bolting	SS 12 Cr, ASTM A193 type 416

4.3.4 Guarantee Parameters of CEP

- Operating point capacity and head
- Pump speed at design capacity and head
- NPSH required ($NPSH_{(R)}$ at design Q – H and single pump operation
- Pump efficiency at MCR
- Power required for single pump operation at MCR
- Max Shut off head
- Parallel operation characteristic curve of pumps
- No critical speed band within ±20% of operating speed.

4.4 Slurry Pumps

Slurry pumps in Power Plant is a vital equipment used for sludge and muck transfer from water treatment plant, slurry transportation ash handling system, sludge pumping from water clarification plant etc

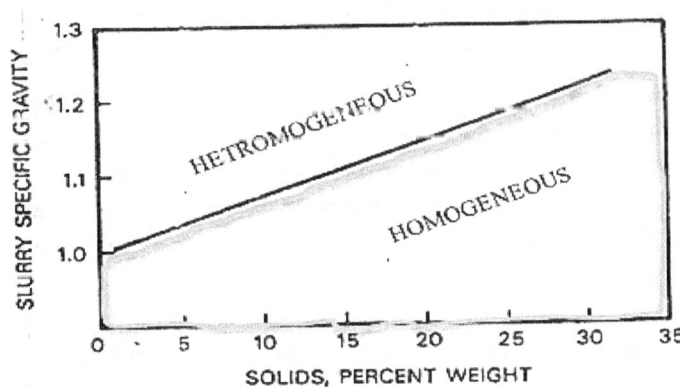

4.4.1 Moc of Slurry Pump Components

Slurry is generated in number of industrial plants and transferred through slurry pump of various design. The design & construction of pump could be centrifugal, reciprocating, diaphragm or jet type. There is popular application of slurry pumps in mining area, paper mills, concrete transfer, municipal solid waste disposal, over and above the power plants. The centrifugal type pumps are most popularly used for slurry transfer in various industries. Discussion is limited to construction and application of horizontal centrifugal slurry pumps.

4.4.2 Moc of Slurry Pumps

SN	Components	Ash Pump	Scrubber Reciprocating pump
		MOC of Slurry Pumps	
1.	Casing	Ductile iron, ASTM A 536Gr 60-45-12	Ductile iron, ASTM A 536Gr 60-45-12
2.	Casing Liners	ASTM A 532, min 550 BHN	Natural rubber
3.	Impeller	ASTM A 532, min 550 BHN	Ductile iron, ASTM A 536Gr 60-45-12
4.	Impeller liner	None	Natural Rubber
5.	Shaft	ASTM A 576Gr 1045, C-Steel	ASTM A 576Gr 1045, C-Steel
6.	Shaft sleeve	ASTM A 276 type 420 harden	ASTM A 276 type 420 harden
7.	Bolting	ASTM A 193 type 410	ASTM A 193 type 410

4.4.3 Constructional Features of Slurry Pumps

Wall thickness of wetted parts i.e. casing and impeller are considered to be higher than conventional centrifugal pumps. Alternatively, internal lining with hard metal or rubber is provided. In first operation of slurry pump, fine slurry particles are penetrated into top layer of rubber and work as stone boxing to resist the erosion from slurry particles being handled.

- Slurry pumps are normally vertical split design for ease of replacement of impeller, liners or repair of other components.
- More flow passage between casing and impeller is intentionally provided to allow non clogging behavior
- If gap between impeller face and suction liner is increased by abrasion, the axial adjustment device is incorporated for adjustment of impeller clearance.
- Extra-large-stuffing box and replaceable shaft sleeve are the special design features in slurry pumps
- Radial and axial thrust bearings are selected for heavy duty applications.

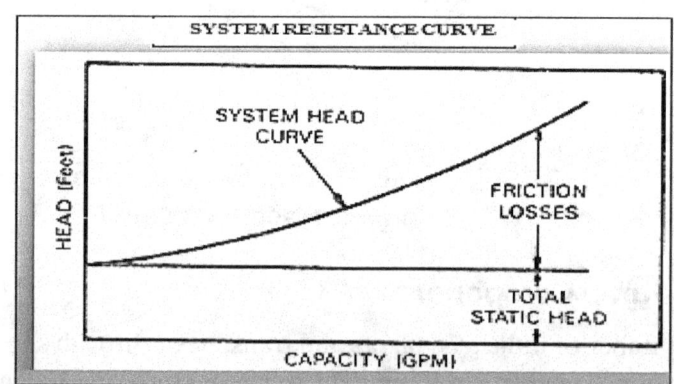

- !0% design margin is necessarily provided to slurry pumps for compensation of friction losses in flow through pipes. Homogeneous slurry (approx.1:5 ratio of solids and water) with velocity 1.5–2.5 M/sec results less problems in slurry transportation otherwise due to non Newtonian flow situation, solids will get settled into pipe and clogging will take place. The lower size of particles (< 2 mils or 50 μM) with higher solids concentration are much more difficult to handle.

- Most of the slurry pumps (eg scrubber re-circulation system) require $NPSH_{(R)}$ ranging from 5 to 10 M.

- Pump wear rate is approx proportional to the pump speed in the third power. Thus, a pump operating at 1000 RPM will wear 8 times, compared to that operating at 500 RPM. Hence, pump speed becomes the vital parameter in slurry pump design & operation. However, by reducing speed, pump efficiency decreases (say 10%) but life of pump is increased appreciably high; thus there will be saving of energy much more than the cost of energy loss due to low efficiency. Hence, lower speed operation is preferred.

- Rubber lined impellers should operate at lower tip velocity to avoid dis-integration of rubber from impeller. The tip velocity should be limited to 17.5–25 M/sec. With hardened or abrasion resistant impeller, tip velocity can be increased to max 37.5 M/sec.

- Beyond 37.5 M/sec impeller tip velocity is selected for specific application of higher head pumping but at the cost of very high abrasion wear.

Chapter 5
BOILER FEED PUMPS

5.1 Boiler Feed Pumps

5.1.1 Boiler Feed Pump application in Power Plant

Boiler Feed Pump is considered the heart of steam generating system. This requires an optimum pressure head and flow rate to perform the specified function (deliver condensate water normally, to boiler drum through heaters and economizer). Therefore, it becomes important to choose suitable head and discharge to inject condensate into boiler against a specific working condition. The criteria of selection, suction & discharge head and flow capacity of BF Pump will be discussed in brief.

5.1.2 Boiler Feed Pump Total Suction Head Determination

Item	Head (M)/ Temp. °C	Reference/comments
De aerator water working temp. max	Say t °C	
De aerator Pressure + Elevation head	x	MCR operating condition heat balance
Static head between De aerator and BF pump impeller I/L line	y	Ref low level of De aerator. Normally Deaerator is kept at 13 M levelfor HP boilers, operating at 70–130 kg/cm²
Frictional loss in pipings, connected between De aerator and BFP suction	z	Design point in negatie value
$NPSH_{(R)}$ of Pump	r	
Vapor pressure of water at working temperature	v	Ref vapor pressure chart at differential temperature.
Recommended margine on $NPSH_{(A)}$	1.0	
Total abs. Head available at suction eye of pump [$NPSH_{(A)}$]	$\{(x+y) - (z+v)\} + 1$	Where a = Atmospheric at altitude of place from sea level
Note: All parameters considered for water at Deaerator working temperature. The suction head NPSH(a) should be more than NPSH(r), design feature of pump at given water temperature being handled		

5.1.3 Determination of Discharge Head of Boiler Feed Pump

Item	Head M	Reference based/comments
Boiler drum pressure (safety blow)	$a = p/\rho$	Ref safety valves setting by manufacturer. Head of boiler drum pressure = Design pressure (kg/cm²g) divided by density of water at working temperature

(Contd.)

Item	Head M	Reference based/comments
Frictional losses in BFP discharge. Line up to Eco I/L, including valves in operating condition	b	Based on MCR flow rate
Eco. Frictional losses	c	As per steam generator's design parameters
Pressure loss at Boiler Feed Control Valve (80% open) at MCR steaming	d	As Per BF Control valve characteristic at 80% valve opening
Static head between Eco I/L to steam drum (water level)	e	As per steam generator's design parameters
Static head between BFP O/L eye to Eco I/L levels	f	As per design parameters of pump piping
Across HP heaters pressure drop	g	Feed haters design references
Total head loss in system (HL)		HL = (b + c + d + e + f + g)
Margin on pressure head (10–15%)	Say 12%	As Per recommendations PP design
BF pump head (H) at given temperature		H = (a-HL) * 112/100 M

a. Thumb rule for BFP flow calculation Q(M³/h) = MCR steam flow * 1.03 * 1.25. Where, CBD & IBD losses are considered 3 to 5%, Margin on flow as per recommendations PP design is considered to be 25 to 30%.

b. BF Pumps are designed normally with specific speed, N = 1000–1300 RPM. The suction specific speed of impellers other than first stage of boiler feed pump should be 8000–9500 and head developed per stage impeller not exceeding more than 670 M to avoid hydraulic instability.

c. The NPSH required by BF Pump at designed operating conditions, based on reduction in first stage head, should not be greater than the calculated NPSH available/1.8.

d. NPSH required during sudden load restriction, based on 3% reduction in first stage total head, should not be greater than calculated NPSH available (lowest-pressure feed water out of service, divided by 1.3)

e. Flow regulation in case of Turbine driven pump, can be adopted on speed of Turbine, but in Electric motor driven pumps–BFP is regulated by discharge valve throttling or by motor speed control through VFD/DC voltage control. In VFD control of motor, power selection is done with +10–15% margin. Motor insulation should be "H" grade.

f. Any critical speed of pump or motor should not fall within 75–125% of operating speed band.

5.1.4 MOC for Boiler Feed Pumps

SN	Components	MOC
1	Casing barrel	Forged carbon steel ASTM A 266, class-I or II or ASTM A 105
2	Diffuser, Volute cases, diaphragm and stage pieces	Ss 12–14% Cr, ASTM A 743 Gr-15 or CA-6 NM
3	Discharge Head	Forged carbon steel ASTM A 266, class-I or II or ASTM A 105 or ASTM A516 (BQ)

SN	Components	MOC
4	Shaft	Ss 12–14% Cr, ASTM A 276 Gr-410 T or ASTM A 479 type 410 II
5	Impeller	Ss 12–14% Cr, ASTM A 743 Gr CA-15, or CA-6 NM
6	Wearing Ring	Ss 12–14% Cr, ASTM A 743 CA-6 NM, CA-40, ASTM A 276 Gr-410, 416, or 420

5.1.5 Guarantee Features of Boiler Feed Pumps (Tested and Conformed)

- Capacity at design operating point of head
- Total head at designed operating point
- Max NPSH required at MCR flow and pressure at given liquid temperature
- Efficiency
- Time required to start the pump from cold to MCR flow and head
- KW input of pump at MCR flow and head
- Min. recommended continuous flow in operation
- Ability of single pump to operate as per characteristic curve at system head curve
- Ability to work with stand by or together running pump in parallel operation
- Max shut off head
- Vibration and sound level within limit specified in ISO standard
- Max pump shaft deflection.

Transient cavitation test- from MCR running condition, throttle suction to reduce suction head to the extent that motor KW is reduced 15–20%. Watch 2–3 minutes for cavitation symptoms.

5.1.6 Thumb Rule for Selection of Boiler Feed Pump

The general guidelines for proper selection:

- Select the pump based on rated conditions.
- The BEP should be between the rated point and the normal operating point.
- The head/capacity characteristic-curve should continuously rise as flow is reduced to shutoff (or zero flow).
- The pump should be capable of a head increase at rated conditions by installing a larger impeller.
- The pump should not be operated below the manufacturer's minimum continuous flow rate. From the pump performance curve, the $NPSH_{(R)}$, varies, with head and flow. Once the specific pump model and size have been determined from the basic process information, then materials of construction is chosen. Selection of pump is based on fluid properties, such as corrosiveness/erossiveness and the presence of dissolved gases. Having all information about the chemical characteristics of condensate, ensure proper material selection of the pump and its components.

5.2 Pre Installation for Boiler Feed Pump

Before starting installation of the equipment, Operation & Maintenance Manual submitted by supplier must be studied thoroughly. Following **drawings** are mainly referred for installation–

√   General Arrangement Drawing

√   Piping & Instrumentation Drawing

√   Foundation Drawing

√   Plant Layout Drawing

5.2.1 Foundation

It must be ready in all respects before installation of pump. Centerlines & elevations must be clearly marked from plant bench mark.

- Proper access to erection site.

- Cranes, hoisting tackles of adequate capacity

- Before starting installation of the equipment, Operation & Maintenance Manual furnished along with the equipment must be studied thoroughly.

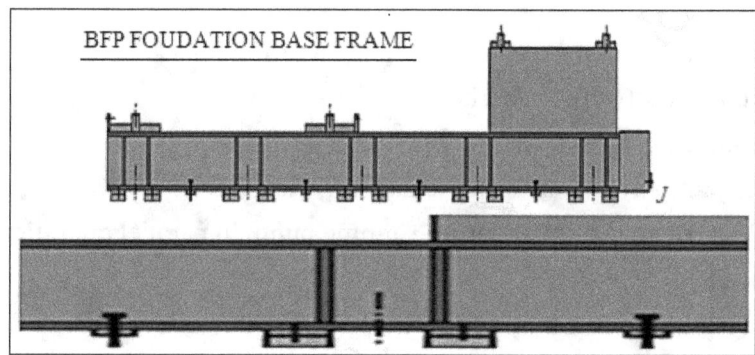

5.2.2 Drawings Mainly Referred for Installation

- Foundation drawings

- BFP OEM drawings

- Piping routing drawings

- De aerator GA and nozzle connection drawings

- Turbine floor drawings in plan and elevations

- BFP Installation drawings and instructions for erection & commissioning

5.2.3 Centrifugal Pumps Base Frame

Base frame of Pump is provided with welded pads of 20 mm thickness at the bottom. They are machined in one setting after welding to the Base frame. Welded & machined pad are in one level.

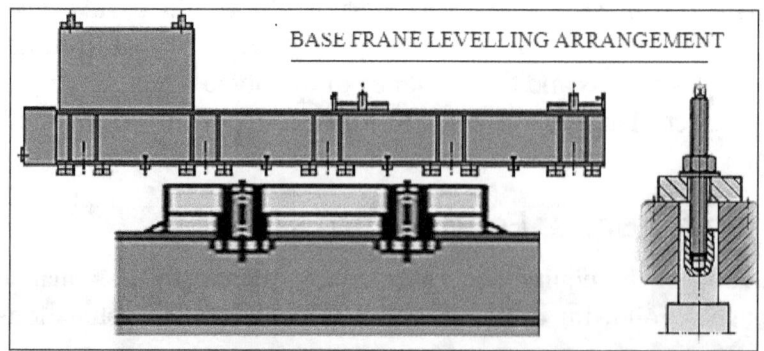

5.2.4 Inspection of the Foundation Block

- Dimension of the foundation
- Elevation with reference plant bench mark
- Position of centre line w.r.t. reference mark

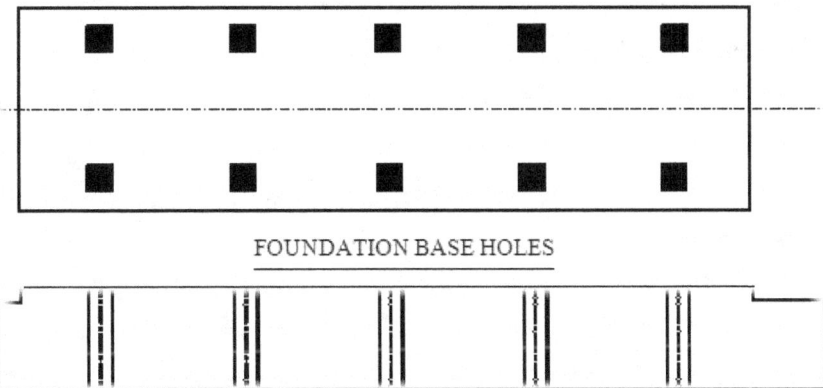

FOUNDATION BASE HOLES

5.2.5 Installation Sequence

If coupling hubs are not factory fitted on Pump & Motor Shafts, then fitting is done at site.

- Suspend the Foundation Bolts into pockets through the base frame
- Mark the positions of the Jack Bolts on the foundation block. Place & grout ordinary metallic plates using Non Shrink Grout. Level the top surfaces of the plates before half grouting the foundation bolts. Allow curing time as recommended by the grout manufacturer.

The pump body expands when subjected to high temperature liquid. To guide the expansion to protect pump from distortion alignment disturbance, locate the pump feet holes with respect to holding down bolts as shown in above sketch. Gaps are so adjusted that they allow sidewise & axial expansion towards discharge end of the pump.

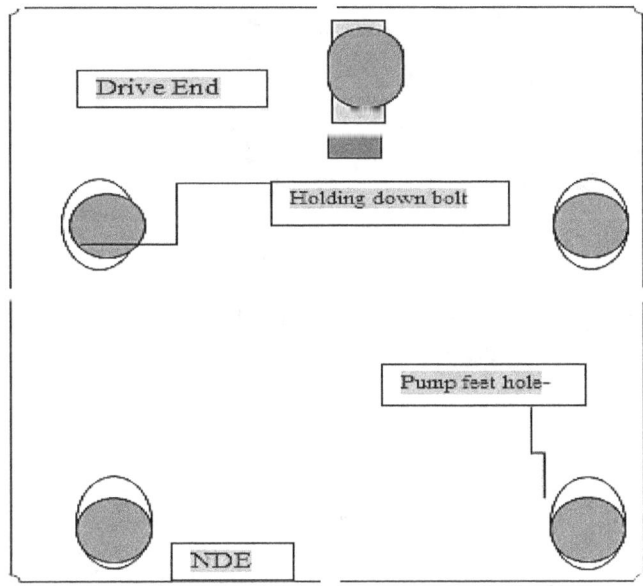

5.2.6 Installation of the Pump Set

- To allow the expansion of the pump body in axial direction towards discharge end, holding down bolts at discharge end are tightened to the extent that freedom of movement of the washer below them are un-disturbed by a light hammer blow. At suction end, tighten the holding down bolts with full torque.

- Remove blind of the discharge flange.

- Level the pump on discharge nozzle **within 0.04 mm/M** in axial & transverse directions

- Check & adjust the elevation of the equipment.

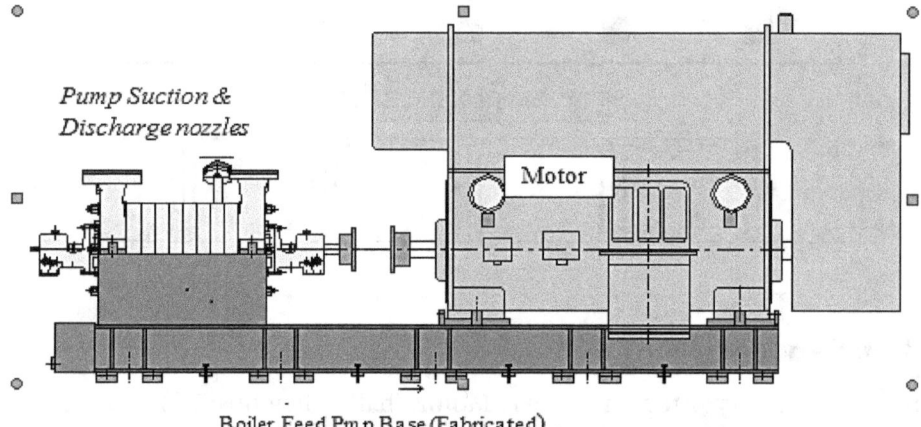

Boiler Feed Pmp Base (Fabricated)

Replace the blind of the discharge casing. Ensure gap of about 80 mm between base frame bottom face & foundation block top surface. Adjust the distance between the Shaft ends (DBSE). Both the Rotors must be at their normal operating axial positions while measuring DBSE

5.2.7 Operating Position of Pump Rotor

In case of pumps with Balancing Disc/Counter Balancing Disc, operating position of the rotor is the position of rotor where both the Discs are closed to touching position.

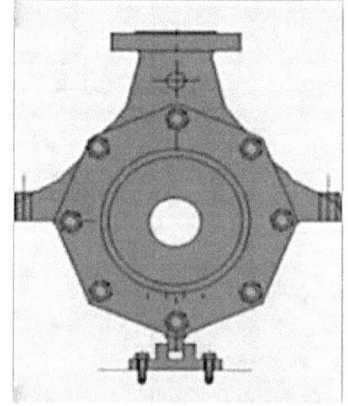

√ In case of pumps with "Lift Off Device," a gap of 1 mm additional is maintained between balancing disc & counter balancing disc at static condition. Hence, DBSE in static condition should be maintained with 1 mm higher from required value.

√ Position of Motor Rotor (applicable only for Motors with Shell bearing) is maintained as specified.

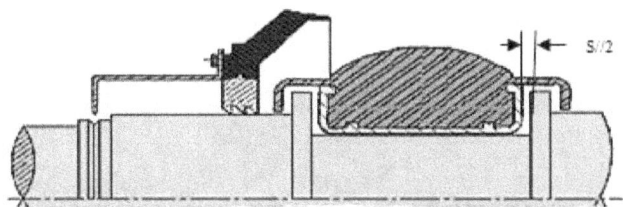

√ Operating position of the rotor of motor is measured by the dial indicator. At correct position, the clearance (S/2) between the rotor and the shell bearing on both ends is equalized.

5.2.8 Leveling and Alignment

– Carry out shuttering around the machined packer plates bolted to the Base frame & grout the pockets, up to half height of the machined packer plates. Use non shrink grout.

– Allow Curing time as recommended by grout manufacturer. Tighten the Foundation Bolts. Recheck & correct the level, if required.

– For correcting level, unbolt the machined packer plates from welded pads & insert suitable S.S. shims between machined packer plates & welded pads.

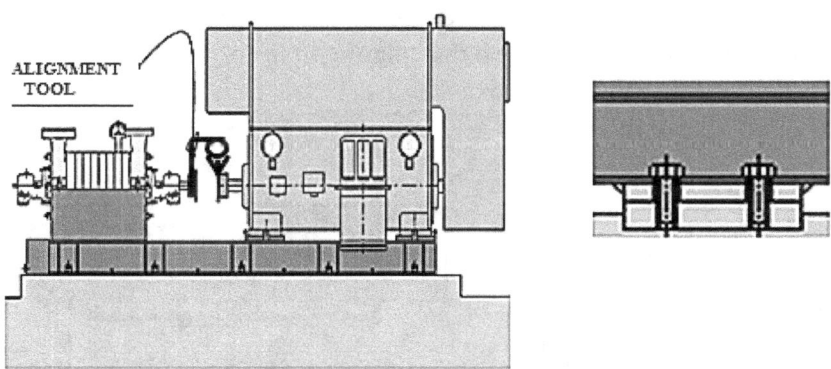

– Grout the Base frame up to its bottom surface with non shrink grout.

– Allow curing time as recommended by grout manufacturer.

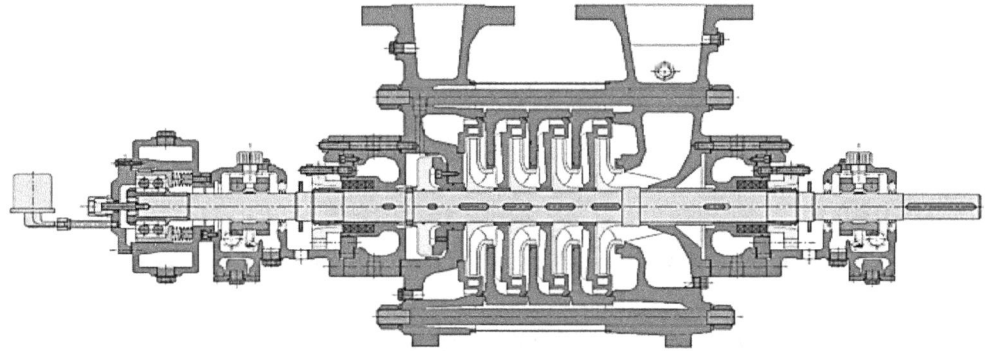

- All cavities in the Base frame like pedestals under the pump feet, hollow cavity under the Motor, etc. must be grouted fully up to the surface minus 2 mm, using conventional grout mix, i.e. Portland cement, sand & aggregate in proportion 1:2:2 & average gravel size not to exceed 16 mm.

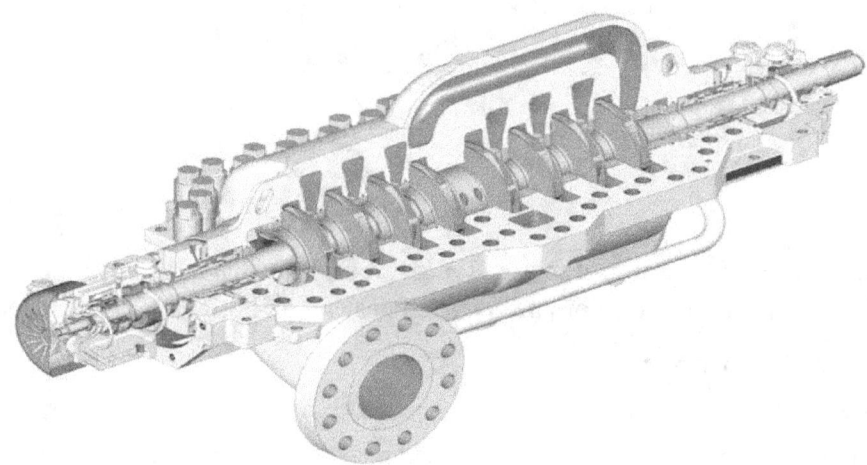

- Align the pump & the motor axially & radially within ± 0.02 mm. Install guide block for thermal expansion of the pump to avoid distortion & mis-alignment. Guide blocks are provided at the bottom of the casing which allows axial expansion without disturbing centerline. Tighten the Guide block bolts after the alignment.
- While alignment, a gap must be ensured between Jack bolts and motor base for alignment movements. Allow thermal expansion so that alignment is not disturbed and there is no distortion in pump components and piping system.

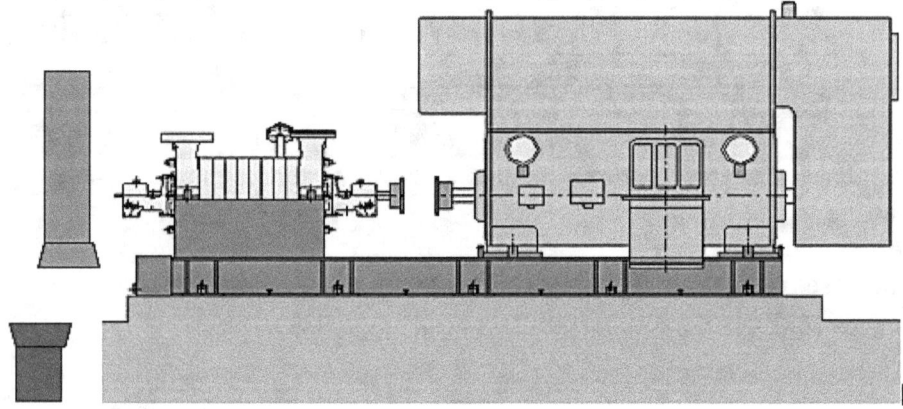

Cast the foundation block & apply oil resistant paint. *Pump set is released for piping load.*

5.3 Pump Pipings

a. Before connecting the piping with the Pump, check radial alignment between companion flanges true within ± 0.15 mm.

b. Make angular alignment of companion flanges true within ± 0.02 mm.

c. Correct gap between companion flanges to maintain angular mis-alignment.

d. Install ARC valve in delivery line & strainer in suction line.

e. Use gaskets of proper sizes & ratings. Tighten all connecting bolts uniformly in sequential order with specified torque.

f. Verify the alignment between the pump & the motor after tightening the piping flanges.

ALIGNED FLANGES   MIS-ALIGNED FLANGES   SUCTION STRAINER

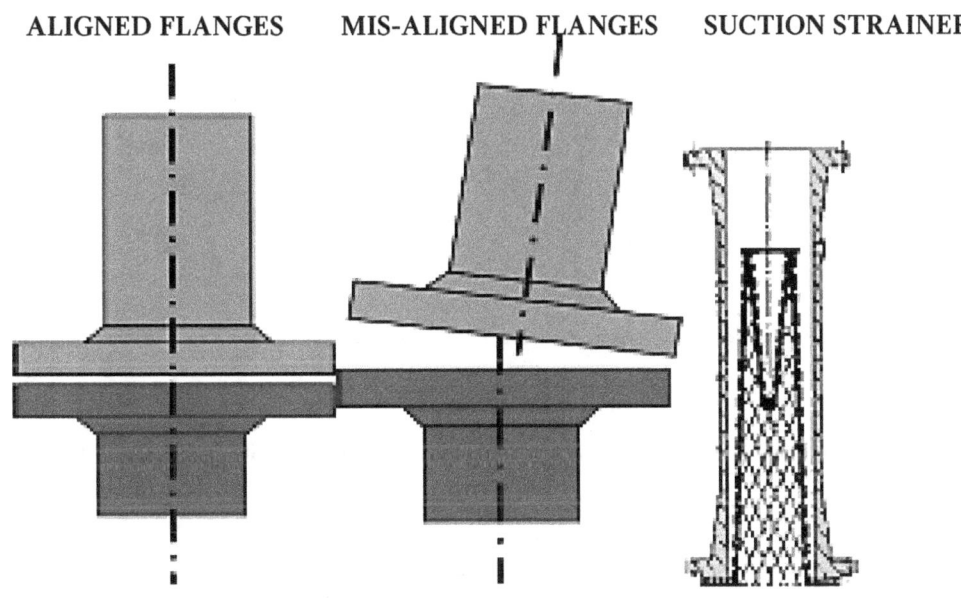

g. Doweling of the pump is carried out only after final checking & correction of the alignment and successful trial run.

h. Only suction feet of the pump are doweled in horizontal plane.

Discharge side is kept free for expansion. Guide blocks are provided at the bottom to take care of expansion in longitudinal and transverse directions.

5.3.1 Pipe Forces

The pipes connected to a pump always impose force and moments on the pump nozzles. Principal causes of such loading are as follows:

- Dead weight of the pipe
- Weight of the pumped liquid
- Steady-state internal pressure
- Pressure surges (e.g. generated by slamming of non-return valve flap or fast closing valves)
- Pressure pulsations
- Thermal expansion
- Seismic forces
- Constraint imposed due to wrong practices of erection.

Steady-state or transient internal pressure transmits full impact on the pump support when axial compensators or expansion sleeves are installed in the line. The torque T developed is given as: $T = 3.14 * p * D^2 * r$ must be taken care by the pump and pipe supports. Don't neglect unless the pressure p is very low.

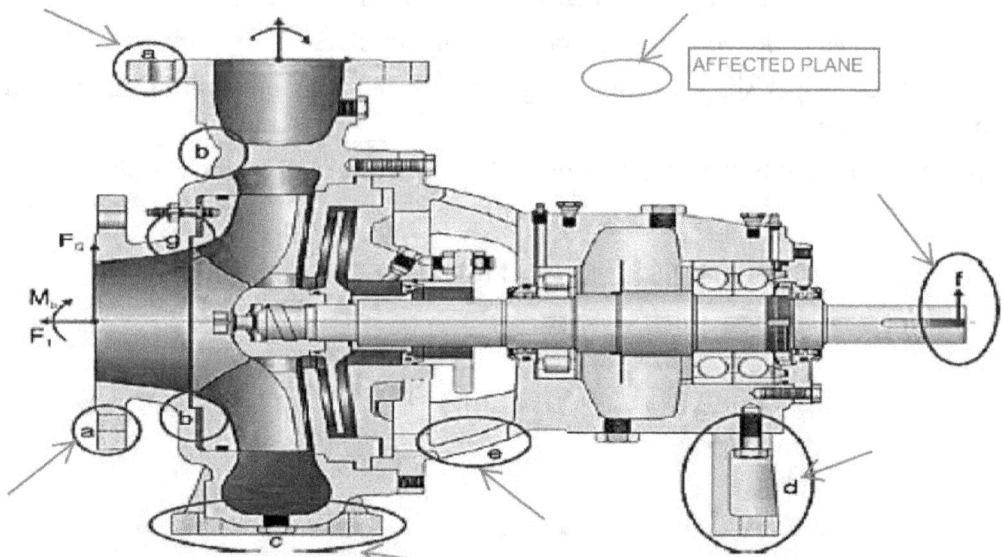

Effect of pipe loads on a pump

It is recommended to use joint compensators or hydrostatically balanced expansion sleeves, unless a sufficiently flexible pipe layout is designed. The effect of pipe loads on a pump is shown in sketches.

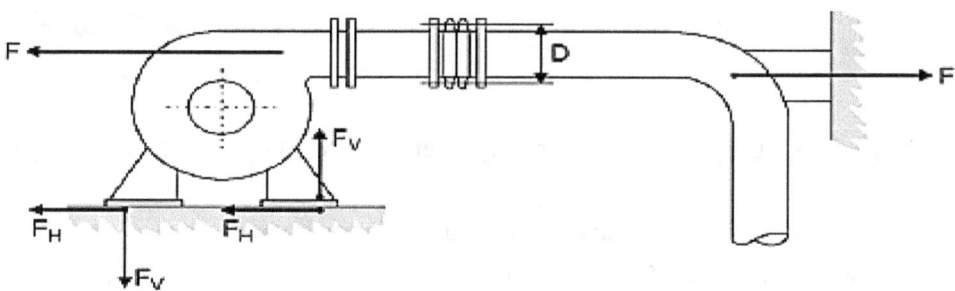

Unbalanced compensator and forces on pump

5.3.2 Stress Affected Areas by Nozzles Force

  a. nozzle flange
  b. casing feet with hold-down bolts
  c. highly loaded areas
  d. support foot with hold-down bolts
  e. radial displacement of coupling
  f. bearing housing deformation
  g. radial displacement of impeller (danger of rubbing). International standards for process pumps provide guidance with respect to allowable forces and moments on pump nozzles.

5.3.3 Points to Be Taken Care during Installation

  1. Selection of the most suitable material for the pipes and seal arrangements;
  2. Correct sizing of the pipe-cross-section to avoid sudden changes in velocity;

Pipe inside diameter (D in M) = $\{Q/(900 * 3.14 * v)\}^{1/2}$

where Q = Disch. (M^3/h); **v** = Velocity (M/sec)

In general the suction pipeline diameter should never be smaller than the pump intake nozzle.

Suction pipelines water velocity = 1.5–2.0 M/sec

Delivery pipelines water velocity = 2.0–3.0 M/sec

3. Correct choice and arrangement of anchor points in pipings layout;
4. Good accessibility of the pipeline to facilitate operation, maintenance and replacement;
5. Compensation for thermal expansion due to alterations in temperature.

5.3.4 Guidelines for Pump Design & Selection

- Preparation of pump performance specification used by the pump manufacturer for particular application.
- Preparation of pump performance specification used by the pump manufacturer for particular application.
- Used hydraulic and pump performance curves and data of manufacture.
- Combination of above to provide an accurate application requirement.
- Need of pumping like transfer of liquid, pressurizing, regulation of flow and head with given system resistance need to be considered.
- Prepare systematic piping flow, considering all equipment in position and operating conditions.
- Determine physical arrangement of piping with clear understanding of operating conditions, instrumentation and control devices.
- Calculate the hydraulic requirement of the pump, determine pump flow, head, NPSHa, suction requirement, system resistance head on which pump will operate, liquid specific gravity temperature and PH.
- MOC of pump components recommended for predicted life of pump.
- Pump trim/construction requirements, pump type, impeller design, shaft seal, shaft sleeves, bearing, coupling and other accessories for successful application.

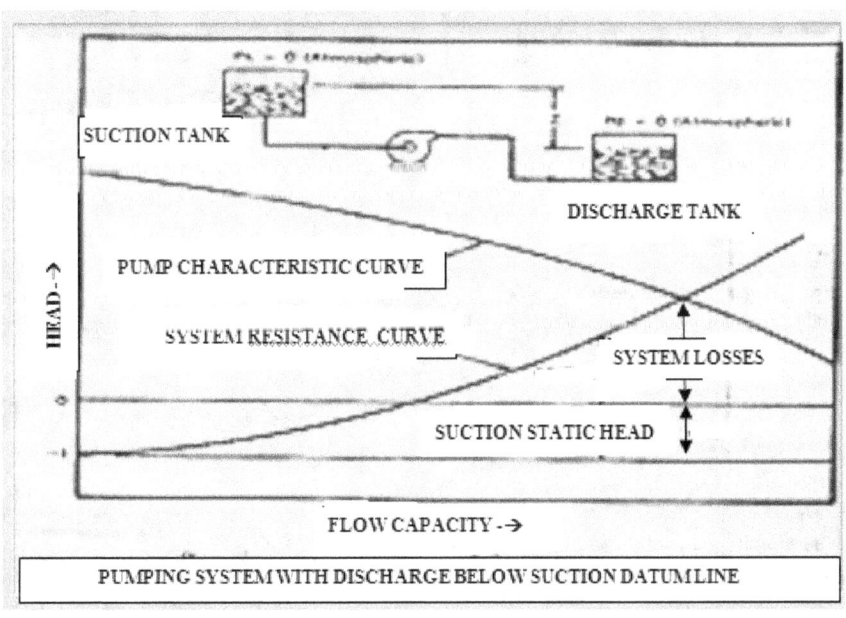

PUMPING SYSTEM WITH DISCHARGE BELOW SUCTION DATUMLINE

- Pump drive arrangement – single speed motor, variable speed motor/synchronous motor, VFD motor, steam turbine drive or diesel engine drive are adopted depending on–
  a. Time/day drive in use.
  b. Acceptable duration on for lost or partial capacity.
  c. Need of variable pumping rates to the piping system.
  d. Importance and type of remote/auto start and stop control.
  e. Pump characteristic.
  f. First cost involvement and annual cost of consumable. and spares in running.

5.3.5 Affinity Laws

Flow: $Q_2 = Q_1 * (N_2/N_1) * (D_2/D_1)$ .... (a)

Head: $H_2 = H_1 * (N_2/N_1)^2 * (D_2/D_1)^2$ .... (b)

Power: $HP_2 = HP_1 * (N_2/N_1)^3 * (D_2/D_1)^3$ .... (c)

Where;

Q = Pump flow (M^3/h), H = Pump head in M,

HP = Horse power, BHP(Kw)

D = Impeller dia (m), N = RPM of Pump in operation

Assumption for efficiency: $\xi p1 = \xi p2$

5.4 Trouble Shooting of Centrifugal Pump

Trouble shooting of centrifugal pumps is presented in tabulated form which is useful to all operating engineers and maintenance crues as a ready reference. This chart in general is prepared to suit almost all type of centrifugal pumps which are popularly used in industries, especially power plants.

Sl. No.	Possible cause of Trouble	Pump does not deliver liquid	Insufficient capacity delivered	Insufficient pressure develop	Pump loses prime after starting	Pump requires excessive power	Stuffing box leaks excessively	Packing has short life	Pump vibrates or is noisy	Bearings have short life	Pump overheats and seizes
	Suction Troubles										
1	Pump not primed	✓									✓
2	Pump or suction pipe not completely filled with liquid	✓	✓		✓				✓		
3	Suction lift too high	✓	✓		✓				✓		
4	Insufficient margin between suction pressure and vapor pressure	✓	✓						✓		✓
5	Excessive amount of air or gas in liquid	✓	✓	✓	✓						
6	Air pocket in suction line	✓	✓		✓						
7	Air leaks into suction line		✓		✓						
8	Air leaks into pup through stuffing boxes		✓		✓						
9	Foot valve too small		✓						✓		
10	Foot valve partially clogged		✓						✓		
11	Inlet of suction pipe insufficiently submerged	✓	✓		✓				✓		
12	Water-seal pipe plugged				✓			✓			

TROUBLE SHOOTING OF CENTRIFUGAL PUMP

Sl. No.	Possible cause of Trouble	Pump does not deliver liquid	Insufficient capacity delivered	Insufficient pressure develop	Pump loses prime after starting	Pump requires excessive power	Stuffing box leaks excessively	Packing has short life	Pump vibrates or is noisy	Bearings have short life	Pump overheats and seizes
13	Seal cage improperly located in stuffing box, preventing sealing fluid from entering space to form the seal				✓		✓	✓			
	System Troubles										
14	Speed too low	✓	✓	✓							
15	Speed too high					✓					
16	Wrong direction of rotation	✓		✓		✓					
17	Total head of sytem higher than design head of pump	✓	✓	✓		✓					
18	Total head of system lower than pump design head					✓					
19	Specific gravity of liquid different from design					✓					
20	Vicosity of liquid differs from that for which designed		✓	✓		✓					
21	Operation at very low capacity								✓		✓
22	Parallel operation of pumps unsuitable for such operation	✓	✓	✓							✓
	Mechanical Troubles										
23	Foreign matter in impeller	✓	✓			✓			✓		
24	Misalignment					✓	✓	✓	✓	✓	✓
25	Foundations not rigid								✓		
26	Shaft bent					✓	✓	✓	✓	✓	
27	Rotating part rubbing on stationary part					✓			✓	✓	✓
28	Bearings worn							✓	✓	✓	✓
29	Wearing rings worn		✓	✓		✓					
30	Impeller damaged		✓	✓					✓		
31	Casing gasket defective, permitting internal leakage		✓	✓							
32	Shaft or shaft sleeves worn or scored at the packing						✓	✓			
33	Packing improperly installed					✓	✓	✓			
34	Incorrect type of packing for operating conditions					✓	✓	✓			
35	Shaft running off-center because of worn bearings or misalignment						✓	✓	✓	✓	✓
36	Rotor out of balance, resulting in vibration						✓	✓	✓	✓	✓
37	Gland too tight, resulting in no flow of liquid to lubricate packing					✓		✓			
38	Failure to provide cooling liquid to water cooled stuffing boxes						✓	✓			
39	Excessive clearance at bottom of stuffing box between shaft and casing					✓	✓	✓			
40	Dirt or grit in sealing liquid, leading to scsoring of shaft or shaft sleeve						✓	✓			
41	Excessive thrust caused by mechanical failure inside the pip or by the failure of the hydraulic balancing device, if any								✓	✓	✓

TROUBLE SHOOTING OF CENTRIFUGAL PUMP

Symptoms →

	TROUBLE SHOOTING OF CENTRIFUGAL PUMP										
Sl. No.	Symptoms ⟶ Possible cause of Trouble	Pump does not deliver liquid	Insufficient capacity delivered	Insufficient pressure develop	Pump loses prime after starting	Pump requires excessive power	Stuffing box leaks excessively	Packing has short life	Pump vibrates or is noisy	Bearings have short life	Pump overheats and seizes
42	Excessive grease or oil in antifriction bearing housing or lack of cooling, causing excessive bearing temperature								✓	✓	
43	Lack of lubrication								✓	✓	
44	Improper installation of antifriction bearings (damage during assembly, incorrect assembly of stacked bearings, use of unmatched bearings as a pair, etc.								✓	✓	
45	Dirt getting into bearings								✓	✓	
46	Rusting of bearings due to water getting into housing								✓	✓	
47	Excessive cooling of waster-cooled bearings, resulting in condensation in the bearing housing of moisture from the atmosphere.								✓	✓	
	Abbreviation						✓ **Probable Reason**				

Chapter 6

SPECIFIC SPEED OF PUMP

The Specific speed (Ns) is a dimension less index which is defined as speed in RPM of a pump- impeller at which a geometrically similar impeller would operate to deliver one GPM flow at one foot head. Specific speed (Ns) is a non-dimensional design index that identifies the geometric similarity of pumps. It is measure of the shape, class, type and proportions of the impellers. The pumps of the same Ns but of different size are considered to be geometrically similar in design.

6.1 Specific Speed and Impeller Configuration

6.1.1 Specific Speed of Impeller

There are verities of pump designs available for performing different task... Pump designers have deviced a way to compare the efficiency of their designs across a large range of pump model & types. Also efficiency expected from a particular pump design is checked. For this purpose pump have been tested and compared using a number or criteria called the specific speed (NS) which helps to do these comparisons. The efficiency of pumps having same specific speed can be compared as a benchmark for improving the design or increase the efficiency... Equation for specific speed is used to compute the value for the pump specific speed, knowing the pump total head (H), the speed of the impeller (n0 and the flow rate Q.

$$N_S = \frac{n(rpm) \times \sqrt{Q(USgpm)}}{H(ft\ fluid)^{0.75}}$$

6.1.2 Suction Specific Speed

Suction specific speed (Nss) is a number that is dimensionally similar to the pump specific speed and is used as a guide to prevent cavitations.

$$N_{SS} = \frac{N\ (rpm) * \sqrt{Q\ (gpm)}}{NPSH_A^{\ 0.75}\ (ft\ fluid)}$$

Specific speed is also used to evaluate the efficiency of standard volute pumps (see following graph). It is to be noted that larger pumps are inherently more efficient and their efficiency drops rapidly at specific speeds of 1000 or less.

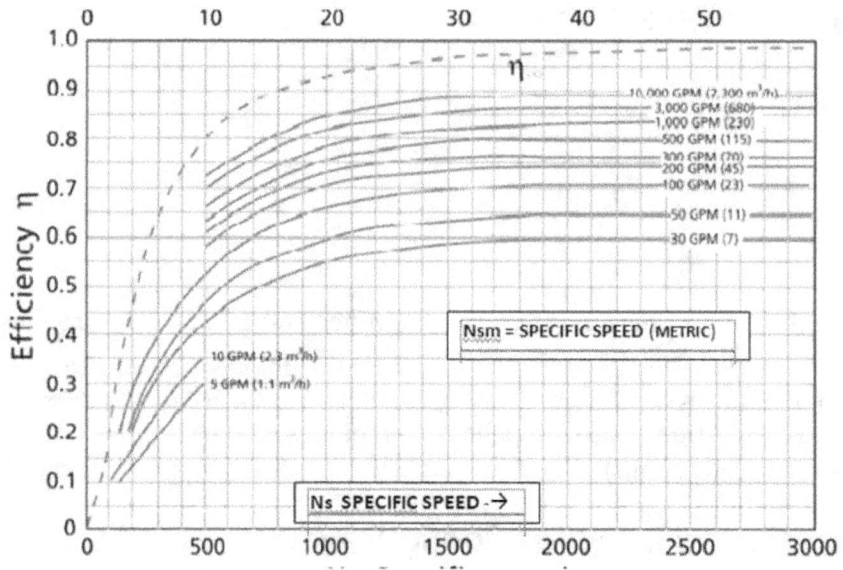

EFFICIENCY OF PUMP WITH DIFFERENT SPECIFIC SPEEDS

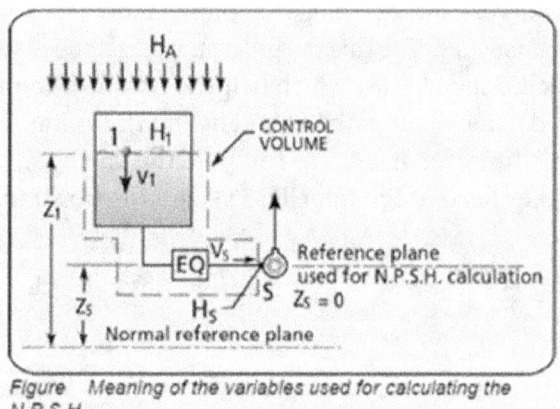

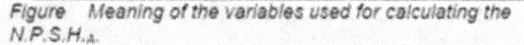

Figure   Meaning of the variables used for calculating the N.P.S.H.$_A$.

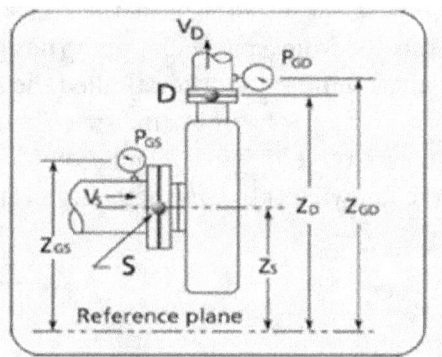

Figure   Location of variables for measuring N.P.S.H.$_A$

We can avoid doing the calculations for equation [4] by measuring the N.P.S.H... The value for the N.P.S.H.A can be computed by taking a pressure measurement at the pump inlet and using equation below–

$$N.P.S.H.\ avail(ft\ fluid\ absol.) = 2.31\ \frac{p_{_{GS}}\ (psig)}{SG} + z_{_{GS}} - z_{_s} + \frac{v_{_S}^2}{2\ g} + H_A + H_{va}$$

If we consider an increase in the pump's speed to increase the flow rate. There will be increase in N.P.S.H (required) also. The suction specific speed is an indication of what the impeller speed limitation for a given N.P.S.H.(A) . The recommendation of Hydraulic Institute states that suction specific speed be limited to 8500 to avoid cavitations. However, some experiments have shown that suction specific speed can be as high as 11000.

Suction specific speed Nss can be obtained from given parameters as–

$$\frac{n(rpm) \times \sqrt{Q(USgpm)}}{N.P.S.H._A(ft\ fluid)^{0.75}} = \frac{1750\ x\ \sqrt{500}}{15^{0.75}} = 5130$$

Thus Suction specific speed Nss is well below 8500... We can easily calculate the new suction specific speed if we have to change the speed of impeller... High pump specific speed indicates that impeller inlet area is large which is reducing the inlet velocity i.e. a low NPSH(R). However, if we continue to increase the impeller inlet area to reduce NPSH(R), a point is reached where the inlet area is too large resulting in suction recirculation (hydraulic instability causing vibration, cavitations, erosion etc). The limitation on Nss value avoids reaching the point. of hydraulic instability.

Sustaining suction specific speed below 8500 is also a way of finding the maximum speed of a pump to avoid cavitations. In case of a double suction pump, in computation of Nss value of Q is considered half of the value of Q_{ACTUAL}.

As per Hydraulic Institute, the efficiency of the pump is maximum when the suction specific speed is between 2000 and 4000. When Nss lies outside this range the efficiency will be de-rated according to the following figure.

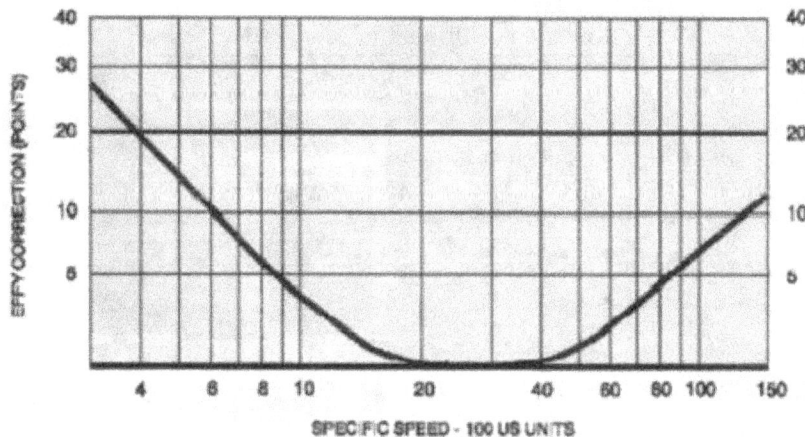

Figure   Pump efficiency correction due to suction specific speed.

As the specific speed increases, the ratio of the impeller outlet eye diameter to that of inlet (or D_s/D_m) will decrease. The ratio approaches to 1.0 for a true axial-flow-impeller. Radial flow impellers develop head principally by centrifugal force whereas axial flow impellers develop head more by axial forces and less by centrifugal force. Radial impellers are generally low flow high head and in contrast to that axial flow impellers are high flow & low head design.

6.2 From Formula of Specific Speed

WE CAN PREDICT–

 i.   A higher specific speed will develop less head and more flow eg. axial flow impeller

 ii. Specific speed identifies the approximate acceptable ratio of the impeller eye diameter (D_s) to the impeller maximum diameter (D_m) in designing efficient impeller.

 iii. Low specific speed impellers (eg. the radial impeller) will develop high head and low flow. Specific speed is normally a design factor in development of new pump. The prototype pumps of the same specific speed are developed for prediction of performance. It is an index number, descriptive of the suction characteristics of given pump design. Each design of impeller has a range of specific speeds in which it will render the best performance.

iv.

Specific speed of impeller	Characteristics of impeller
Ns: ≤ 1000	Radial vane area impellers
Ns: 500 to 5000; D1/D2 > 1.5	radial flow impellers
Ns: 4,000 to 4,500	Francis vane area impellers
Ns: 5000 to 10000; D1/D2 < 1.5	mixed flow impellers
Ns: 9,500 to 10,000	Mixed flow area impellers
Ns: 10000 to 15000; D1/D2 = 1	Axial flow impellers

6.2.1 Recommended Specific Speed of Impellers in Power Plants

Applications	Typ. Impeller profile	Spec. speed Ns
Boiler feed pumps	Radial vane or Francis vane area of impellers	1000 to 1800
Condensate pumps	Francis vane area impellers	1500 to 2500
Circulating pumps	Mixed flow area of impeller	4500 to 7500

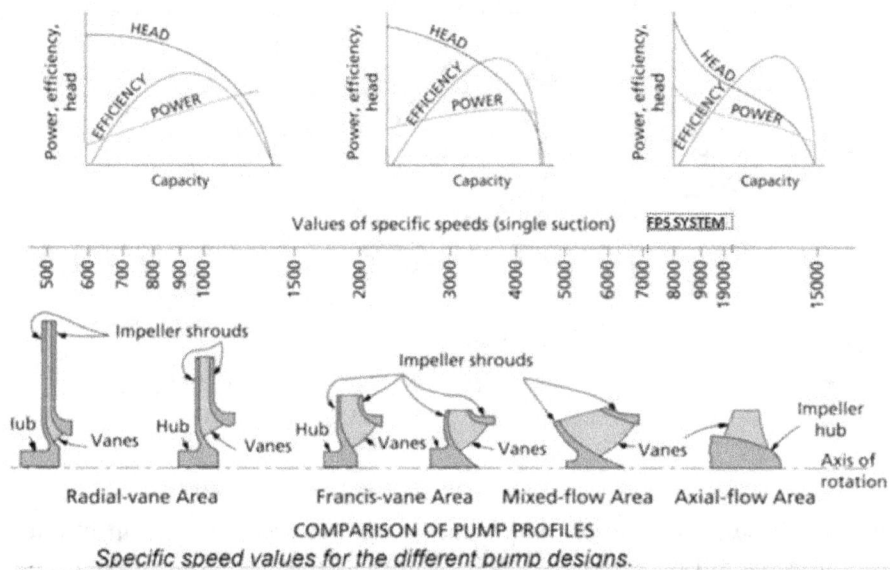

Specific speed values for the different pump designs.

6.2.2 Affinity Laws

The affinity laws are derived from a dimensionless analysis of three important parameters twhich describe pump performance (flow, total head and power). The reduced impeller are geometrically similar and operate at dynamically similar specific speed and the affinity laws may predict the performance of the pump at different diameters for the same speed or different speed but at same diameter... Since, in practice impellers of different diameters are not geometrically identical hence, change of impeller diameter more than 10 to 20% are recommended. To avoid over cutting the impeller, trimming should be done in steps with careful measurements and at each step of trimming, compare predicted performance with the measured one and adjust accordingly. The affinity laws were developed using the law of similitude as follows–

AFFINITY LAWS -				
Flow vs diameter	$\dfrac{Q}{nD^3} = K$	or		$\dfrac{Q_1}{Q_2} = \dfrac{n_1}{n_2}\dfrac{D_1^{3}}{D_2^{3}}$
Total head vs diameter	$\dfrac{gH}{n^2 D^2} = K$	or		$\dfrac{H_1}{H_2} = \dfrac{n_1^{2}}{n_2^{2}}\dfrac{D_1^{2}}{D_2^{2}}$
power vs diameter and speed	$\dfrac{P}{\dfrac{\gamma}{g} n^3 D^5} = K$	or		$\dfrac{P_1}{P_2} = \dfrac{n_1^{3}}{n_2^{3}}\dfrac{D_1^{5}}{D_2^{5}}$
Subscript 1 *2 denotes value before and after change. H= total head, P = power, n = speed (RPM), D = Impeller diameter,				
When n is constant, we can write affinity laws as –	$\dfrac{Q_1}{Q_2} = \dfrac{D_1^{3}}{D_2^{3}}$	$\dfrac{H_1}{H_2} = \dfrac{D_1^{2}}{D_2^{2}}$		$\dfrac{P_1}{P_2} = \dfrac{D_1^{5}}{D_2^{5}}$
When D is constant, we can write affinity laws as -	$\dfrac{Q_1}{Q_2} = \dfrac{n_1}{n_2}$	$\dfrac{H_1}{H_2} = \dfrac{n_1^{2}}{n_2^{2}}$		$\dfrac{P_1}{P_2} = \dfrac{n_1^{3}}{n_2^{3}}$
Affinity law do not apply to points which belongs to system curves				

It is assumed that the two operating points are compared at the same efficiency. The two operating points, say 1 and 2, depends on the shape of the system curve (see Figure).

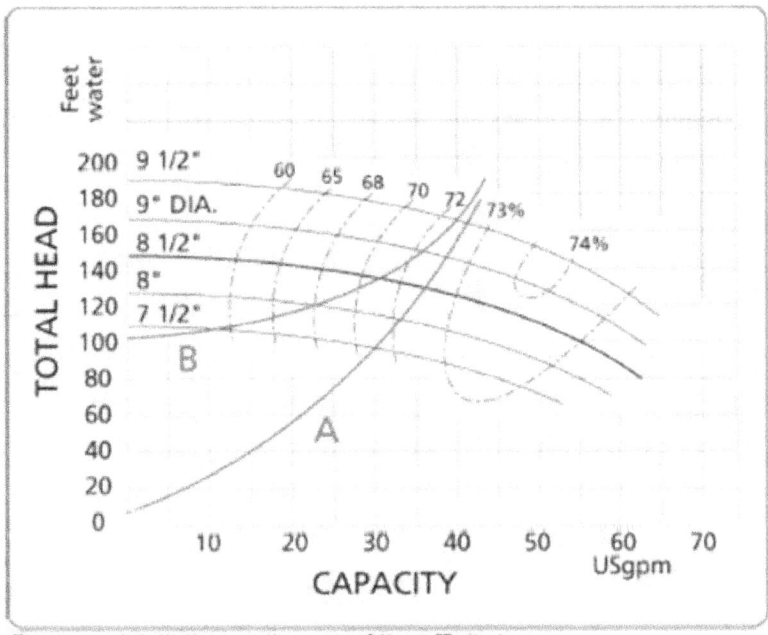

Figure   Limitation on the use of the affinity laws.

The points that lie on system curve A will all be approximately at the same efficiency. whereas the points that lie on system curve B are not. The affinity laws do not apply to points that belong to system curve B. which describes a system with a relatively high static head compared to. System curve A.

6.2.3 Orifice Effect & Water Flow Through Pipes

• The necessary hole-diameter **dBi** of the orifice is calculated from the head difference to be throttled ΔH, using the equation:

$$dBi = f * Q/g * \Delta H$$

where

dBi = Hole diameter of the orifice in mm   **f** = Throttling or pressure drop coefficient.

Q = Flow rate in m3/h;   **g** = Gravitational constant 9.81 m/s²

ΔH = Head difference to be throttled in

Since, the area ratio (dBi/d)² must be estimated in advance, an iterative calculation is necessary.

• **PROPERTIES OF WATE`R AT DIFFERENT TEMPERATURES**

PROPERTIES OF WATER AT DIFFERENT TEMPERATURES						
Density of water at 20 deg C = 1.0 (or in FPS system@ 68 deg F density = 62.32 lb/cuft)						
Temp. F	Temp. C	pv g psi	pv g (kg/cm²)	pv abs ft	pv abs M	Spec. Gr
50	10			0.41	0.123	1.002
60	15.55556			0.59	0.177	1.001
70	21.11111			0.84	0.252	1
80	26.66667			1.17	0.351	0.998
90	32.22222			1.62	0.486	0.997
100	37.77778			2.2	0.66	0.995
110	43.33333			2.96	0.888	0.993
120	48.88889			3.95	1.185	0.99
130	54.44444			5.2	1.56	0.988
140	60			6.78	2.034	0.985
150	65.55556			8.74	2.622	0.982
160	71.11111			11.2	3.36	0.979
170	76.66667			14.2	4.26	0.975
180	82.22222			17.85	5.355	0.972
190	87.77778			22.3	6.69	0.968
200	93.33333			27.6	8.28	0.965
210	98.88889			34	10.2	0.961
220	104.4444	2.49	0.169387755	41.45	12.435	0.957
230	110	6.07	0.41292517	50.35	15.105	0.953
240	115.5556	10.27	0.698639456	60.75	18.225	0.948
250	121.1111	15.12	1.028571429	73	21.9	0.944
260	126.6667	20.72	1.40952381	87.35	26.205	0.939
270	132.2222	27.15	1.846938776	103.3	30.99	0.935
280	137.7778	34.48	2.345578231	122	36.6	0.93
290	143.3333	42.85	2.914965986	144	43.2	0.925
300	148.8889	52.3	3.557823129	169	50.7	0.92

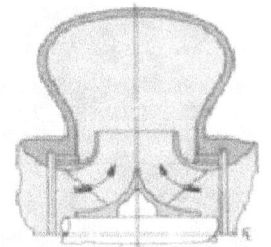

Figure a. Radial flow
pump cross-section.

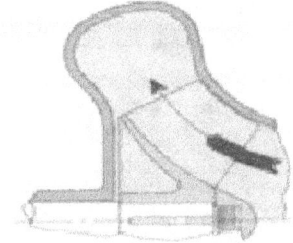

Figure b. Mixed flow
pump cross-section.

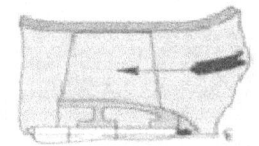

Figure c. Axial flow
pump cross-section.

When Nss lies beyond 2000 – 4000, the efficiency is de-rated as per graph given below–

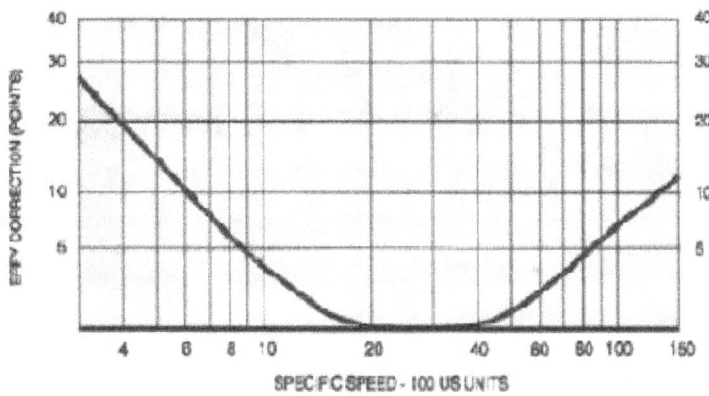

Pump efficiency correction

Velocity Diagram of an Axial Flow Pump

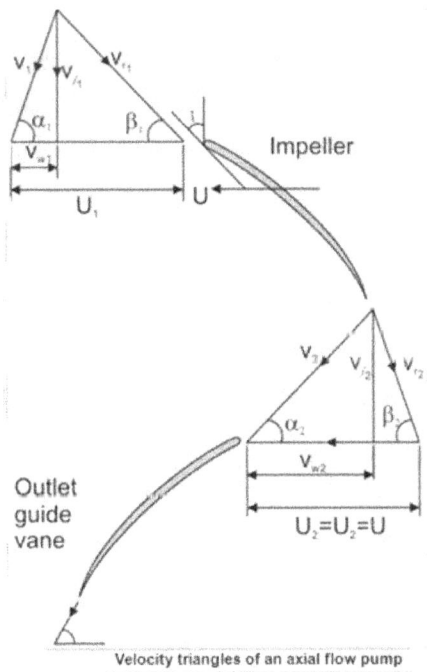

Velocity triangles of an axial flow pump

Maximum energy transfer to the fluid per unit weight $= u(u - V_f \cot \beta_2) \big/ g$

Velocity Diagram of Centrifugal Pump

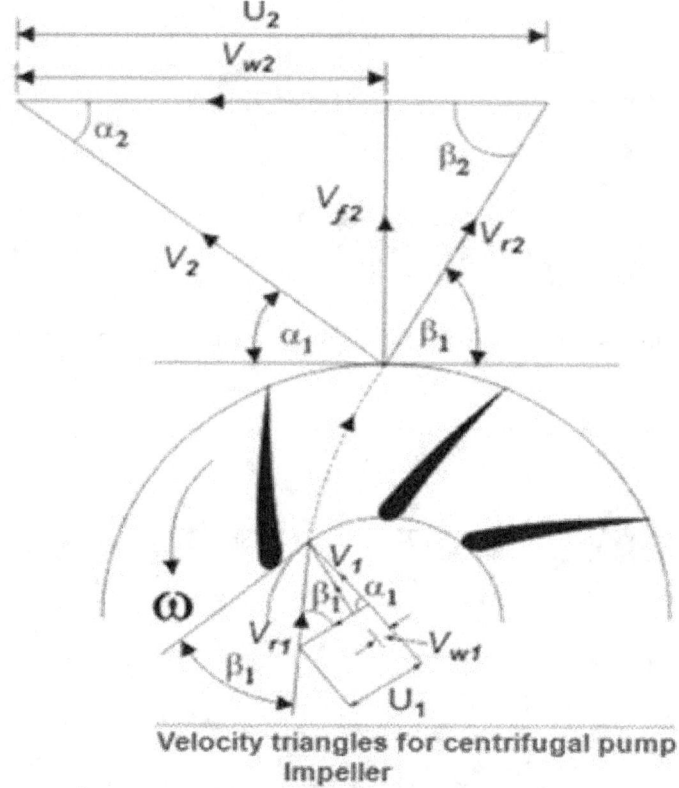

Velocity triangles for centrifugal pump Impeller

Work done on the fluid per unit weight = $V_{w2} U_2 / g$

Chapter 7
CHARACTERISTIC CURVES

7.1 Characteristic Curve for Centrifugal Pumps

The characteristic cure of centrifugal pump (Q vrs H) at low rage of discharge is unstable. The streep characteristics curve is avoided for pump operation as it renders unstability in parellel operation or in transfer of liquid into accumulator, filled with gas or steam as shown in following graphs–

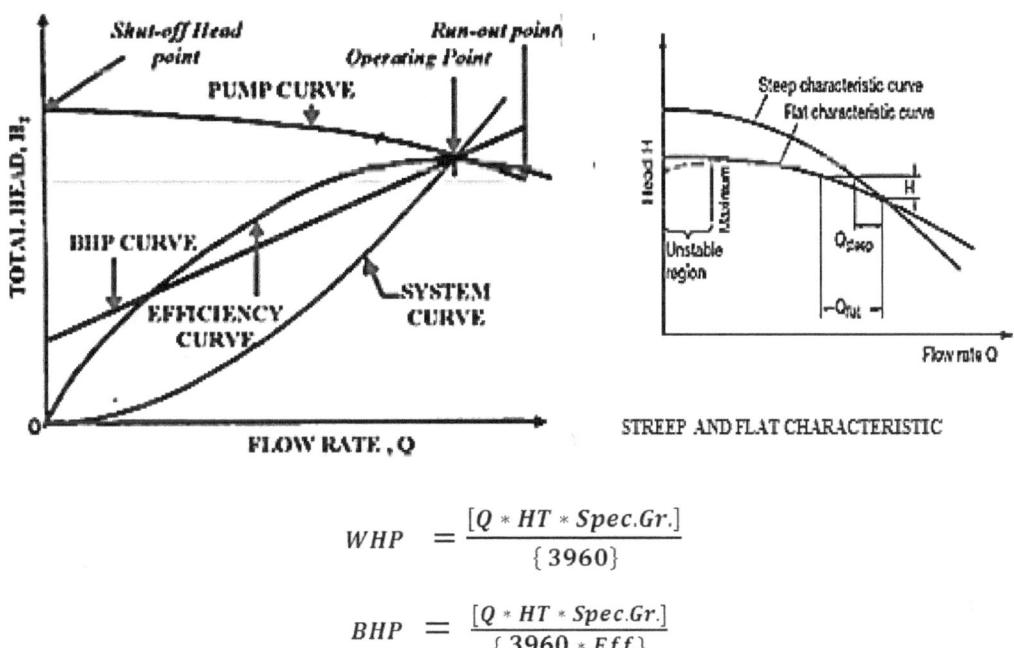

STREEP AND FLAT CHARACTERISTIC

$$WHP = \frac{[Q * HT * Spec.Gr.]}{\{3960\}}$$

$$BHP = \frac{[Q * HT * Spec.Gr.]}{\{3960 * Eff.\}}$$

where–

- Q = Capacity (GPM); H_T = Total differential head (ft)
- Spec. Gr. = Specific gravity of liquid; Eff. = Pump Efficiency
- Pump Efficiency = WHP/BHP

7.2. Understanding Different Heads of Centrifugal Pump

7.2.1 Total Head

Total head and flow are the main criteria which are used to compare performance of one pump with other. Secondly selection of a centrifugal pump for desired application is mostly based on these parameters.

Total head is related to the discharge pressure of the pump. The pump manufacturers normally do not know the exact criteria of pump applications except discharge and head while selecting the pump. The discharge pressure to some extent depends on the pressure available at suction of the pump. The discharge pressure will be different for negative and positive suction head in the same pump. Therefore, to eliminate such confusions, the standard practice, is adopted to define pump pressure in terms of Total head (i.e. the difference of pressure at the outlet and inlet eyes of the pump).

Further, the pressure produced by pump will also depend on liquid density (higher pressure for high fluid density keeping same flow rate). *The fluid in the measuring vertical tube at discharge or suction eye of the pump will rise to the same height for all fluids regardless of the density.*

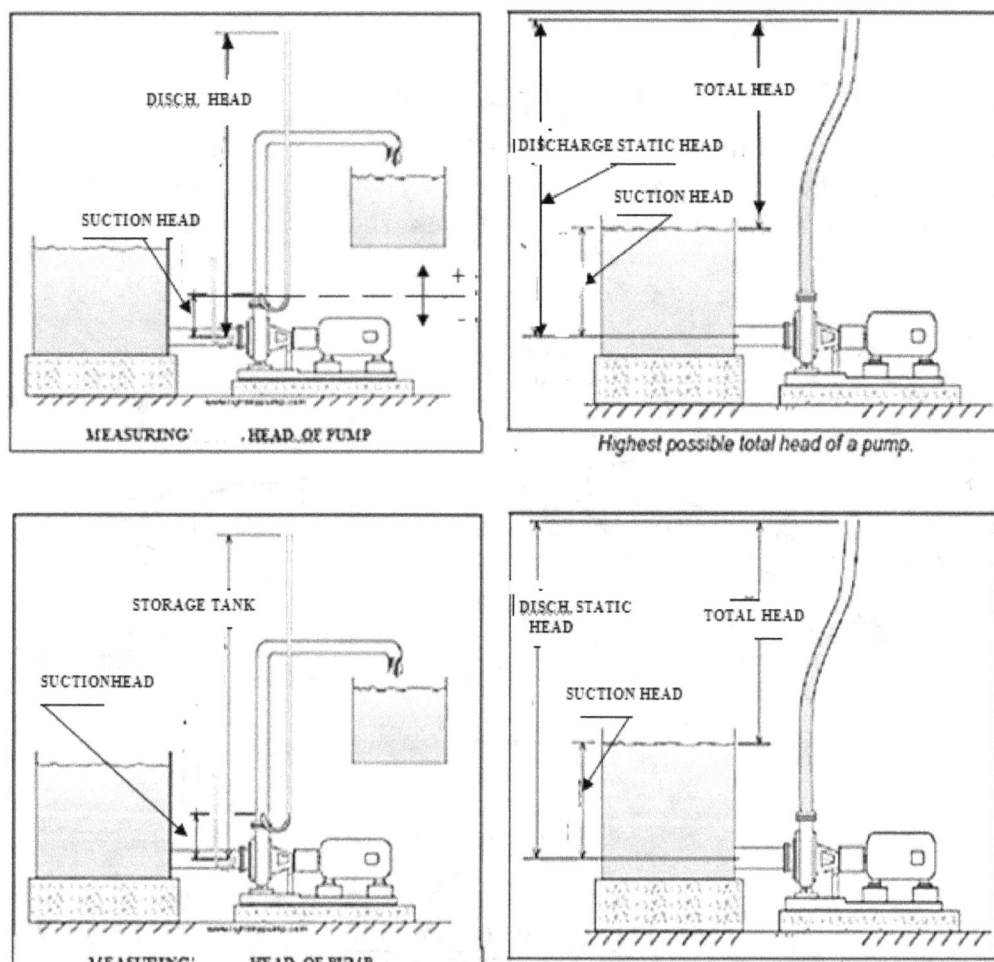

Highest possible total head of a pump.

Highest possible total head of a pump.

7.2.2 Total Head Is the Height That the Liquid Is Raised by the Pump with Zero Discharge Less the Height of the Liquid, at the Suction Side Which Can Be Positive or Negative

Energy is expressed in Kg.M or foot-pounds which is the amount of force required to lift an object up multiplied by the vertical distance, similar to weight lifting.

Assume if 100 pounds (445 Newtons) weight is lifted upto 6 feet (1.83 m), the energy required is 6 * 100 = 600 ft-lbf (814 N-m). Therefore, head can be defined as energy divided by the liquid weight displaced.

Similarly, in the case of weight lifting, the energy divided by the weight displaced will render the weight lifted to the height 'h' = force (600)/100 = 6 feet (1.83 m). This is not useful to the weight lifter but in pumping system, the displacement height of liquid is an important parameter.

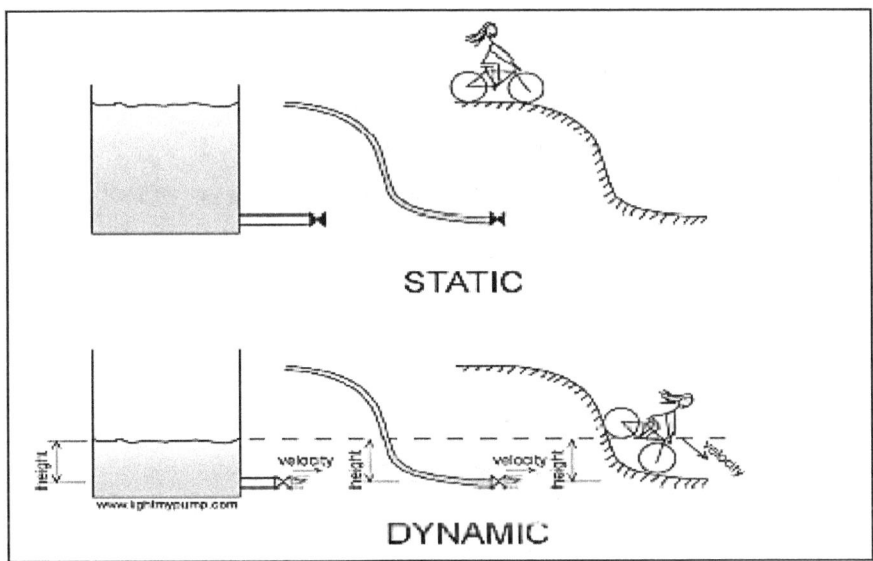

The relationship between height, pressure and velocity.

In the figure above we see a tank full of water, a tube full of water and a cyclist at the top of a hill. The tank produces pressure at the bottom and so does the tube. The cyclist has elevation energy that he will be using as soon as he moves.

The energy that the pump must supply is the frictional energy plus energy to lift the liquid to height. The liquid lifted to a height contains elevation/potential energy. If the liquid escapes the system at high velocity then we would have to consider the velocity energy also but normally velocity is considerably low hence, it can be neglected.

Therefore,   **PUMP ENERGY = FRICTION ENERGY + ELEVATION or POTENTIAL ENERGY**

For example we consider tanks of different sizes from small to big, the pressure will be the same everywhere at the bottom, no matter how big the surface of tank is made as long as the surfaces are at the same level. It is often expressed in kg/cm² (psi).

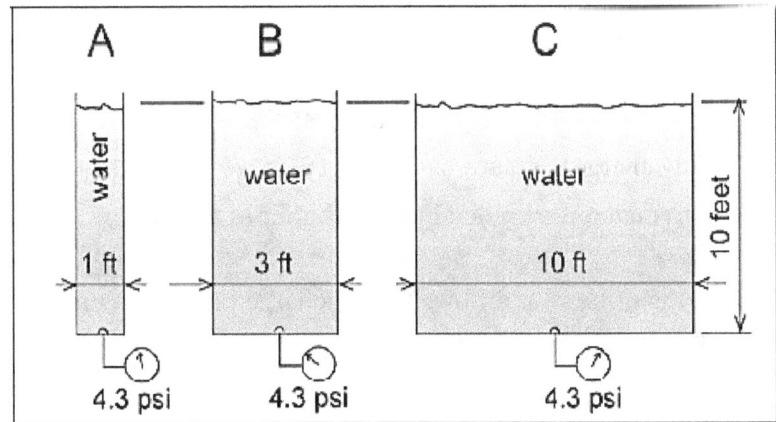

Pressure at the bottom of tanks of various sizes.

The force (weight of water) per unit area is the pressure. The density of water is taken as 1.0 gm/cc or 62.3 pounds per cubic foot. Thus pressure is force acting on unit surface area.

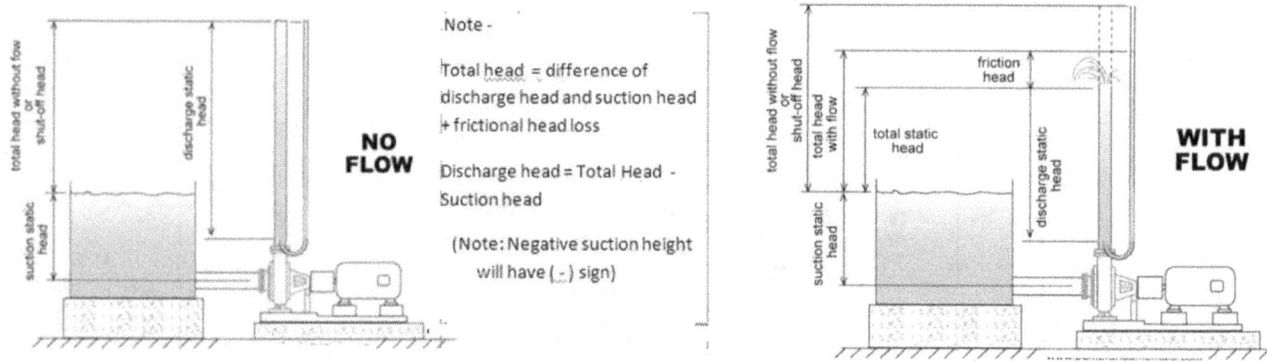

7.2.3 Negative Suction Pumping System

The sketch is self explaiatory for different terms used in the system.

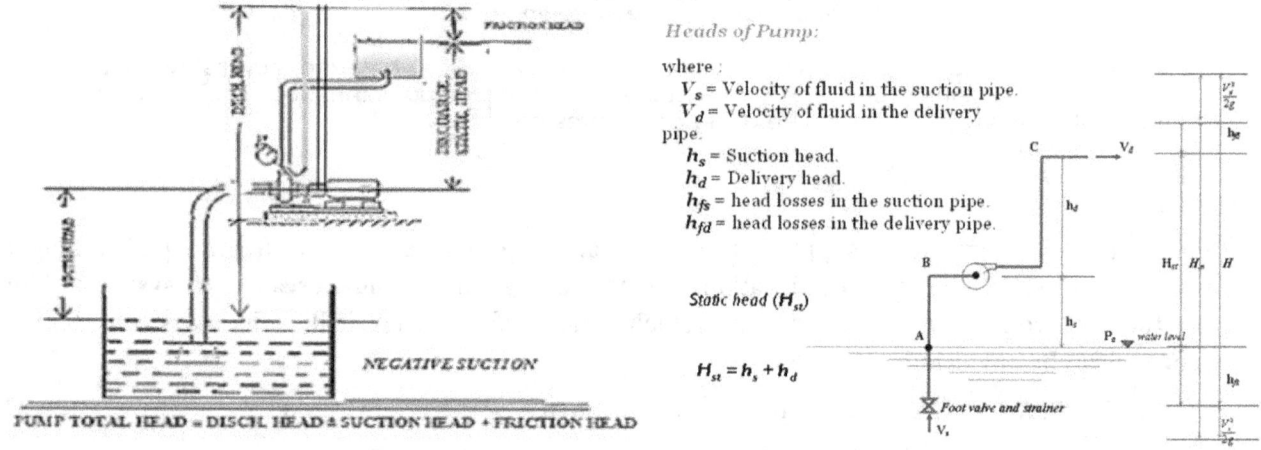

7.2.4 The Total Head

THE TOTAL HEAD is equal to the difference between the pressure head at the discharge Hd and the pressure head at the suction Hs. Theoretically Hs equals to −32 ft but practically −15 feet approx (due to frictional head loss in suction pipings). Total head = Hd − Hs. Therefore: Hd = Total Head + (± Suction head)

$$Hd = 100 + (-15) = 85 \text{ feet}$$

$$\text{The discharge pressure} = p = 1.0 * (85 \ 37/2.31) = 37 \text{ psig}$$

Measure discharge pressure to match the preded head. If that does not conform, there may be something wrong with the pump.

Pumps are often rated in terms of head and flow. The discharge pipe end is raised to a height at which flow is just stopped. We measure the height from pump shaft centre to static column top at zero discharge.

7.2.5 The Static Head

The head depends on flow rate but in this case since there is no flow and hence no friction. Thus, head of the pump is *THE MAXIMUM HEIGHT THAT THE FLUID CAN BE LIFTED REFERENCE TO THE SURFACE OF THE SUCTION TANK.*

No flow → no frictional head loss, the total head produced by pump will equal to the STATIC HEAD.

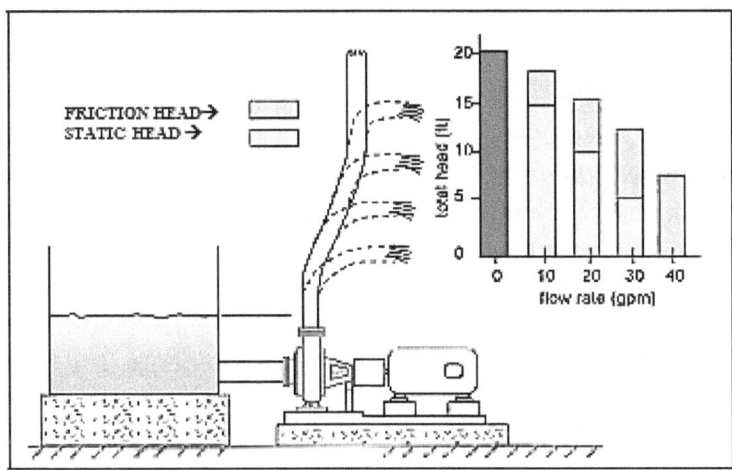

Variation of total head vs. pipe end elevation

THE STATIC HEAD – the discharge pipe end is raised vertically until the flow stops, the pump cannot raise the fluid higher than this point and the discharge head is maximum. The total head analogy of pump can be compared with the cyclist who applies maximum force to the pedals without climbing any distance uphill.

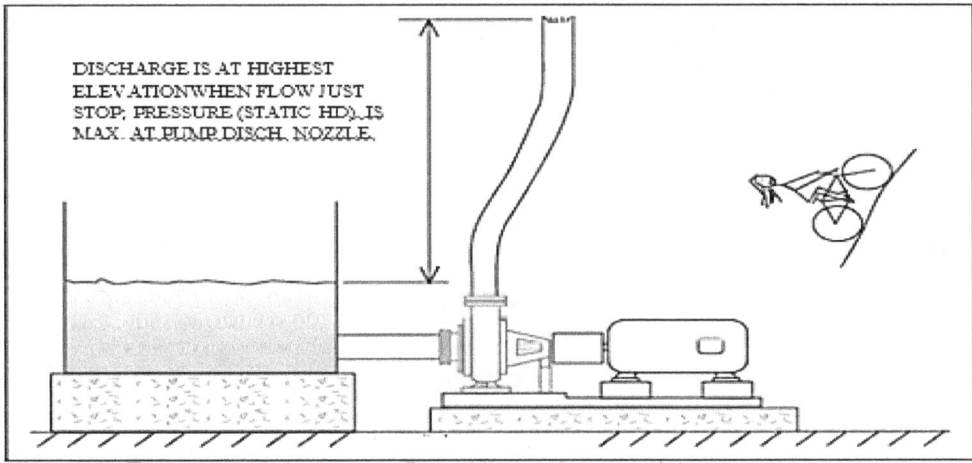

The pump produces zero flow at its maximum outlet pressure.

If the liquid surface of the suction tank is at the same elevation as that at discharge end of the pipe then the static head will be zero and the flow rate will be limited by the friction in the system. The analogy of pump in this case can be compared with a cyclist on a flat road, where velocity depends on the amount of friction between the wheels and the road and the air resistance.

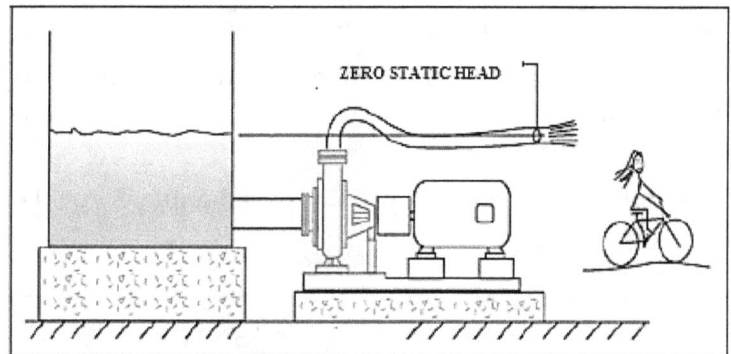

Flow rate is limited by friction in the system when the static head is zero.

If the discharge pipe end is lower than the liquid surface of the suction tank then the static head will be negative and the flow rate high. If the negative static head is large then it is possible that a pump is not required to transfer energy to move liquid since the energy provided by the difference in elevation may be sufficient to move the liquid and cover the frictional losses. The pump in this case will function as siphon. The analogy of cyclist can be visualized from down the hill movement of cycle where the stored elevation energy less the looses is transformed progressively into velocity energy. The streeper the slope (higher potential energy), the faster will be cylist speed.

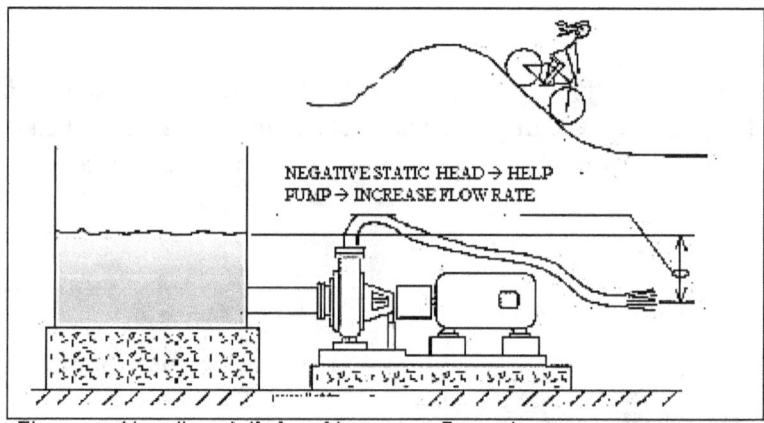

Negative static head increases flow rate.

For identical systems, the flow rate will vary with the static head. If the pipe end elevation is high, the flow rate will be low. Compare this to a cyclist on a hill with a slight upward slope where velocity will be moderate and correspond to the amount of energy supplied to overcome the wheel friction with road and the change in elevation (air resistance neglected).

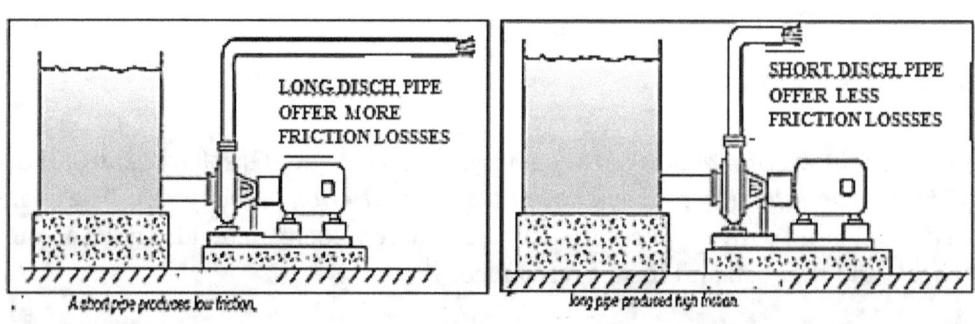

A short pipe produces low friction. *long pipe produced high friction.*

If pump is supplying water to a bath tub placed on 2nd floor, it will need enough head to reach that level (static head plus friction loss in pipes and fittings). If pump is assigned to fill the bath quickly, then the taps on the bath should be fully opened to offer a little resistance or frictional. In contrast to this, if water flow through a slower filling of bath tub is required then the pump has to develop more head & same flow maintained by closing bathtub taps to create more resistance.

The pump can deliver more head by some other means like increasing speed of pump motor or increase of impeller diameter or both. In practice,these changes may not be preferred and if cost of new pump with a higher head installation is permitted.

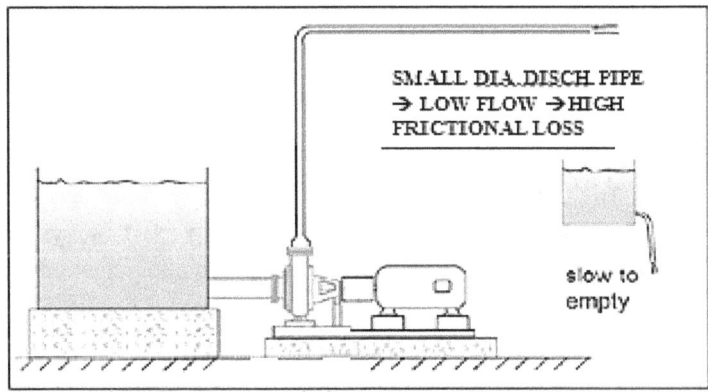

The smaller the discharge pipe cross section, lesser the flow. The pump adjust itself according to the diameter of the pipe following the characteristic curve. The pump is designed to produce a certain average flow for systems and accordingly pipes are sized. The impeller size and its speed dectates the liquid flow rate. If attempted to push for rated flow through a smaller pipe the discharge pressure will increase and the flow will decrease; similar to example of emptying a tank with a smaller drain pipe will take a longer time to empty out the tank. In pumping system, longer the pipe the friction losses will be more.

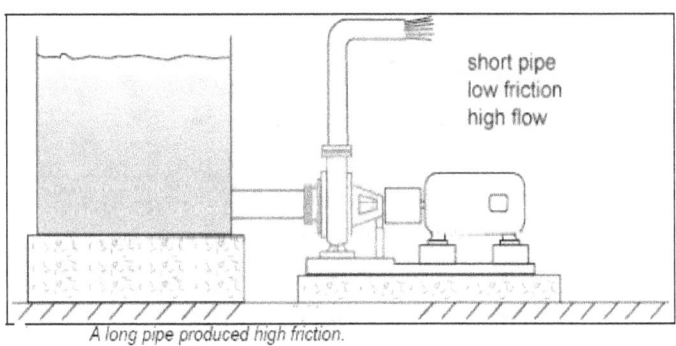

A long pipe produced high friction.

Manometric Head (H$_m$)

$$H_m = \frac{p_d - p_s}{\gamma} + (z_d - z_s) \qquad \text{but} \quad \frac{p_d}{\gamma} = h_d + h_{fd} \qquad \text{and} \quad \frac{p_s}{\gamma} = -(h_s + h_{fs})$$

$$H_m = \frac{v_s^2}{2g} + h_s + h_{fs} + h_d + h_{fd} : = H_{st} + h_f + \frac{v_s^2}{2g}$$

Where $h_{fd} = f\frac{L}{D}(V_d^2/2g)$ and $h_f = h_{fs} + h_{fd}$

$$H_m = h' - H_L = \frac{V_{w2}U_2}{g} - H_L$$ (where H_L = impeller losses)

Total Head (H)

$$H = \frac{p_d - p_s}{\gamma} + (z_d - z_s) + \frac{V_d^2 - V_s^2}{2g} = H_{st} + h_f + \frac{V_d^2}{2g}$$

$$H_m = H + \frac{1}{2g}(V_s^2 - V_d^2)$$

When $V_s = V_d$
Hence $H_m = H$

7.3 Selection of A Centrifugal Pump

It is unlikely to buy a centrifugal pump off the shelf, install it in an existing system and expect to deliver exactly the same flow rate as required. The flow rate depends on the physical characteristics of system such as friction, length, size of pipes and elevation difference to transfer liquid. The pump selection require system situation and desired parameters of pump.

7.3.1 The Main Factors That Affect The Flow Rate Of A Centrifugal Pump:

√ friction, depending on the length and construction of pipe and the diameter
√ static head depending on the difference of the pipe end discharge height vs. the suction tank fluid surface height
√ fluid viscosity, if the fluid is different than water.

Head Measurement in pumping system–

Pressure to Head Conversion formula is given as–

$$\text{HEAD in ft} = \frac{\text{PRESSURE (psi)} * 2.31}{\text{SPECIFIC GRAVITY}}$$

7.3.2 Significance of the Best Efficiency Point (BEP)

a. **B.E.P as a measure of optimum energy conversion**

The efficiency of centrifugal pumps is expressed in percentage and it describes the ratio of change of centrifugal or velocity energy to pressure energy. in sizing, selecting and designing the centrifugal pump for a given application, the best efficiency point (B.E.P) is considered in initial design stage. The B.E.P. is the area on the charactristic curve where the change of velocity energy into pressure energy at a given flow rae is maximum or the pump function is most efficient.

b. **Departure from B.E.P and Unstable operation**

The impeller is subjected to non-symmetrical forces while pump operating at lower or higher range of the B.E.P. These forces render mechanically unstable conditions like excessive hydraulic thrust, temperature rise, vibration, erosion, separation and cavitation. Thus, the operation of a centrifugal pump away from B.E.P, (left or right on efficiency curve) render comparatively poor performance, and faces premature failures of bearings & mechanical seal, shaft deflection related problems, increase in process fluid temperature in pump may cause cavitations and seizure of close tolerance components.

c. **B.E.P as an important parameter in calculations**

Capacity at BEP is an important parameter in calculation of various parameters of pump viz... specific speed, hydrodynamic sizing, viscosity correction, head rise at shutoff, etc. Many users prefer to operate pump within 90% to 110% of B.E.P/operating point for optimum performance and smooth operation.

The operating point of a centrifugal pump, known as duty point, is obtained by the intersection of the pump characteristic curve with the system resistance curve where efficiency of pump should be maximum. The flow rate Q and the developed head H are determined at intersection point (duty point/operating point). To change the operating point either the system resistance curve or the pump characteristic curve is required to be altered.

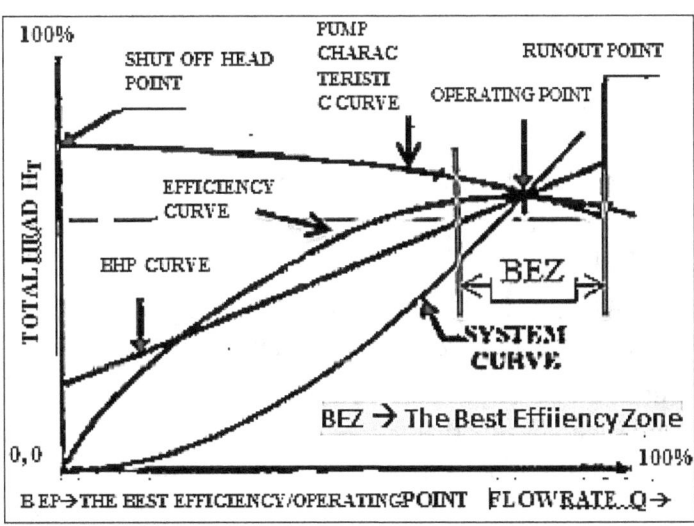

d. **Change of characteristic curves & System resistance curve**

i. The **system characteristic curve** can be altered by–

√ changing the flow resistance (for example, by changing the setting of a throttling device, by installing an orifice or a bypass line, by rebuilding the piping or by its becoming incrusted) and/or

√ changing the static head component (for example, with a different water level or tank pressure).

ii. **Axial flow (propeller)** pumps, by changing the blade pitch setting

Please note that effect of these measures for changing the characteristic curve can only be predicted for non- cavitating operation

iii. A **pump characteristic curve** can also be changed by–

√ changing the speed of rotation of pump

√ starting or stopping standby pump operation in series or parallel

√ changing the impeller's outside diameter for pumps with radial impeller(s), size restricted by casing design

√ installing or changing the setting of installed pre swirl control equipment for pumps in mixed flow impellers,

7.4 Selection of a Centrifugal Pump

a. **Determine the flow rate**

To select a centrifugal pump, first determine the flow rate and head. For various applications, flow rate will often depend on process plant production variations. For determining the required flow rate, several flow tests are taken and averaged for fair accuracy. The seasonal requirements and change in process demand should be considered while conducting the flow tests.

b. **Determine the static head**

This a matter of taking measurements of the height between the suction tank fluid surface and the discharge pipe end height from reference point (pump's shaft center line or the discharge tank fluid surface elevation).

c. **Determine the friction head**

The friction head depends on the flow rate, the pipe size and the pipe length and material. For fluids different than water the viscosity will be an important factor to consider.

7.4.1 Calculate the Total Head

The total head is the sum of the static head (remember that the static head can be positive or negative) and the friction head.

We can select the pump based on the pump manufacturer's catalogue and other information, using total head and flow required as well as suitability to the application.

When discharge pipe end is lowered in elevation, the pump flow is established and the total head will decrease to a value that corresponds to the flow at characteristic curve. Let's start from the point of zero flow with the pipe end at its maximum elevation and gradually pipe end is lowered. At certain elevation the flow begins. If there is flow there must be friction loss. The frictional head is consumed from the maximum total head and actual head is reduced. At the same time, the static head is reduced which lead to further reduction of total head.

7.4.2 Pumping System

- **SYSTEM CHARACTERISTICS AND PUMP HEAD**

  The piping and equipment through which the liquid flows from the pump comprise the pumping system. A pumping system should deliver a liquid efficiently at specified rate of flow and head through a particular system. While procuring a pump, the required flow of liquid and total head, the system's resistance must be specified. The total head of a centrifugal pump is usually measured in M but in case of a positive displacement pump the differential pressure is specified.

  Pump pressure is usually measured in kg/cm². The pressure M-head and kg/cm², can be expressed in kg.M i.e. Newton of energy added to each kg of liquid pumped at related flow and head. The purchaser should furnish the system- resistance of piping installation so that correct design & selection can be made by pump manufacturer.

  In a positive displacement Pump, the system resistance will dictate the generated head & power consumption. Any underestimating of differential pressure can result more power and pumping head.

- **PARALLEL OPERATION PIPING SYSTEM**

  The piping system through which the liquid is pumped, offers resistance to flow due to friction. If the liquid discharges to a pressurized system, additional resistance is encountered, requiring pressure or head energy. The pump must therefore overcome the total system resistance, that is, friction plus elevation and pressure heads for desired rate of flow.

  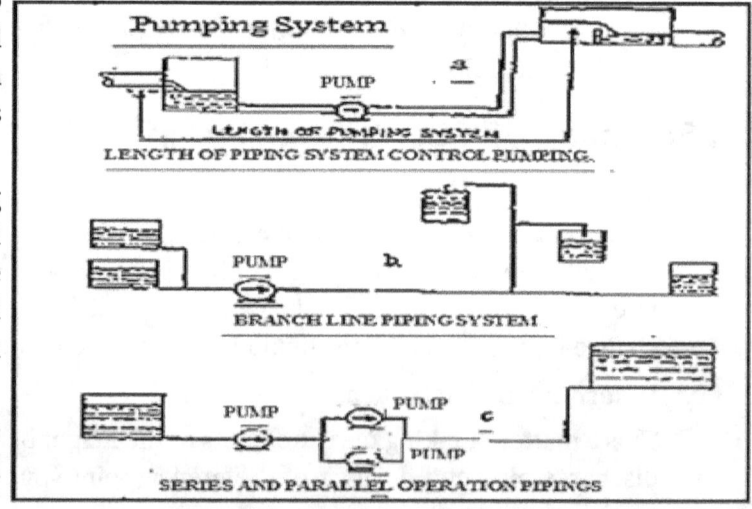

  a. Only the length of the piping containing liquid & controlled by the pump is a part of the pumping system. A pump and the limit of the length is shown in fig (a).

  b. The pump suction and discharge piping can consist of branch lines, as shown in fig (b).

c. There can be one or more pumps in a pumping system. Several pumps can be piped together in series or in parallel or both, shown in fig. (c). When there is more than one pumps, the flow through the system will be determined by the combined performance of all pumps.

7.5 Installation of Pumps

While installing pump, following points should be taken care:

a. Location of pump should be accessible for operation and maintenance.

b. Prefer to install pump with positive suction ie. Pump center line below the suction liquid level (conforming minimum suction height of liquid above suction centre line).

c. Pump location should be well illuminated, free from dust & moisture as for as possible.

d. Provide ample head room for assembly & dis-assembly of pump with proper handling facility for pump components, valves and pipings.

e. Pipings installation should not obstruct pump and motor dismantling & assembly. Free space is provided for leveling & alignment of machine.

f. Pipings installation should be properly designed for supports, expansion compensation and hot alignment so that pipings do not transmit any force or torque on pump's flanges in running or stationary stage of the pump. Provide expansion bellows in suction and discharge pipings to compensate expansion as well as to facilitate removal & re fixing of stop valves, NRV or piping's alignment with pump.

g. Pipings should be supported independently so that it does not transfer load on pump in any working or non-working condition of the pump.

h. Take care of hot and cold alignments of machine. Refer OEM manual for alignment.

i. Connect piping only after leveling, alignment, half-grouting & curing of pump foundation. However, after alignment of pipings with pump, coupling should be disconnected for final alignment checking and complete grouting work.

j. Dowelling of pump should be done only after trial run of the pump. However, the pump and motors should be locked in position with side screws/machined blocks at feet.

k. Suction pipings should be designed for minimum head loss and negative suction height should not be more than 15–20 ft from the liquid level to pump centre line in case of suction tank opened to atmosphere.

l. Other installation cares, discussed separately should be taken care.

7.5.1 Design Guidelines

Flow approach either from a pump bay/suction well or an inlet bend to the impeller should be undisturbed. With careful design, both solutions may be regarded as hydraulically equivalent. For larger capacities (big stations involving heavy construction costs) the decision for one or the other must be reached individually on the strength of an economic assessment taking the entire installation costs into account. In case of doubt, model tests should be carried out for both arrangements, to achieve the highest possible plant reliability.

7.5.2 Suction Piping

The suction pipe should be as short as possible and run with a gentle ascending slope towards the pump. If necessary, eccentric suction piping as shown in figure should be provided (with a sufficient straight length of pipe upstream of the pump L ≥ d) to prevent the air pockets formation. If, on account of the site conditions, fitting an elbow immediately upstream of the pump cannot be avoided, then provide an accelerating elbow for a smooth flow of liquid. For the same reason, an elbow with multiple turning vanes (see Fig. 63) is required in front of double-entry pumps or pumps with mixed flow (or axial flow) impellers unless this is impossible because of the nature of the medium handled (no stringy),

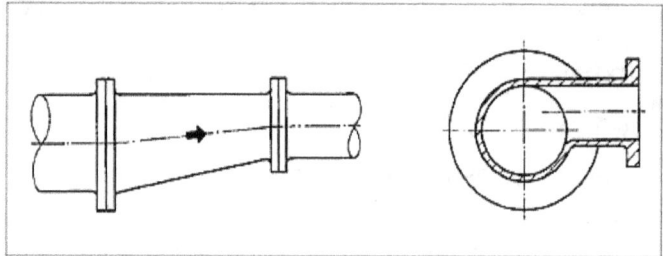

Eccentric reducer and branch fitting to avoid air pockets

The suction and inlet pipes in the suction tank or pump sump must be sufficiently wide apart to prevent air entraining in the suction line. At places where sufficient space between suction pipes is restricted a positive deflectors should be provided. The mouth of the inlet pipe must always laid below the minimum submergence height of liquid. If the suction pipe in the tank or sump is not submerged adequately due to too low liquid level, rotational movement of the liquid air entraining hollow vortex will develop – a funnel-shaped depression at the surface, a tube-shaped air cavity in a short duration of time, extending from the surface to the depth of suction pipe inlet. This will result an unsteady operation of pump and the flow is decreased. Therefore, the required minimum submergence (minimum suction depth) should be maintained in installation.

7.5.3 Use of Proper Elbows in Suction to Reduce Frictional Losses

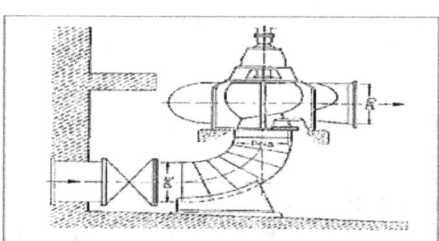

Flow-accelerating elbow upstream of a vertical volute casing pump with high specific speed

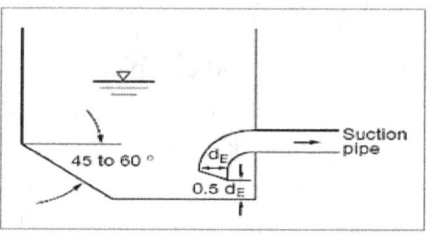

Inclined sump walls to prevent deposits and accumulation of solids

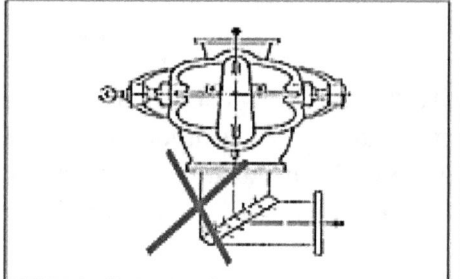

Intake elbow with multiple turning vanes upstream of a double-entry, horizontal volute casing pump (plan view)

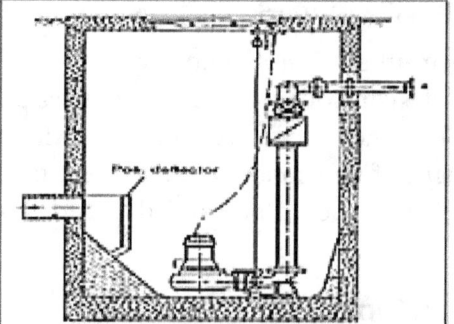

Installation of a positive deflector in the intake chamber of a submersible motor pump

7.5.4 Submergence of Suction Point

a. **Minimum submergence of suction point to avoid hollow vortices are shown in graphs below–**

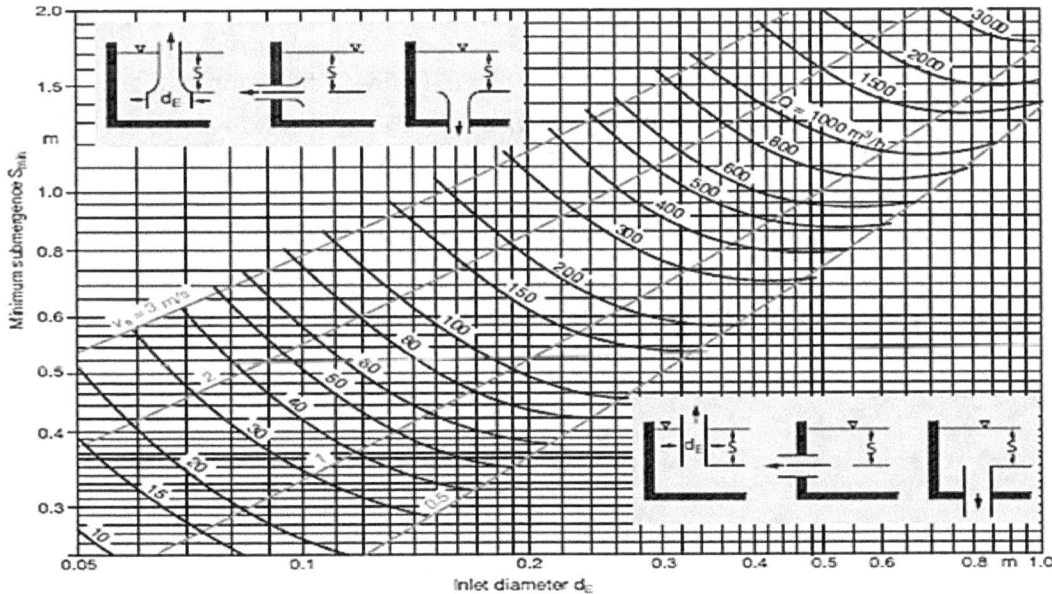

Minimum submergence S_{min} of horizontal and vertical suction pipes
required for suction tanks to avoid hollow vortices

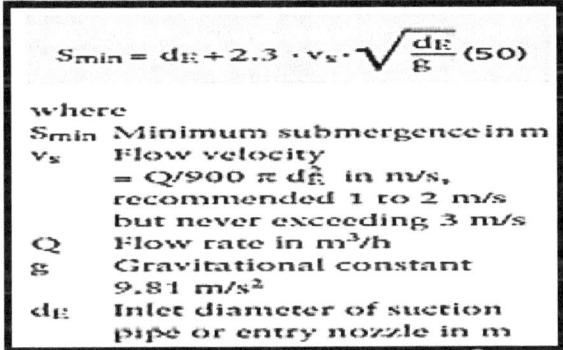

$$S_{min} = d_E + 2.3 \cdot v_s \cdot \sqrt{\frac{d_E}{g}} \quad (50)$$

where
S_{min} Minimum submergence in m
v_s Flow velocity
   $= Q/900 \, \pi \, d_E^2$ in m/s,
   recommended 1 to 2 m/s
   but never exceeding 3 m/s
Q Flow rate in m³/h
g Gravitational constant
   9.81 m/s²
d_E Inlet diameter of suction
   pipe or entry nozzle in m

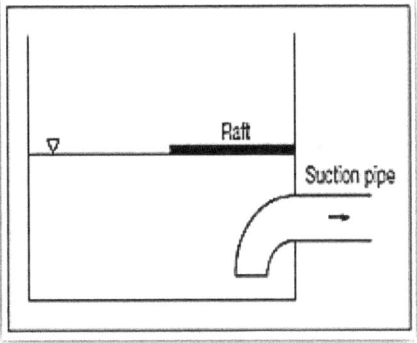

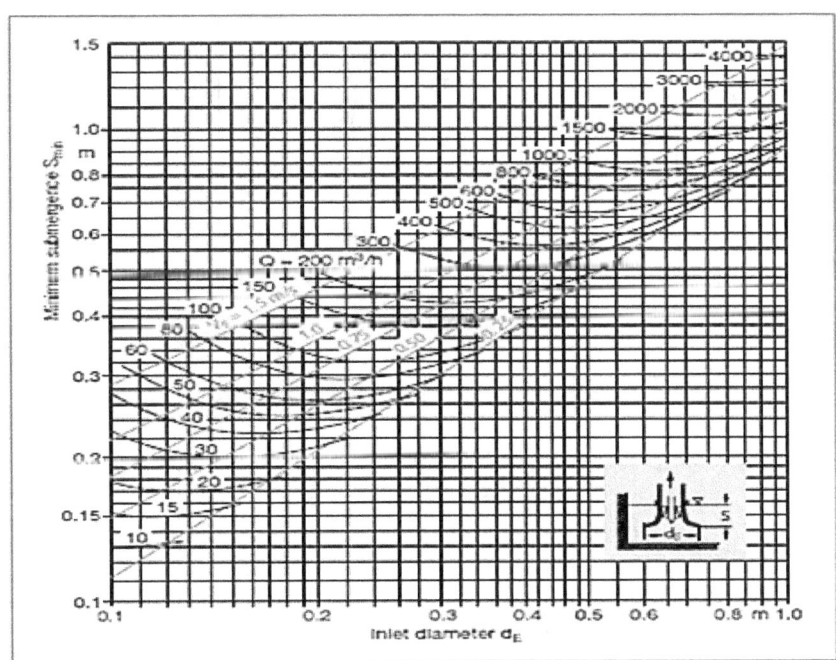

Minimum submergence S_{min} of tubular casing pump suction
pipe to avoid hollow vortices

7.6 Intake Chamber Construction

The intake structure is constituted of approach flow zone, followed by pump bay or inlet bend for the dry pit arrangement or wet pit installation. $NPSH_{(A)}$ is the water level which is required above pump suction ($NPSH_{(A)} = NPSH_{(R)} + 1.5\ M$) to avoid cavitation. The approach flow section distributes the water uniformly for variety of services to the pump bay.

- If series of installation is made, maximum flow velocity in the duct width at the first pump is allowed = 0.3 m/s max. and sufficient spacing between pump to pump is required to be maintained. The baffle cylinder duct width W = 3.5 D whereas-vortex device duct width W = 3 D.

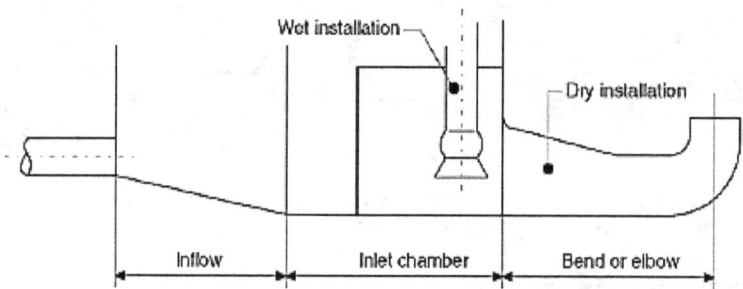

- In conjunction with pump $NPSH_{(A)}$ the minimum submergence S for the inlet bend becomes necessary to avoid vortex formation. The intake geometry and sound hydraulic design must be considered for trouble free operation of installations.

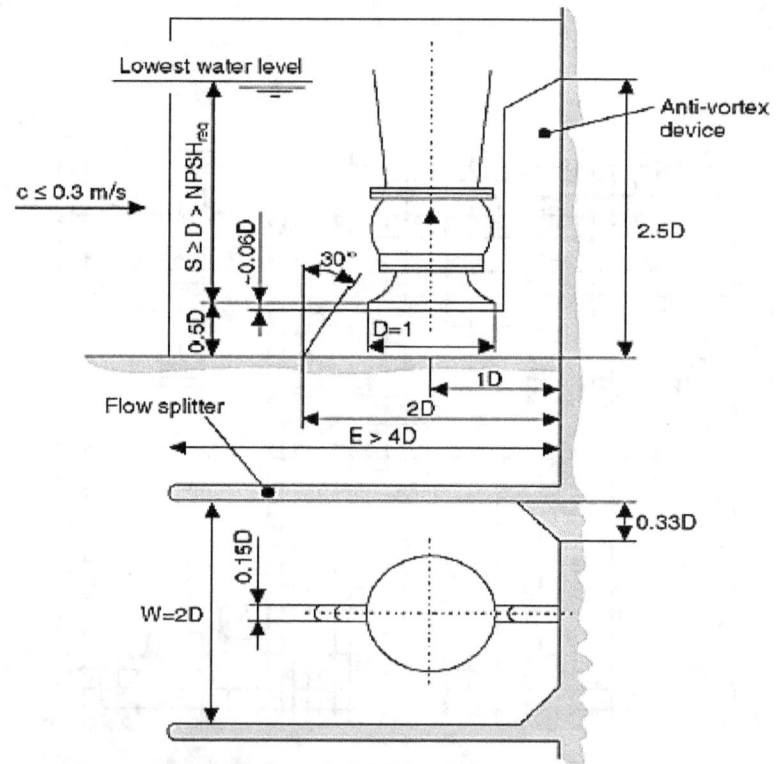

Open pump bay, normal installation for flow rates ≤20 m³/s. submergence S ≥ 2D ≥ NPSH$_A$

7.6.1 Illustration

A typ. Dimension of suction piping and pump inlet eye is laid in following sketch for further study purpose.

- **D1 = impeller entry diameter,** S = minimum submergence (observing the necessary NPSHA)
- g = gravitational acceleration (9.81 m/s²)
- **S = minimum submergence ≥ 2 * D ≥ NPSH(A)**

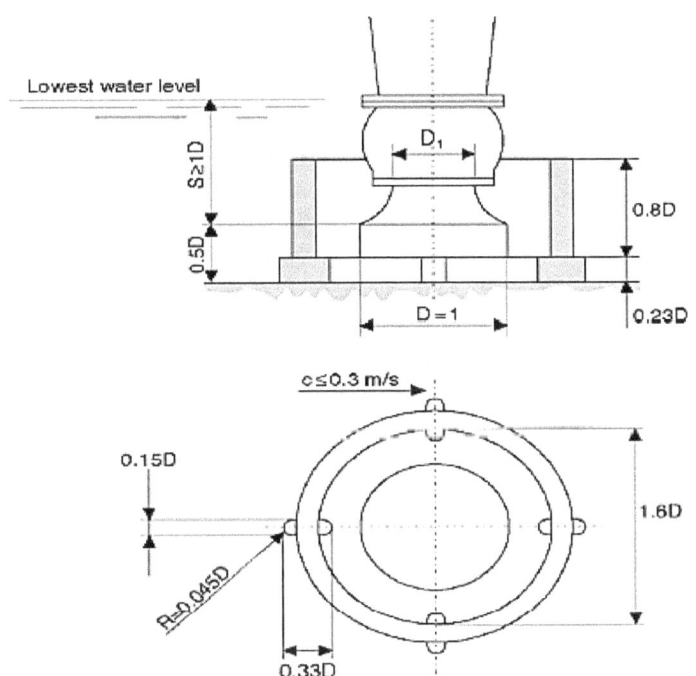

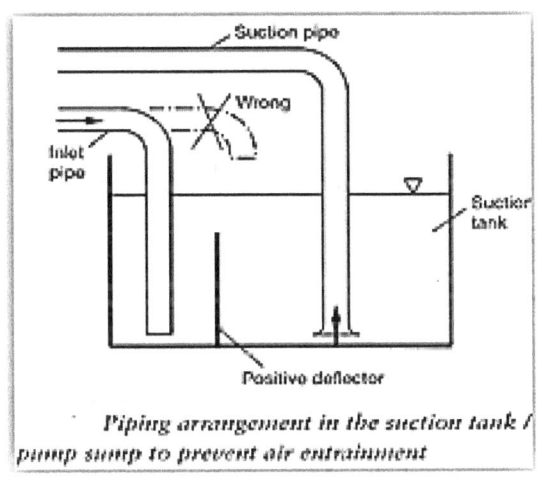

Piping arrangement in the suction tank / pump sump to prevent air entrainment

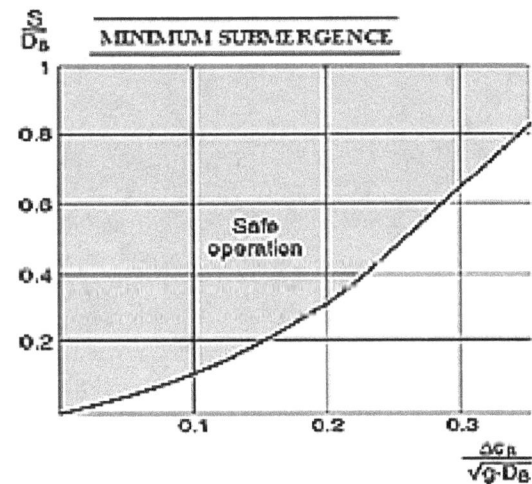

a. **Calculation of submergence S:**

B = pump spacing minus partition thickness; pump spacing in station and bend inlet with B.

The inlet height Hc (inlet bend/connection to pump bay)

Flow velocity v = $Q_{(pump)}/(B * H_c)$

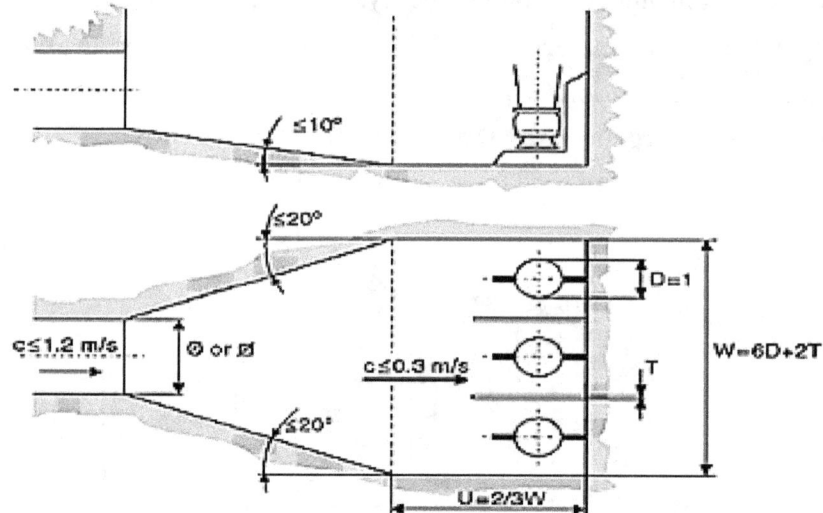

PUMP BAY ARRANGEMENT, DIFFUSER
ANGLE ≤ 20 DEG; GRID AT DISTANCE U.

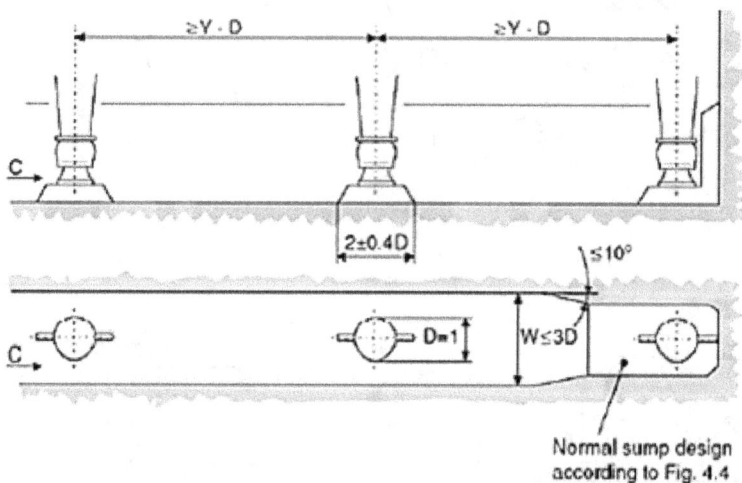

Normal sump design
according to Fig. 4.4

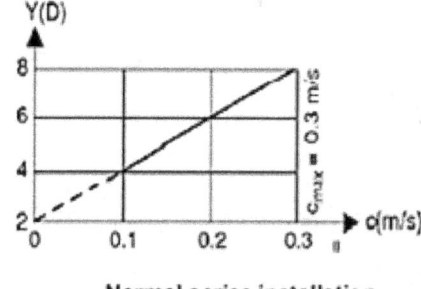

Normal series installation

$V_s = 0.7$ m/s with screen
before the bend

$V_s = 1.6$ m/s without screen
before the bend.

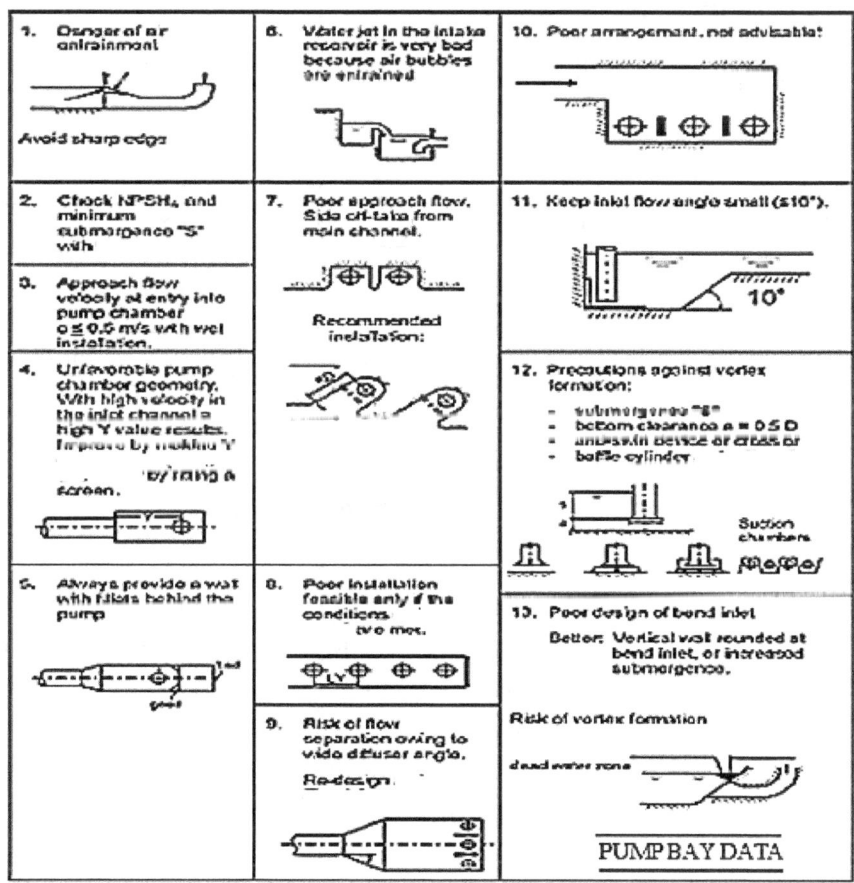

Bend geometry

INLET BEND TO PUMP

7.6.2 Recommended Parameters

Inlet velocity in chamber –	0.5 m/s
Inlet velocity in suction bell (range 1.2 to 2.1 m/s) –	1.7 m/s
Velocity in pipeline DL to pump –	4.0 m/s
Dimensions: D –	1.75 DL

Straight pipe sections of adequate length (L = 7 D) before the suction nozzle.

The velocity at the branch connection must not exceed about 2.5 m/s, to keep the submergence "S" within acceptable limits.

The approach flow pipe should be engineered so that sources of disturbance like bends, branches or valves are a as possible. Such components may cause uneven velocity distribution and vortex formation.

PUMP BAY DATA

7.6.3 End Suction Pumps

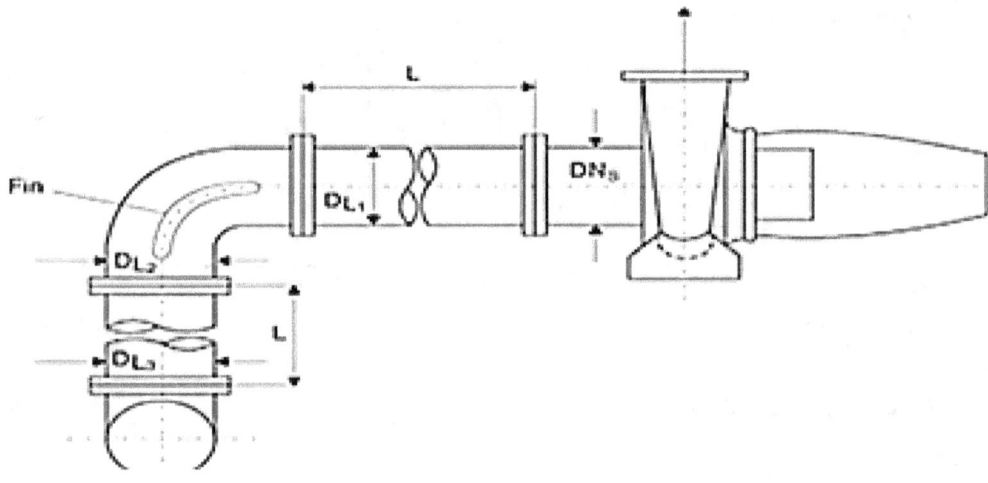

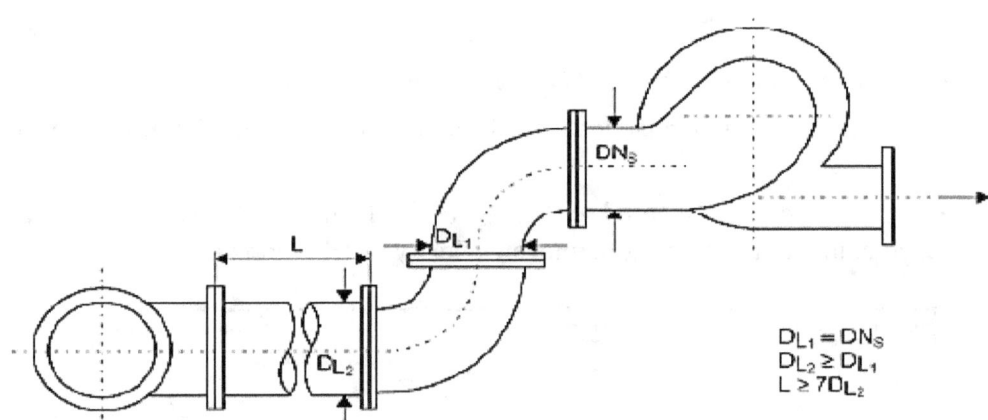

$DL_1 = DN_S$
$DL_2 \geq DL_1$
$L \geq 7DL_2$

SPEC SPEED	MODERATE NS	NS ≥ 55
Design conditions	DL2 = DL1 = DNs	Acceleration = 2.25
	DL3 ≥ DL2	DL1/DNs = 1.25–1.5
	L ≥ 7 * DNs	Provide acceleration nozzle before suction eye
		L ≥ 2 DL2

7.6.4 Some Important Tips for Pump and Piping Installation

- $NPSH_{(available)}$ should be always greater than $NPSH_{(required)}$ {not less than + 0.6 to +1.5 M}
- Keep suction pipe short and directly orientated suction pipings from liquid to pump flange, i.e. pipe's upper dia tip in horiz. position pipe should not be above pump suction eye top position.
- Use suction pipe in one step higher size to that of pump suction eye.
- Keep suction pipe slopping down from pump suction eye to suction liquid. Avoid any installation which will give rise to air pockets.
- Suction pipes should always be provided with long radius bends (2 R). No short and sharp bends should be permitted.

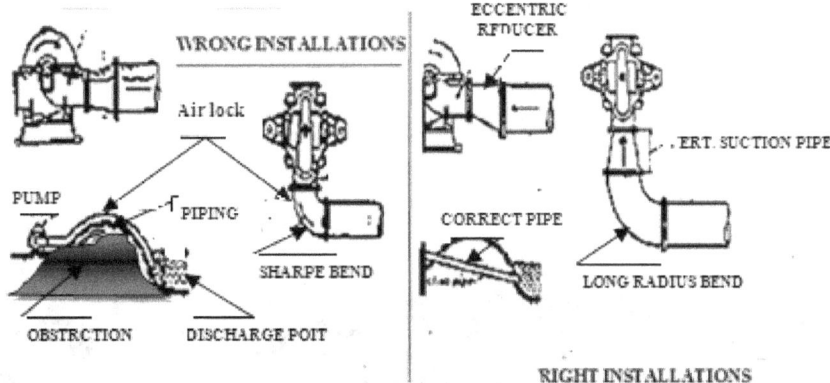

- Lay suction pipe under the construction rather than above construction to avoid air pockets.
- Use eccentric reducer having flat portion at top to connect horizontal pump suction wyw to avoid air pockets.
- Do not use valve in negative suction pipes. However, in positive suction pipes it can be installed if necessary.
- Horizontal elbow should be used as far as possible sufficiently away from suction flange.
- Always use separate suction pipe for each pump for parallel running pump installations. Don't use common suction unless proper designe of suction header is available.
- Suction pipe should be below submergence height from liquid surface (minimum 1.0 M as standard practice).
- Testing of suction pipings- hydro-test at 1.5 * working or design pressure (refer ASME viii).
- Avoid foot valve if other priming facility like positive head, ejector or vacuum pump priming is available.
- Opening of foot valve should be 2 to 4 * suction pipe area. Provide foot valve back wash system from pressurized source of water.
- Provide suction strainer with total opening area, in case of positive suction line, as follows-
- Strainer opening area = 2.5 to 4.0 * suction pipe cross sectional area. In case of high sludge or foreign matters in liquid, use rotating type strainer with scrapping device or back flushing arrangement.
- Install NRV at all discharge pipes of pumps, connected with same header(pumps running in parallel) and pumping against liquid head.
- In suction chamber of liquid, ensure that returning liquid does not make churning effect near suction line. Use separate chamber for return water which should over flow to suction chamber of pumping system. For dirty water pumping, a settling and suction chamber should be provided. Settling chamber will be connected to sludge pit from which settled sludge could be pumped out.

$$\text{LOSSES IN PIPE LINE } \Delta P = \lambda * \frac{L}{d1} * v^2 * \frac{N}{m^2} \quad \text{Or} \quad Hv = \lambda * \frac{L}{d1} * \frac{v^2}{2g} \text{ (m)}$$

where:

Δp = pressure loss (N/m²)

Hv = head loss (m)

λ = pipe friction loss coefficient (_)

L = pipe length (m)

di = inside diameter of pipe (m)

ρ = density (kg/m²)

g = acceleration due to gravity (m/s²)

v = mean flow velocity (m/s)

k = pipe roughness value (mm)

Laminar flow	Turbulent flow		
	Hydraulically smooth pipes	**Hydraulically rough pipes**	**Pipes in transient range**
	Limits: $R_e \cdot \dfrac{k}{d_i} < 65$	Limits: $R_e \cdot \dfrac{k}{d_i} > 1300$	Limits: $65 < R_e \cdot \dfrac{k}{d_i} < 1300$
Formula for λ $\lambda = \dfrac{64}{R_e}$ $\lambda = \dfrac{64 \cdot \nu}{c \cdot d_i}$	Formula for λ: (a) Blasius formula for the range $2320 < R_e < 10^5$ $\lambda = 0.3164 \cdot R_e^{-0.25}$ (b) Nikuradse formula for the range $10^5 < R_e < 5 \cdot 10^6$ $\lambda = 0.0032 + 0.221 \cdot R_e^{-0.237}$ (c) Prandtl and von Kármán formula for the range $R_e > 10^6$ $\dfrac{1}{\sqrt{\lambda}} = 2\lg\left(R_e \cdot \sqrt{\lambda}\right) - 0.8$	Formula for λ: Nikuradse formula $\dfrac{1}{\sqrt{\lambda}} = 2\lg\dfrac{d_i}{k} + 1.14$	Formula for λ: Prandtl-Colebrook formula $\dfrac{1}{\sqrt{\lambda}} = 2\lg\left[\dfrac{2.51}{R_e\sqrt{\lambda}} + \dfrac{k}{d_i} \cdot 0.269\right]$

7.6.5 Guide Lines in Branch Looping of Pipes

- **T-Connections**

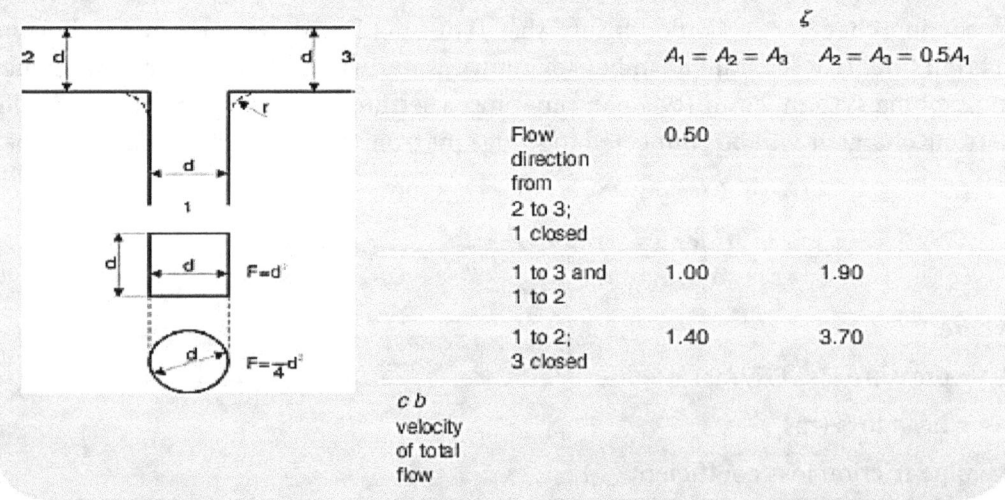

Table   T-Pieces, 90° Angle Round or Square Cross-section, Sharp Edges. By rounding off the edges the ζ values are slightly reduced, by 10 to 30% if strongly rounded

		ζ	
		$A_1 = A_2 = A_3$	$A_2 = A_3 = 0.5A_1$
	Flow direction from 2 to 3; 1 closed	0.50	
	1 to 3 and 1 to 2	1.00	1.90
	1 to 2; 3 closed	1.40	3.70

c b velocity of total flow

- **Branch Looping**

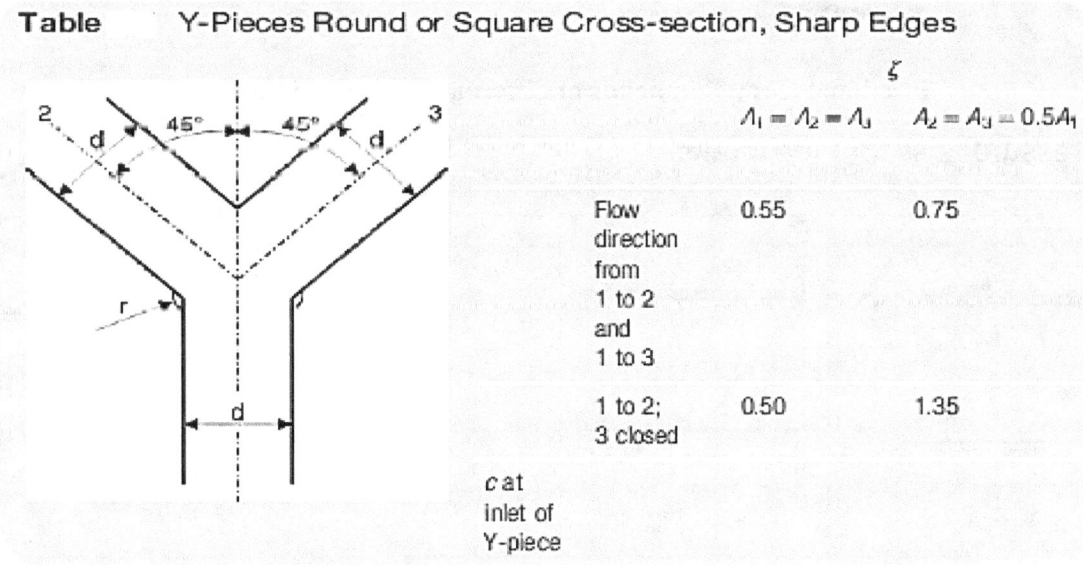

Branch Loops (Circular Cross-section)

	Q_z/Q	0.0	0.2	0.4	0.6	0.8	1.0
ζ_z		0.96	0.88	0.89	0.96	1.10	1.29
ζ_d		0.05	−0.08	−0.04	0.07	0.21	0.35
ζ_z		−1.04	−0.40	0.10	0.47	0.73	0.92
ζ_d		0.06	0.18	0.30	0.40	0.50	0.60
ζ_z		0.98	0.79	0.64	0.57	0.60	0.75
ζ_d		0.05	−0.05	−0.02	0.07	0.20	0.34
ζ_z		−0.92	−0.30	0.13	0.40	0.57	0.66
ζ_d		0.04	0.24	0.30	0.25	0.10	−0.19
ζ_z		0.90	0.68	0.50	0.38	0.35	0.48
ζ_d		0.04	−0.06	−0.04	0.07	0.20	0.33

- **Y – Type Looping**

Table   Y-Pieces Round or Square Cross-section, Sharp Edges

		ζ	
		$A_1 = A_2 = A_3$	$A_2 = A_3 = 0.5A_1$
Flow direction from 1 to 2 and 1 to 3		0.55	0.75
1 to 2; 3 closed		0.50	1.35

c at inlet of Y-piece

- **Pipe & Bend Resistance Loss Co efficient–**

Table   Pipe Bend, Resistance Loss Coefficient ζ

$\frac{r}{d_i}$						
α	1	1.5	2	4	6	
15°	0.03	0.03	0.03	0.03	0.03	
30°	0.07	0.07	0.07	0.07	0.07	Interior
45°	0.14	0.11	0.09	0.08	0.075	pipe wall
60°	0.19	0.16	0.12	0.10	0.09	*smooth*
90°	0.21	0.18	0.14	0.11	0.09	
15°	0.10	0.08	0.06	0.05	0.04	
30°	0.23	0.19	0.14	0.11	0.08	Interior
45°	0.34	0.27	0.20	0.15	0.12	pipe wall
60°	0.41	0.33	0.24	0.19	0.15	*rough*
90°	0.51	0.41	0.30	0.23	0.18	

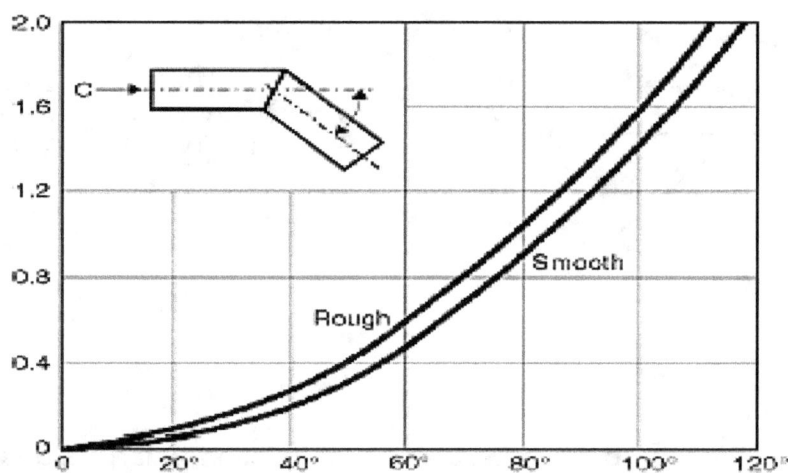

Frictional loss (Ç) in bends of different degree and roughness

7.6.6 Pressure Loss for Pipe Intakes

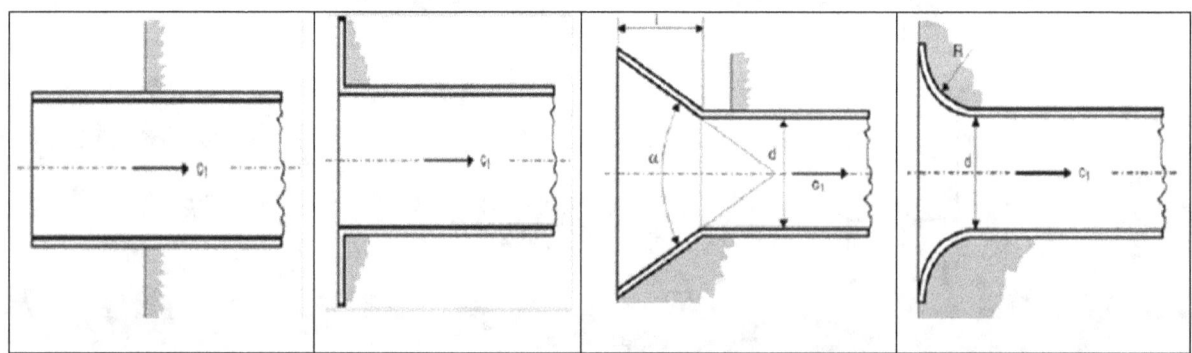

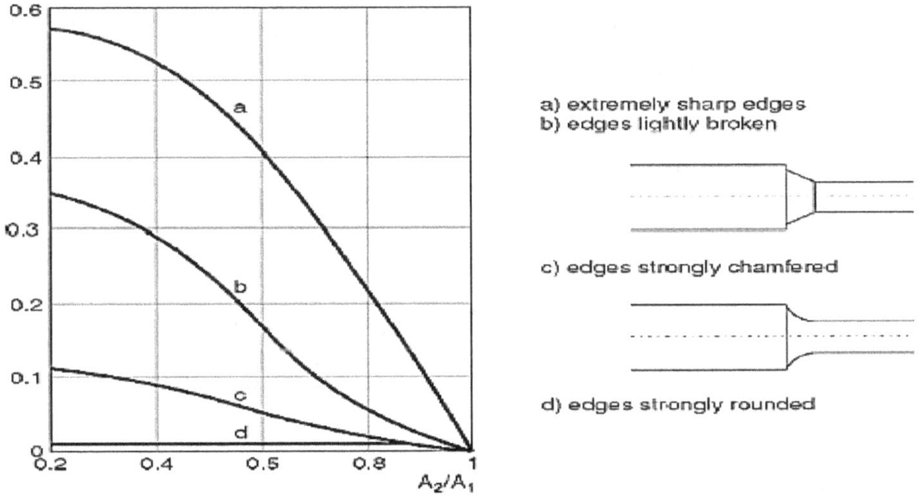
0.6
0.5
0.4
0.3
0.2
0.1
0
0.2
0.4
0.5
0.8
1
A2/A1
a
b
c
d
a) extremely sharp edges
b) edges lightly broken
c) edges strongly chamfered
d) edges strongly rounded

Resistance loss coefficient ζ for constrictions

7.6.7 Diffuser with Continuous Divergence

Diffusers with continuous divergence should have a cone angle of a ≤ 8°.

A_1/A_2	0.25	0.3	0.4	0.5	0.6
ζ-values $\alpha = 8°$	0.06	0.05	0.04	0.03	0.02
$\alpha = 16°$	0.15	0.13	0.10	0.07	0.05
$\alpha = 24°$	0.26	0.23	0.17	0.12	0.00

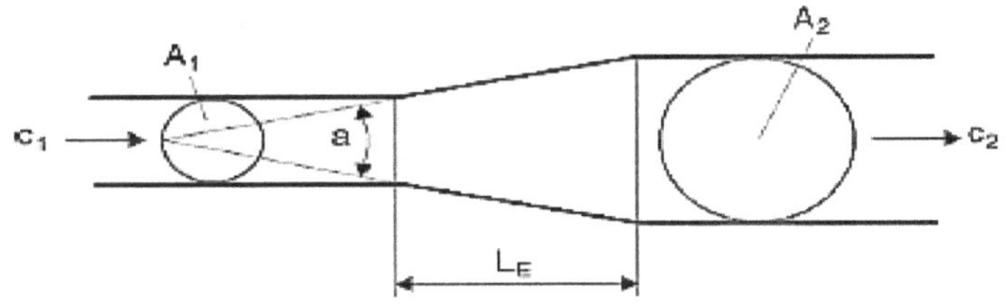
A1
A2
c1
c2
a
LE

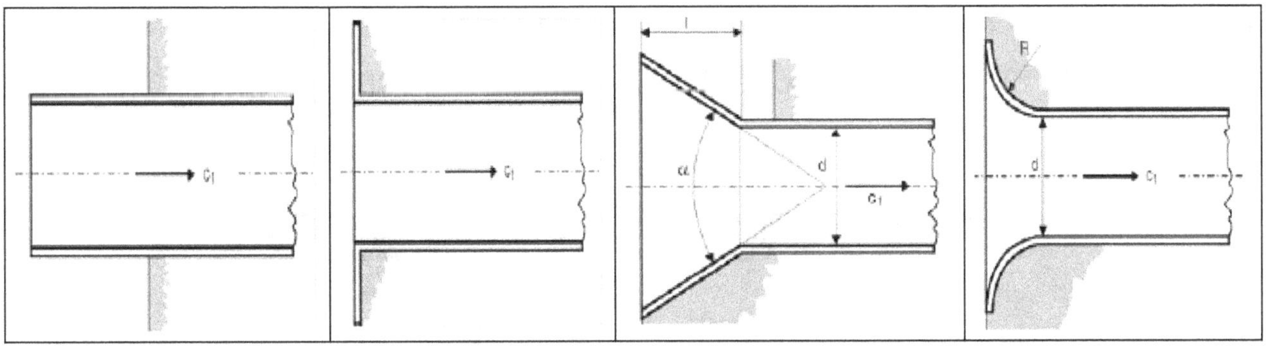
c1
l
α
d
R

– projecting: sharp-edged: 2.573 broken: 0.671	– even with wall: sharp-edged: z = 0.5 rounded: z = 0.25	Enlarged pipe intakes: – conical shape: z depends on 1/d and a Optimum for a = 40 to 60		– rounded shape: For R/d = 0.2 z = 0.03
1/D	0.1	1.0	0.1	0.6
á	60	40–60	40–60	40
ξ	0.4	0.1	0.2	0.1

7.6.8 Pressure Losses in Valves

The resistance loss coefficients z are approximate values. For exact calculations precise values must be obtained from the valve manufacturers.

- Straightway valves (completely open) (according to I.E. Idel Chik)
- Inclined seat valves

Cast valves, DN 25 up to 200	$\zeta = 2.5$
Wrought iron valves, DN 25 up to 50	$\zeta = 6.5$

7.6.9 Angle Valves (Valve Completely Open), DN 25 To 200; z = 2.0

DN	25	32	40	50	65	80	100	125 up to 200
ζ	1.7	1.4	1.2	1	0.9	0.8	0.7	0.6

7.6.10 Non-Return Valves Linear Seat Valves

NON-RETURN VALVES LINEAR SEAT VALVES, DN 25 to 200; z = 3.5; Inclined seat valves, DN 50 to 200; z = 2.0

- Foot valves with suction strainer (according to VAG valves)
- Foot valves arranged in groups (according to VAG valves)

	DN	50 up to 80	100 up to 350
ζ-values	c = 1 m/s	4.1	3.0
	c = 2 m/s	3.0	2.25

7.6.11 Metallic Seal Valves, Short

DN	400	500	600	700	800	1000	1200
ζ-values	7.0	6.1	5.45	4.95	4.55	4.05	3.9

DN	50	100	150	200	300	400	500	600	800	1000
ζ-values	0.23	0.2	0.18	0.17	0.15	0.14	0.12	0.11	0.10	0.09

7.6.12 Metallic Seal Valves, Medium Length

DN	50	100	150	200	300	400	500	600	800	1000
ζ-values	0.29	0.26	0.23	0.19	0.17	0.16	0.15	0.13	0.12	0.10

7.6.13 Metallic Seal Valves, Long Length

DFN	50	100	150	200	300	400	500	600	800	1000
ζ-values	0.32	0.29	0.26	0.23	0.20	0.19	0.18	0.16	0.15	0.13

7.6.14 Shut-Off Valves with Locking Parts Open

DN	150	200	250	300	400	500	600	700	800	900	1000	1200	1400	1600	2000
PN 6	–	–	–	–	–	–	–	–	–	–	–	0.24	0.24	0.22	0.22
PN 10	–	0.75	0.65	0.60	0.50	0.45	0.40	0.38	0.36	0.34	0.32	0.32	0.30	0.28	0.28
PN 16	1.50	1.00	0.80	0.70	0.66	0.64	0.62	0.60	0.55	0.55	0.50	0.45	0.40	0.38	0.38
PN 25	1.50	1.00	0.80	0.75	0.70	0.70	0.70	0.65	0.60	0.60	0.55	0.52	0.45	0.42	0.42
PN 40	1.50	1.00	0.85	0.80	0.80	0.75	0.75	0.70	0.65	0.65	0.60	0.60	0.50	0.45	0.45

(ζ-values)

7.6.15 Non-Return Valves without Lever and Weight

DN	50	100	150	200	300	400	500	600	800	1000
$c = 1$ m/s	3.1	3.0	3.0	2.9	2.9	2.9	2.8	2.8	2.6	2.4
$c = 2$ m/s	1.4	1.4	1.3	1.3	1.3	1.2	1.1	1.0	0.9	0.9
$c = 3$ m/s	0.9	0.9	0.8	0.8	0.8	0.7	0.6	0.5	0.4	0.4
$c = 1$ m/s	0.8	0.8	0.7	0.7	0.6	0.6	05	0.4	0.4	0.3

(ζ-values)

7.6.16 Frictional Head Loss in Flow through Bends and Valves

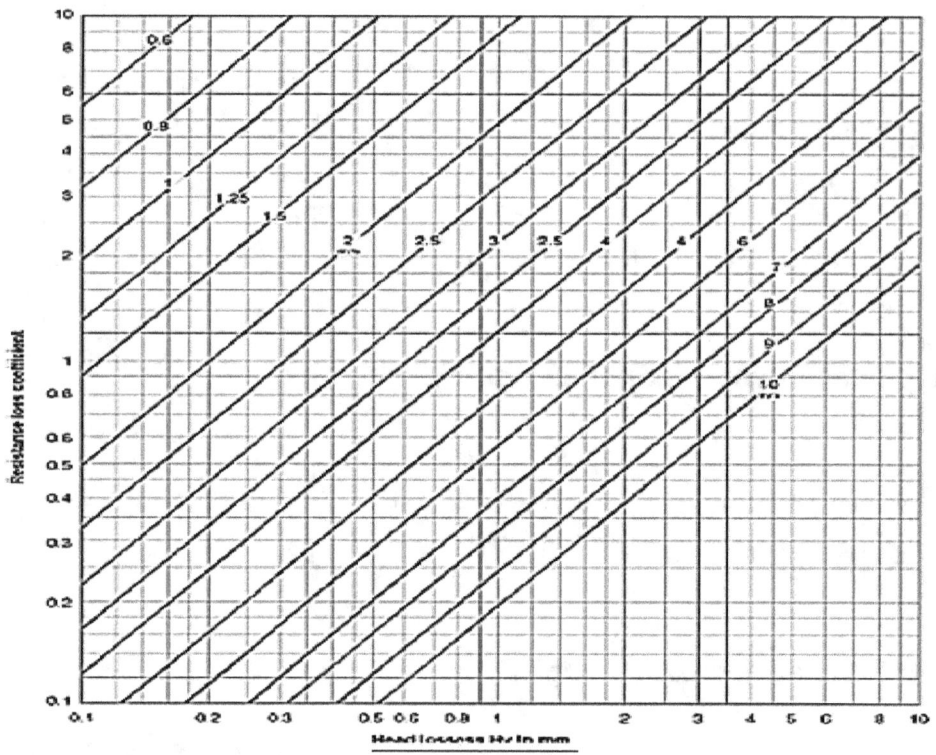

NOMOGRAM TO DETERMINE THE FRICTIONAL HEAD LOSSES OF BENDS & VALVES

7.6.17 Frictional Head Loss in Flow through Old Pipes

LOSS OF HEAD IN FEET, DUE TO FRICTION, PER 100 FEET OF 17 YEAR OLD STEEL PIPE

For New Pipe Multiply Readings by 0.6—For 25 Year Old Pipe Multiply Reading by 1.2

U.S. G.P.M.	PIPE SIZE								
	1	1¼	1½	2	2½	3	4	5	6
10	11.7	3.1	1.4						
15	25.0	6.6	3.0	1.1					
20	42.0	11.2	5.2	1.9					
25	64.0	16.6	7.9	2.7					
30	89.0	23.5	11.0	3.8	1.3				
35	119	31.2	14.7	5.1	1.7				
40	152	40.0	18.8	6.6	2.2				
50		60.0	28.4	10.0	3.3	1.4			
60		85.0	39.6	13.9	4.7	1.9			
70		113	53.0	18.4	6.2	2.6			
80		145	68.0	23.7	7.9	3.3			
90		180	84.0	29.4	9.8	4.1	1.0		
100			102	35.8	12.0	5.0	1.2		
120			143	50.0	16.8	7.0	1.7		
140			190	67.0	22.3	9.2	3.3		
160				86.0	29.0	11.8	3.9		
180				107	35.7	14.6	3.6		
200				129	43.1	17.6	4.4	1.5	
220				154	52.0	21.3	5.2	1.6	
240				182	61.0	25.1	6.2	2.1	
260					70.0	29.1	7.2	2.4	
280					81.0	33.4	8.3	2.8	1.1
300					92.0	38.0	9.3	3.1	1.3
325					107	44.0	10.7	3.6	1.5
350					122	50.3	12.4	4.2	1.7
375						57.7	13.7	4.6	1.9
400						65.0	16.0	5.4	2.2
425						73.0	17.6	6.0	2.4
450						80.0	19.8	6.7	2.7
475						89.0	22.0	7.4	3.0
500						98.0	24.0	8.1	3.3
550							28.7	9.6	4.0
600							33.7	11.3	4.7
650							39.0	13.2	5.5
700							44.9	15.1	6.3
750							51.0	17.2	7.1
800							57.0	19.4	8.0
850								21.7	9.0
900								24.0	10.0
1000								26.7	12.0

Chapter 8

OPERATION OF CENTRIFUGAL PUMPS

In centrifugal pump, when a certain mass of fluid is rotated by an external driving source, it is thrown away from the central axis of rotation and a centrifugal energy is imparted to liquid which get converted into pressre head in casing, enabling to raise liquid to the higher elevation.

8.1 Starting Procedure in Centrifugal Pump

1. Close the delivery valve and prime the pump.
2. Start the motor connected to the pump shaft.
3. Open the delivery valve gradually when motor RPM reaches 100% to enable liquid to starts flowing to the discharge pipe.
4. A partial vacuum is created at the suction eye by centrifugal action. The liquid rush from sump to the pump suction due to pressure difference created at the two ends of suction pipe.
5. As the impeller continues to run, liquid movement will continue to the pump suction eye and spinning of impeller increases the energy of the liquid to deliver to the reservoir or application place.
6. While stopping the pump, the delivery valve is closed first, otherwise there may be back flow from the reservoir or pressurised system. If NRV is installed in delivery pipe, closing of discharge valve may not be mandatory.

An uniform velocity & flow is maintained in the delivery pipe by special design of the casing. As the flow proceeds from casing to the delivery pipe, the area of the casing is fully occupied by fresh liquid resulting continuous flow of liquid from the impeller amd casing. Thus, an uniform flow is maintained in the delivery pipe.

8.1.1 Cavitation in Pumps

Cavitations begins to appear in centrifugal pump when the pressure at the suction eye is equal to or start falling below the vapour pressure of the liquid at operating liquid temperature. Observation of vpour pressure at suction eye may not be practical hence, observation of cavitations symptom in initial stage is prefered (e.g. a sudden drop in flow, efficiency, head or motor power reduction.

8.1.2 A Pump Characteristic Curve Can Be Modified by

- changing the speed of rotation
- starting or stopping one pump operated in series or parallel
- changing the impeller's outside diameter in radial impeller(s) pump. with radial impeller(s)
- installing or changing the setting of installed pre-swirl control in mixed flow impellers.

8.1.3 Effects of Cavitation

When cavitation begins to appear in centrifugal pump, following harmful effects may appear sooner or later:

i. Pitting and erosion of casing & impeller surfaces

ii. sudden drop in head, efficiency and power requirement.

iii. Noise and vibration can be produced by collapse of vapor bubbles.

iv. The factors which promote the cavitations are–

    √ High runner speed

    √ Restricted flow in suction

    √ Too high specific speed against design parameters

    √ Too high temperature of the liquid being pumped.

8.1.4 Preventive Actions to Avoid Cavitations

- It is necessary that $NPSH_{(A)}$ head is at least 0.8–1.5 m higher in any worst condition of operation compared to $NPSH_{(R)}$ designed. i.e. $\mathbf{NPSH_{(A)} - NPSH_{(R)} \geq 0.8–1.5\ M}$

- Variation in flow rate can be used for control of cavitations by increasing $NPSH_{(R)}$ to some extent in smaller condensate pumps. The variation of flow alters $NPSH_{(R)}$ value by change of working point.

- The head per stage in design stage should be limited to 50 M, otherwise the pump may run unevenly and may cause excessive wear of vanes.

- If condensate flow in the hot well is below operating quantity, the condensate level will drop resulting depression of $NPSH_{(A)}$ and pump may cavitate. The break-off characteristic curve will now intersects system curve, away from working point for a lower flow rate.

- When pressure at inlet suction nozzle of pump approaches to liquid vapor pressure, then tendency of cavitation will commence. The total energy at any point of pumping system constitute of kinetic and pressure energy in pumping system.

8.1.5 Cavitation is Two Stage Phenomenon

a. Formation of vapor cavities as a result of suction pressure approaching to liquid vapor pressure.

b. Collapses of bubbles when moves out from low pressure zone to high pressure zone of pump. The vapor cavity collapses in HP zone and simultaneously formation of vapor down stream (suction side) continues.

Depending on suction pressure, the formation and collapse of vapor bubbles happen at very short interval which render to damages (formation of pits on metal surfaces) in impeller and or casing. Manufacturer normally specify the cavitation parameters of pump at design liquid temperature. Variation in pressure, lower power requirement and reduces the pump flow and sound like stone breaking are produced along with some other symptoms of cavitations.

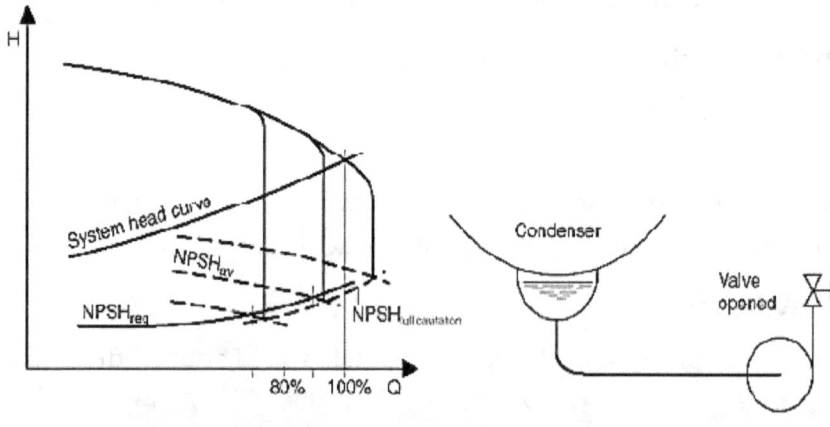

Cavitation control

When flow of condensate hot well is increased due to reduction in pump flow, $NPSH_{(A)}$ will increase and cavitations will be diminished, resulting restoration of normal flow of the pump (see Fig.) and consequently with increased flow the level will be reduced rendering to low $NPSH_{(R)}$. Thus, pulsation between cavitations and normal operation will exist till correction of $NPSH_{(R)}$. Therefore, intersection of variable $NPSH_{(A)}$ with pump $NPSH_{(R)}$ at full cavitation determines the flow rate.

8.1.6 Factors which Promote Cavitations

- Restricted suction
- Higher impeller speed
- Higher specific speed at design parameters
- Higher temperature of the liquid being pumped.

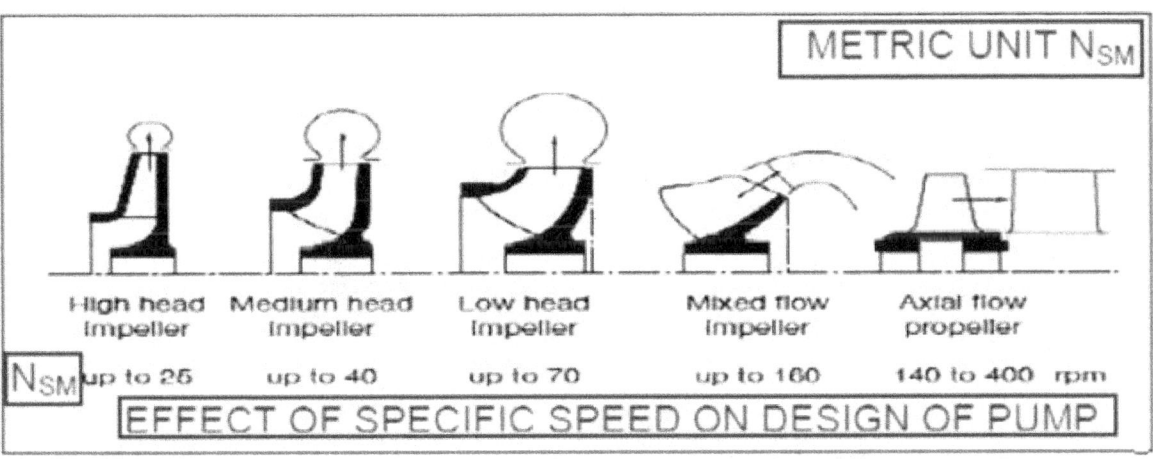

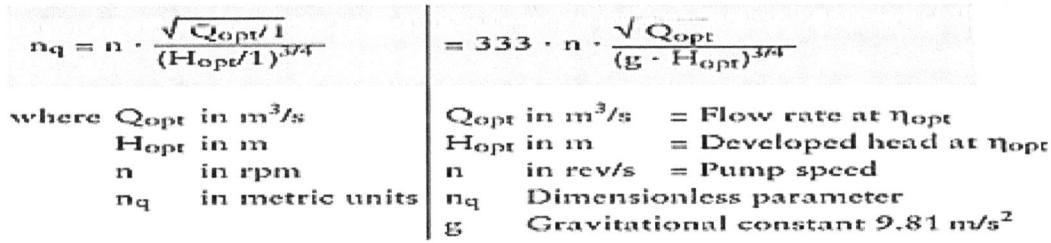

$$ n_q = n \cdot \frac{\sqrt{Q_{opt}/1}}{(H_{opt}/1)^{3/4}} \qquad = 333 \cdot n \cdot \frac{\sqrt{Q_{opt}}}{(g \cdot H_{opt})^{3/4}} $$

where
Q_{opt} in m^3/s
H_{opt} in m
n in rpm
n_q in metric units

Q_{opt} in m^3/s = Flow rate at n_{opt}
H_{opt} in m = Developed head at n_{opt}
n in rev/s = Pump speed
n_q Dimensionless parameter
g Gravitational constant 9.81 m/s^2

8.2 Regulation/Controls in Centrifugal Pumps

8.2.1 Pump Capacity Control Methods

1. Throttling
2. Switching pumps on or off
   a. pumps operating in parallel;
   b. or pumps operating in series
3. Bypass control
4. Speed control

5. Impeller vane adjustment
6. Pre-rotation control
7. Cavitations control.

8.2.2 Flow Control by Throttling

Closing a throttle valve in the discharge line increases the system resistance and it becomes steeper. Thus, intersection of the pump characteristic curve with new system resistance curve will form a new working point which is obvious from graph below. $NPSH_{(A)}$, of pump installation is reduced by throttling in suction line. Hence, it is not admissible.

Changing the flow rate Q by operating a throttle valve is the simplest flow control method not only for an instant flow adjustment but also for continuous flow control, with no investment. However, throttling through valve or orifice, is an energy wasting method, since the flow energy is converted irreversibly to heat by increasing the system resistance.

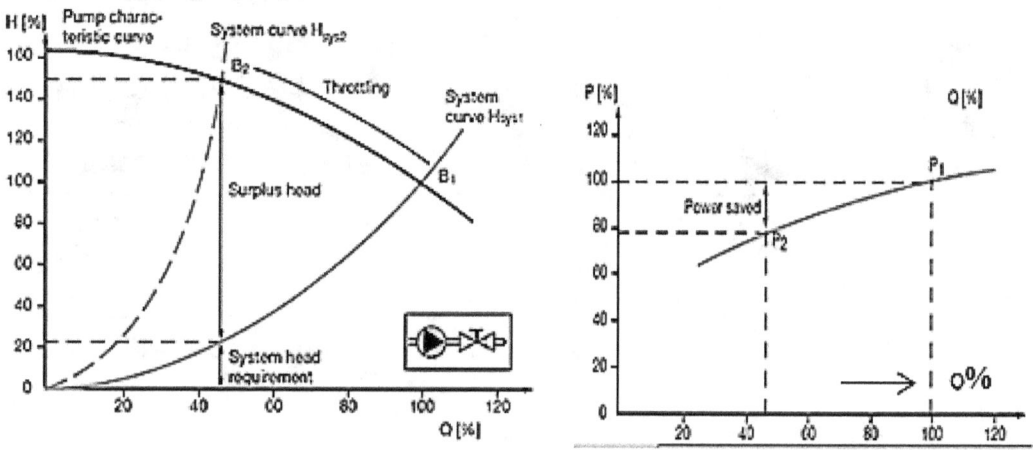

8.2.3 Variable Speed Flow Control

At various speeds of rotation n, a centrifugal pump has different characteristic curves, which are related to each other by the affinity laws. If the characteristics H and P as functions of Q are known for a speed n1, then all points on the characteristic curve for N_2 can be calculated by the following equations:

$$Q_2 = Q_1 * (N_2/N_1) \qquad \text{(a)}$$
$$H_2 = H_1 * (N_2/N_1) \qquad \text{(b)}$$
$$P_2 = P_1 * (N_2/N_1) \qquad \text{(c)}$$

These are true when pump is operating at BEP

Eq. (c) is valid as long as the efficiency η does not decrease with the speed 'n' variation. With a change of speed, the operating point is shifted on; H-Q curve. For different speeds the intersection of the system curve & characteristic curve (working point) will be different. Therefore, operating point will vary by change of speed resulting variation in flow. Thus, by changing operating speed of impeller, flow control can be exercised very accurately. However, the pump should not operate beyond BEP zone.

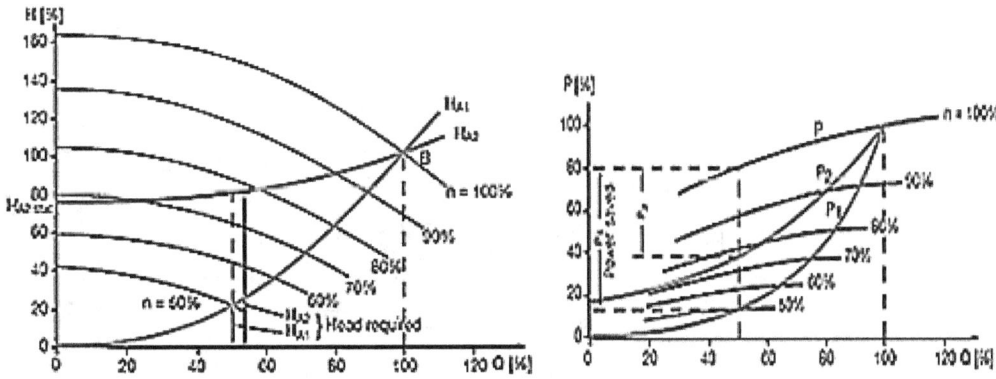

Operation of a variable speed pump for different system characteristic curves H_{sys1} and H_{sys2}
(Power savings ΔP_1 and ΔP_2 at half load each compared with simple throttling)

8.2.4 Recommended Flow Velocities for Cold Water

Inlet suction piping = 0.7–1.5 m/s

Discharge piping   = 1.0–2.0 m/s

8.2.5 Recommended Flow Velocities for Hot Water

Inlet suction piping = 0.5–1.0 m/s

Discharge piping   = 1.5–3.5 m/s

The pipe friction factor λ varies with the flow conditions of the liquid and the relative roughness d/k of the pipe surface. The flow conditions are expressed according to the affinity laws.

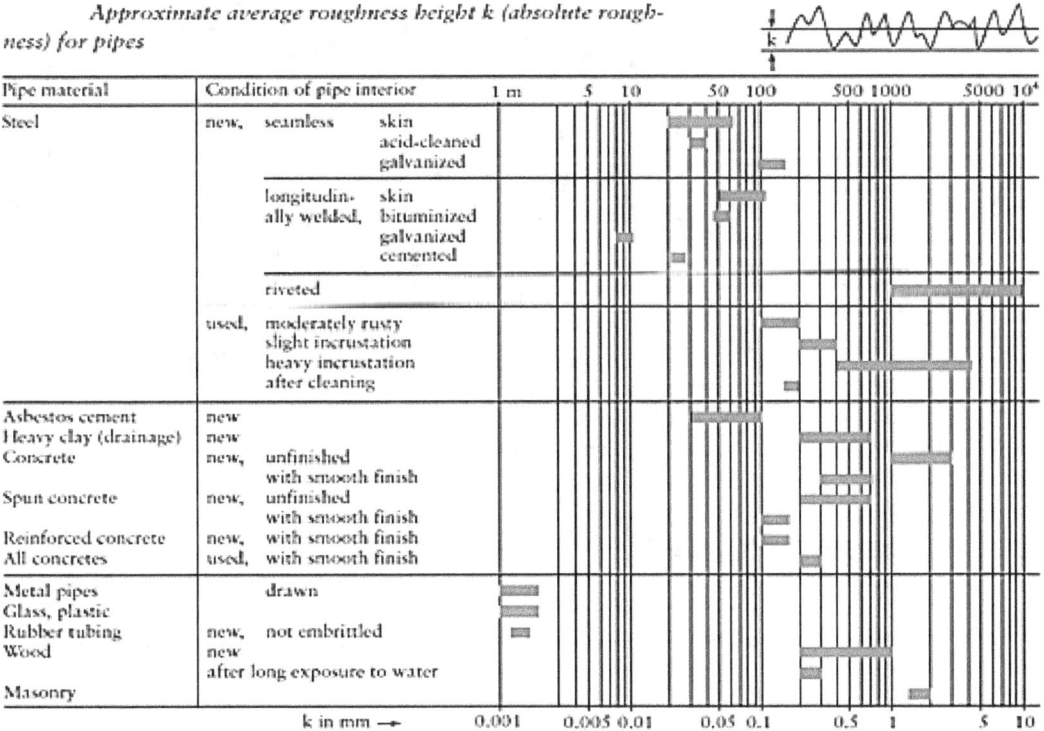

Approximate average roughness height k (absolute roughness) for pipes

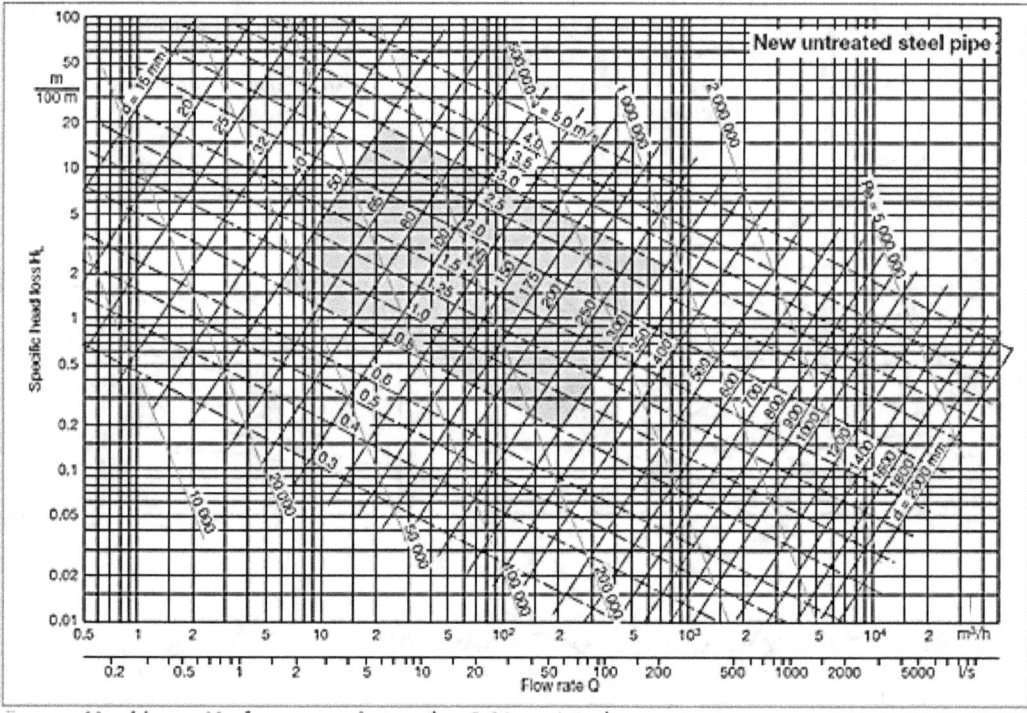

Head losses H_L for new steel pipes (k = 0.05 mm)

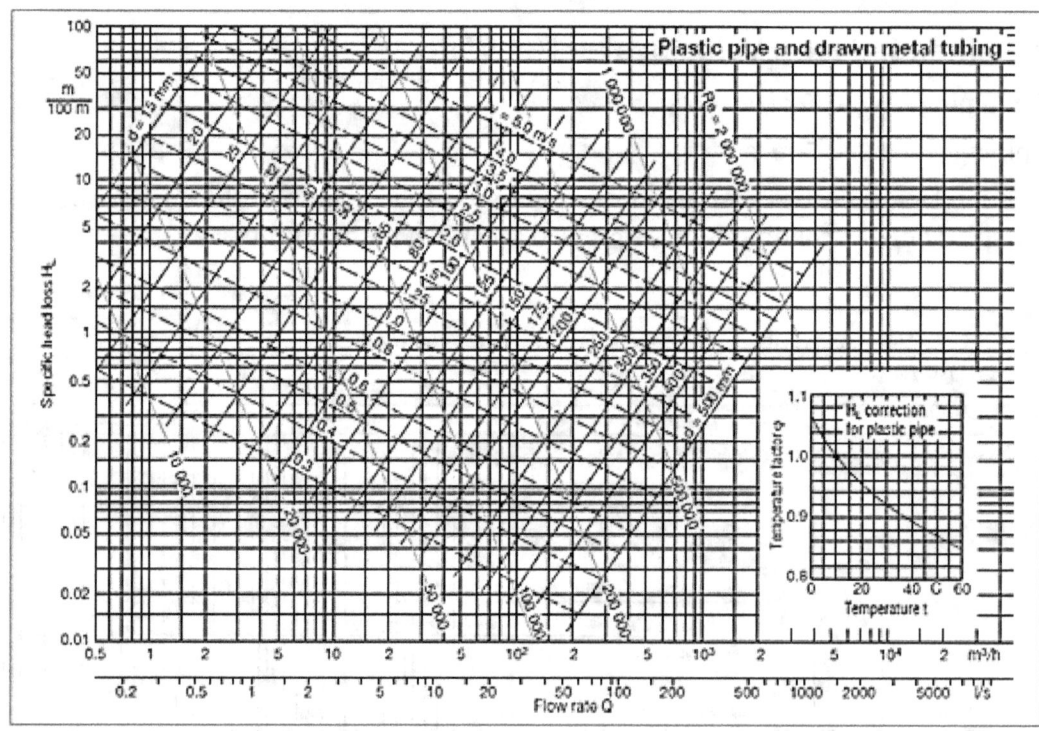

Head losses H_L for hydraulically smooth pipes (k = 0)
when t ≠ 10 °C multiply by the temperature factor φ.

8.3 Parallel Operation of Centrifugal Pumps

Where one pump is unable to deliver the required flow Q at the operating point, it is possible to have two or more pumps working in parallel in the same piping system, each with its own non-return valve.

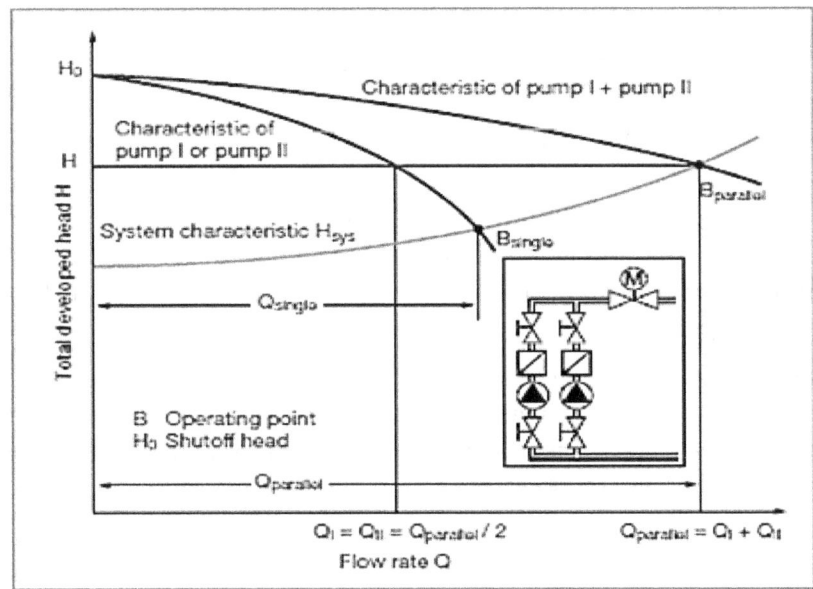

Parallel operation of 2 identical centrifugal pumps with stable characteristic curves

- All the pumps to run in parallel must have same characteristics, (shown in characteristic curves).
- System resistance for all pumps should remain same during operation. Inter section of combined characteristic curve (pump#1 + pump#2) and system resistance curve is known as parallel operation working point which should be unaltered for constant flow ($Q_{PARALLEL} = Q_1 + Q_2$).
- $Q_{PARALLEL}$ is slightly less than $Q_1 + Q_2$
- For frequent variable control, at least one of the pumps should be provided with a variable speed drive or a control valve in the common discharge piping.
- If frequent pumps running at fixed speeds and having unstable characteristics are run in parallel, difficulties can arise while taking another pump in line. The problems arises when the developed head H1 of the pump running is larger than the shutoff head of second pump (head at Q = 0) In such situation the second pump becomes unable to overcome the pressure on its non-return valve (System curve as shown in the following figure).

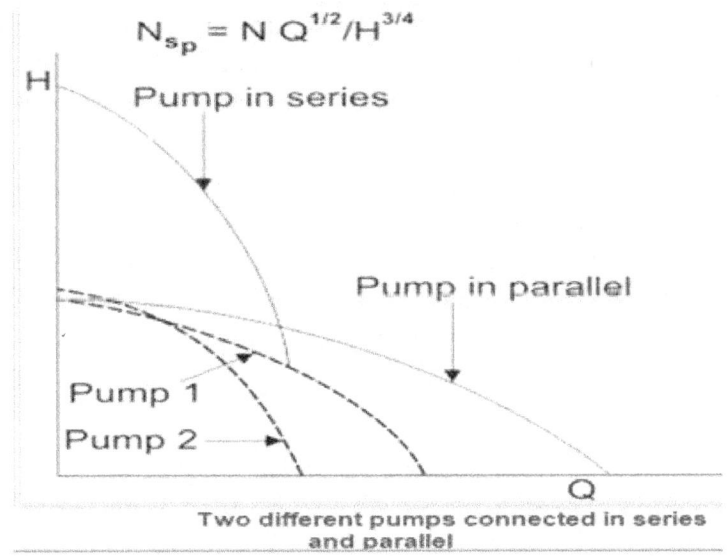

Two different pumps connected in series and parallel

- Pumps with unstable characteristics are not suitable for very low flow operation. For a lower system curve such pumps are able to operate properly without much of problem since the developed head of the pump running is lower than the shutoff head of the pump to be started.

  In case of difference in working point of any pump may lead to hunting between two working points or one pump will be over loaded and other pump will have tendency of sleeping.

- Intersection points of pump characteristic curve and system resistance curve (working point) for all pumps should be at same point. In case of difference in working pointed of any pump, will result hunting between two working points or one pump will be over loaded and other one will be in sleeping tendency.

Flow in piping system of two or more pumps before common header should be kept laminar with Renold's number closed to each other. If flow changes from laminar to turbulent in one of the pump, it may render change of system resistance curve and working point of that pump. Thus, there will be tendency to operate both pumps at different working points and hunting will take place.

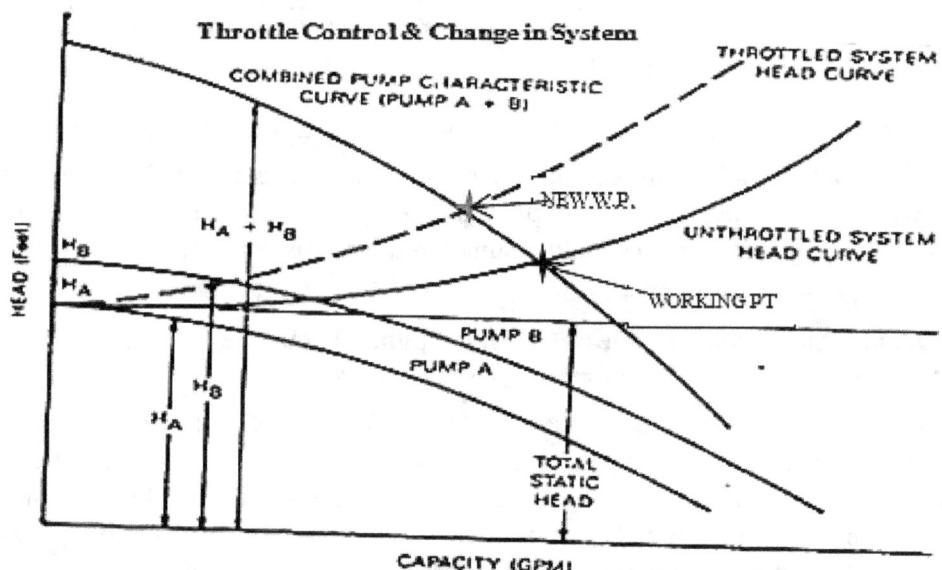

Where one pump is unable but delivers the required flow Q at the operating point, it is possible to have two or more pumps with its own non-return valve & working in parallel in common piping system. Parallel operation of pumps is easier when their shutoff heads H_0 are equal & identical. If the shutoff heads H_0 differ, the lowest shutoff head marks the point on the common H/Q curve for the minimum flow rate $Q_{(min)}$, below which no parallel operation is possible, since the non-return valve of the pump with smaller shutoff head will be held shut by the other pump(s).

During parallel pumping it must be kept in mind that after stopping one of the two identical centrifugal pumps the flow rate $Q_{(single)}$ of the second pump does not fall to half of $Q_{(parallel)}$. Then the second pump should immediately maintain the minimum flow of $0.5 * Q_{(parallel)}$ or operate at working point $B_{(single)}$ above. This must be conformed while checking the NPSH values and the drive power during performance test. The reason for this behavior is the parabolic shape of the system characteristic. That is the reason why on taking a second identical pump in line does not double the flow rate $Q_{(single)}$ of the pumps. The behavior of flow while pumps in parallel operation can be given as–

$$Q_{(parallel)} < 2 \cdot Q_{(single)}$$

This effect when starting or stopping one additional pump is more intense when the system curve is steeper or when the pump characteristic is flatter. As long as pumps I and II are running in parallel, practically flow rate $Q_{(parallel)}$ is the algebraic sum of Q_1 and Q_2, i.e.: $Q_{(parallel)} = Q_1 + Q_2$

Starting or stopping individual pumps, operated in parallel does save substantial energy, but it allows only a stepped control of the flow rate.

For frequent variable control, at least one of the pumps must be fitted with a variable speed drive or a control valve in the common discharge piping.

Parallel operation of reciprocating pump and centrifugal pump: It is important to choose equal head leaving discharge to be different as shown in following graph. The working point for both pumps will be at different locations maintaining equal heads.

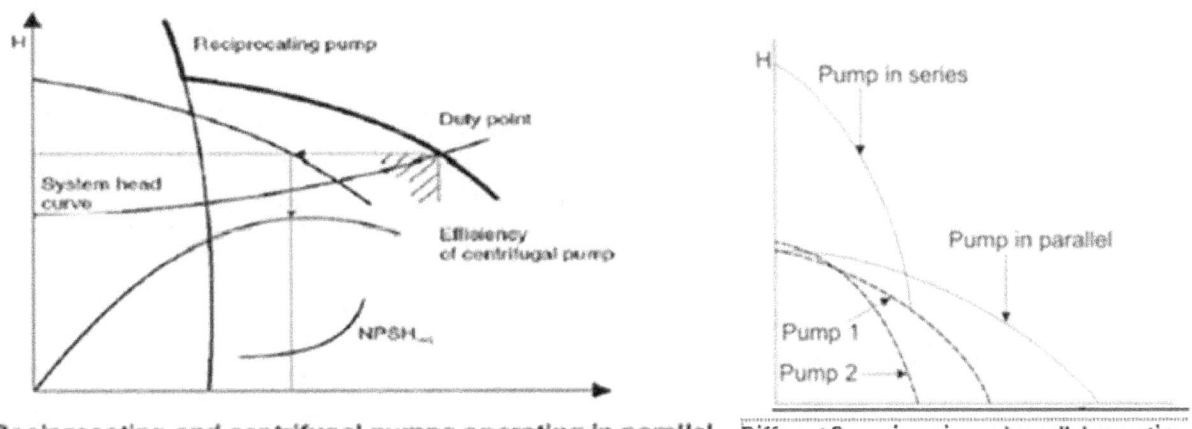

Reciprocating and centrifugal pumps operating in parallel   Different Pump in series and parallel operation

The flow developed by a centrifugal pump varies with the system resistance head, while that of a positive-displacement pump is independent of the system head.

8.3.1 In Series Operation

The pumps are connected one after the other so that the developed heads can be added for a given flow rate. This means that the discharge of the first pump is the inlet for the second pump. In high pressure pump suitable diameter and strength of the casing becomes the prime consideration. Therefore, multistage pumps are usually preferred where shaft sealing is not a problem except for the hydraulic transport of solids.

Where flow dependent losses are dominating, (i.e. H **geo** _ H **dyn**) pumps may be operated in series. With predominantly dynamic losses, series operation is advisable because it is still possible to deliver about 70% of the original flow if one out of 2 pumps is shut down (see figure below). Moreover the pump operates at better NPSH and good efficiency.

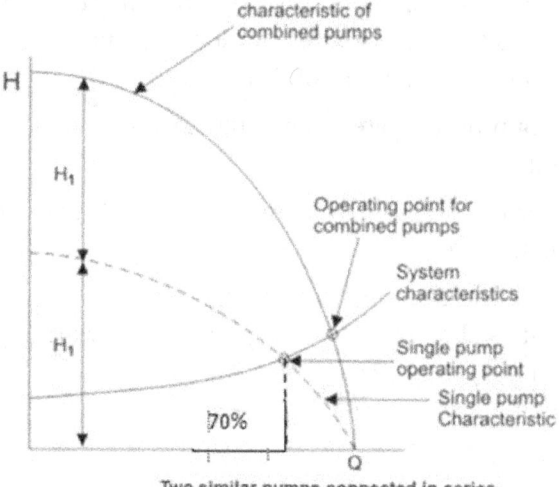

Two similar pumps connected in series

G_{hee} & H_{dyn} are mainly static system hence, series operation is not recommended. When one pump fails in series operation, the other runs at lower flow and lower efficiency (i.e. uneven performance). Such pumps can run in parallel and deliver greater flow at higher efficiency. When pumps operate in series, it must be noted that the seals and casings on the downstream pump are designed for higher pressure.

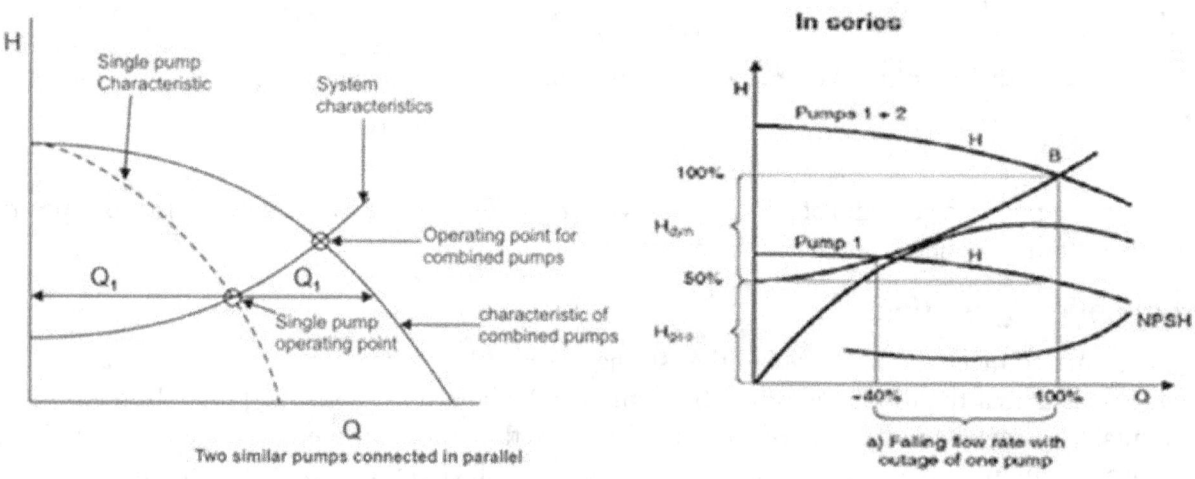

Two similar pumps connected in parallel

The shaft seal may have to be balanced with suction pressure of the upstream pump. Recirculation losses may occur from this, which must be taken into account in the overall efficiency calculation.

8.3.2 Regulation of Centrifugal Pump

At various speeds of rotation, a centrifugal pump has different characteristic curves, which are related to each other by the affinity laws. If the parameters H and P as functions of Q are known for a speed N_1, then all points on the characteristic curve for speed N_2 can be calculated from following equations:

$$Q_2 = Q_1 * (N_2/N_1) \quad \text{............ (a)}$$

$$H_2 = H_1 * (N_2/N_1)^2 \quad \text{............ (b)}$$

$$P_2 = P_1 * (N_2/N_1)^3 \quad \text{............ (c)}$$

Eq. (c) is valid as long as the efficiency η does not decrease on reduction of speed. With a change of speed, the operating point get shifted. In H/Q curves for different speeds of rotation, each curve has an intersection with the system resistance curve. Thus operating point moves along the system curve for smaller flow rates when the speed of rotation is reduced.

8.3.3 Speed Regulation

The speed regulation is adopted for capacity variation, to meet the process demand. On speed regulation of pump, the position of pump's characteristic curve is changed and intersection of system resistance curve with pump characteristic curve forms a working point. When speed is raised, the working point shifts right to the original working point, resulting increased discharge. Contrary to this, if speed is decreased the discharge is reduced. This type of speed regulation is economical and popularly adopted in power plant eg. BF pumps, CE Pumps or CW where we can achieve an appreciable saving of energy. The *equation of similarity* in speed control can be used as given here–

$$\frac{Q_1}{Q_2} = \frac{n_1}{n_2} \quad \frac{H_1}{H_2} = \left(\frac{n_1}{n_2}\right)^2 \quad \frac{NPSH_1}{NPSH_2} = \left(\frac{n_1}{n_2}\right)^2$$

It is clear that increasing the speed increases the $NPSH_{(R)}$, hence, higher $NPSH_{(A)}$ must be provided in design of the system to operate the pump at variable speed. Moreover, increasing speed increases power input approximately with third power of the speed. Consequently, the strength of the shaft and motor rating must always be verified by designer before adopting the speed variation. With small speed changes (up to 10%) efficiency remains virtually unchanged but for higher speed that is lowered.

Larger speed change, the velocity and consequently the Reynolds number alters in the channels. The efficiency is degraded at higher speeds.

8.3.4 Minimum Flow in Regulation

Efficiency decreases as the pump runs beyond working point along the characteristic curve. When regulation is applied at low flow rate, the energy difference of BHP and Water HP of pump will be converted into irreversible heat and transferred to liquid & components of pump. Therefore, pump's minimum flow requirement, recommended by manufacturer should be maintained in variable flow demands.

$$\text{Temperature rise } \Delta t = \frac{\{(1 - \xi p) \cdot h\}}{(778 \cdot Cp \cdot \xi p)}$$

Where:

ξp = pump efficiency,

Cp = specific heat of liquid (for water = 1,0)

II – Pump total head ft

Under-rated flow not only induces heat but also produces noise, hydraulic pulsation and rapid deterioration of impeller due to interior circulation and churning of liquid in un-directional & un-controlled manner.

8.3.5 Impeller Blade Adjustment

For axial pump or even semi-axial pumps (high specific speed) impeller is provided with adjustable blades, an economical flow control system which enable a wide flow rate variation. For increase in flow rate, the impeller blade angle (turning the blades around an axis essentially perpendicular to the hub) is increased. Impeller blade adjustment depends on the shape of the system head curve which can allow flow variation 70 to 50% of its rated value in the best efficiency zone.

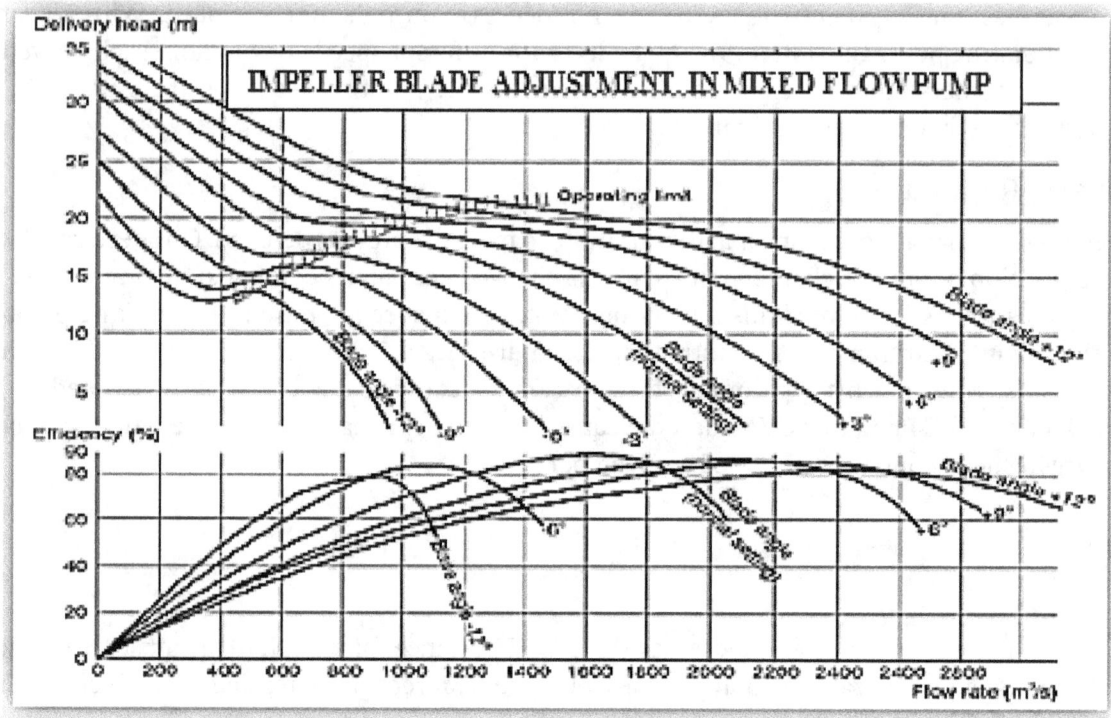

8.3.6 Behavior of Guide Vane Control

The impeller preceded with controllable inlet guide vanes (IGV), will have an altered characteristic due to varied inflow to the impeller. The change in guide setting,imparts a circumferential movement to the inflowing media, altering the energy conversion process inside the pump and consequently change of the characteristics curve. The inlet guides can be used on any pump but its effect on the characteristic results increase in specific speed of the pump. *The inlet guides are employed primarily in mixed-flow or axial-flow pumps but* efficiency of the pump drops much streeper in the part loads and overload operation compared to no control by inlet vane or impeller blade control system in the same situation.

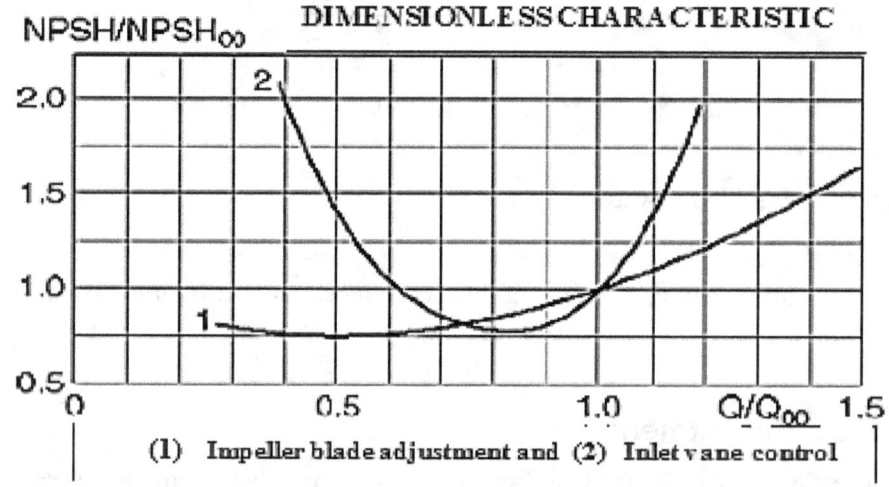

(1) Impeller blade adjustment and (2) Inlet vane control

Moreover, guide vane control/pre-rotation control is more susceptible to cavitations as compared to impeller blade control which is obvious from characteristics graphs above.

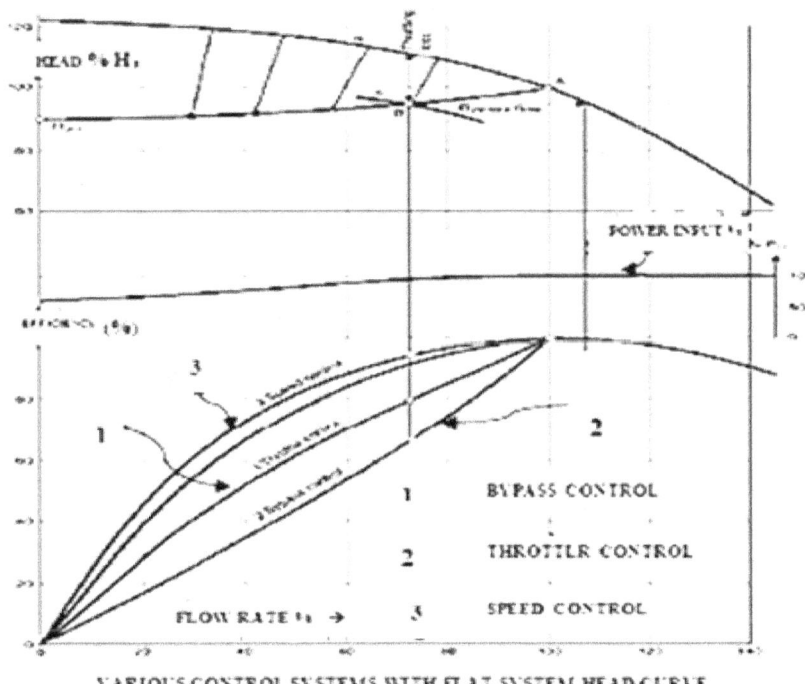

VARIOUS CONTROL SYSTEMS WITH FLAT SYSTEM HEAD-CURVE

SPECIAL CONTROLS

a. VFD drive with pump motor

b. Modulating recirculation

c. Auto On-Off

d. Manual On-Off

8.3.7 Permissible Speed of Pump or Capacity value:

From given formula of Ns we calculate RPM (N), given with Ns, Q, and NPSH$_{(A)}$ and head H–

Ns = [N * $Q^{0.5}$]/$H^{0.75}$ ....for single stage and single suction impeller pumps

The results are presented in graphical form for different conditions i.e. capacity, of pump suction, impeller- single/double/multi stages. The suction NPSH required. NPSH$_{(A)}$ must be ≥ suction NPSH available NPSH$_{(R)}$. Difference of (NPSH$_{(A)}$ and NPSH$_{(R)}$) is the safety margin (0.8 to 1.5 M) provided for eventualities of pump factors (eg. Size, power consumption, intake design, range and mode of operation, type of service etc,

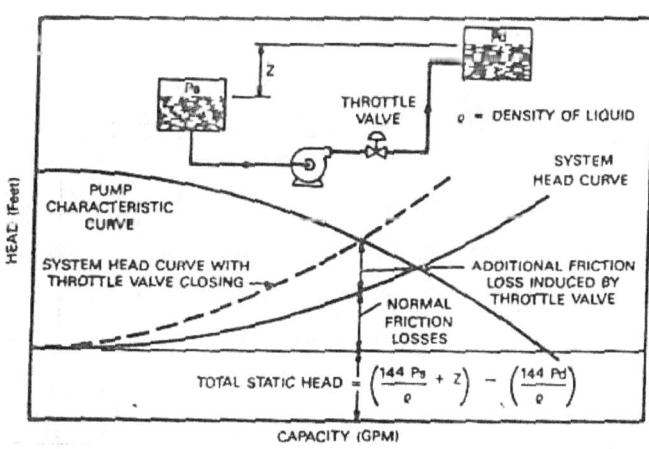

Pumping system with throttle flow,

8.3.8 Flow Reduction by Reducing Impeller DIA

In a radial or mixed flow centrifugal pump *permanent* flow control by trimming and turning down the outside diameter D of the impeller. The trimming should be limited to the diameter at which the impeller vanes are still overlapping in radial view. This technique can not apply to Impellers, made of hard materials, e.g. Ni-hard metal etc. For multistage pumps, only the vanes but not the shrouds of the impellers are cut.

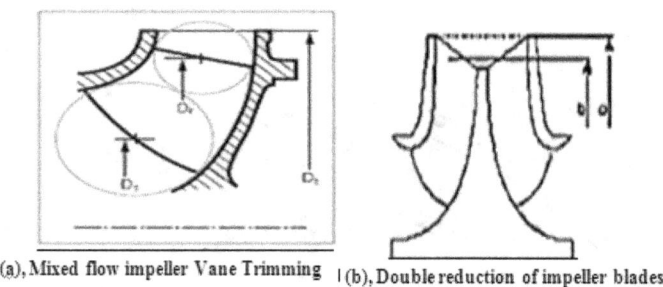

(a), Mixed flow impeller Vane Trimming | (b), Double reduction of impeller blades

- Instead of trimming the impeller, sometimes the impeller and diffuser of one stage from multistage pump are removed and replaced with a blind impeller. The impellers with a non cylindrical exit are either trimmed or only blades are cut as shown in fig (b) above.

- If the impeller diameter is required to be reduced according to a thumb rule presented below. The exact calculation cannot be made, since the geometrical similarity of the vane angle and exit width are not preserved while turning down the impeller. Following approximate relationship can be applied between Q, H and the impeller diameter D–

$$(D_t/D_r)^2 \approx Q_t/Q_r \approx H_t/H_r$$

Where: subscript "**t**" designates the condition before reduction of the impeller outer diameter and "**r**" is the condition after the reduction. The required (average) reduced diameter is given as:

$$D_r \approx D_t * (Q_r/Q_t)^{1/2} \approx D_t * (H_r/H_t)^{1/2}$$

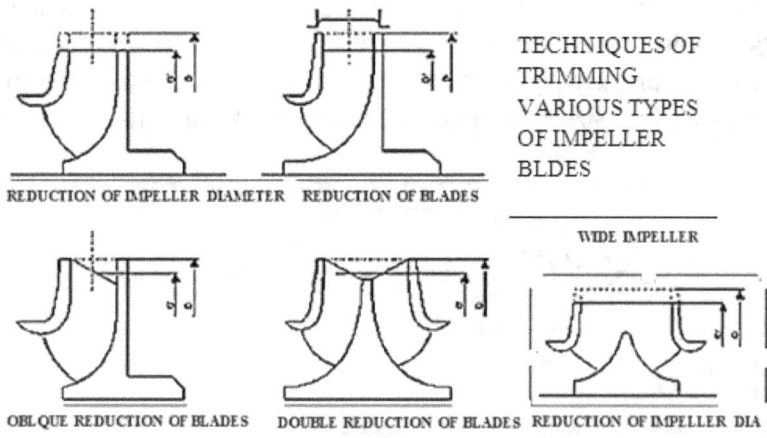

TECHNIQUES OF TRIMMING VARIOUS TYPES OF IMPELLER BLDES

REDUCTION OF IMPELLER DIAMETER   REDUCTION OF BLADES

WIDE IMPELLER

OBLQUE REDUCTION OF BLADES   DOUBLE REDUCTION OF BLADES   REDUCTION OF IMPELLER DIA |

Larger corrections, increases the $NPSH_{(R)}$ of pump in overload conditions, because specific vane loading is increased by the reduced vane length & also affects velocity distribution at the impeller inlet. If a small correction (say 5%) to impeller Vane diameter is necessary, it may be assumed that the required $NPSH_{(R)}$ will increase negligibly.

When impeller diameter of pump is trimmed and reduced, outlet width, blade angle and blade length are also reduced. The effect of trimming depends on design of the impeller (i.e. specific speed and NPSH). Consequently, only approximate effects of reducing the impeller diameter on characteristics curve can be predicted.

The parameters needed to determine the reduction in diameter can be estimated from figs. shown; in the H/Q curve a line is drawn connecting the origin and the new operating point Br is established. The extension of the line intersects the characteristic curve for full diameter D_t at the point B_t. In this way the values of Q and H with the subscripts t and r can be found, which is used in Eq. to find the desired reduced diameter D_r. The ISO 9906 method is more accurate but also more involved by consideration of the average diameter D1 of the impeller leading edge (sub-script 1), valid for N_q < 79 and for a change of diameter < 5%, as long as the vane angle and the impeller width remains.

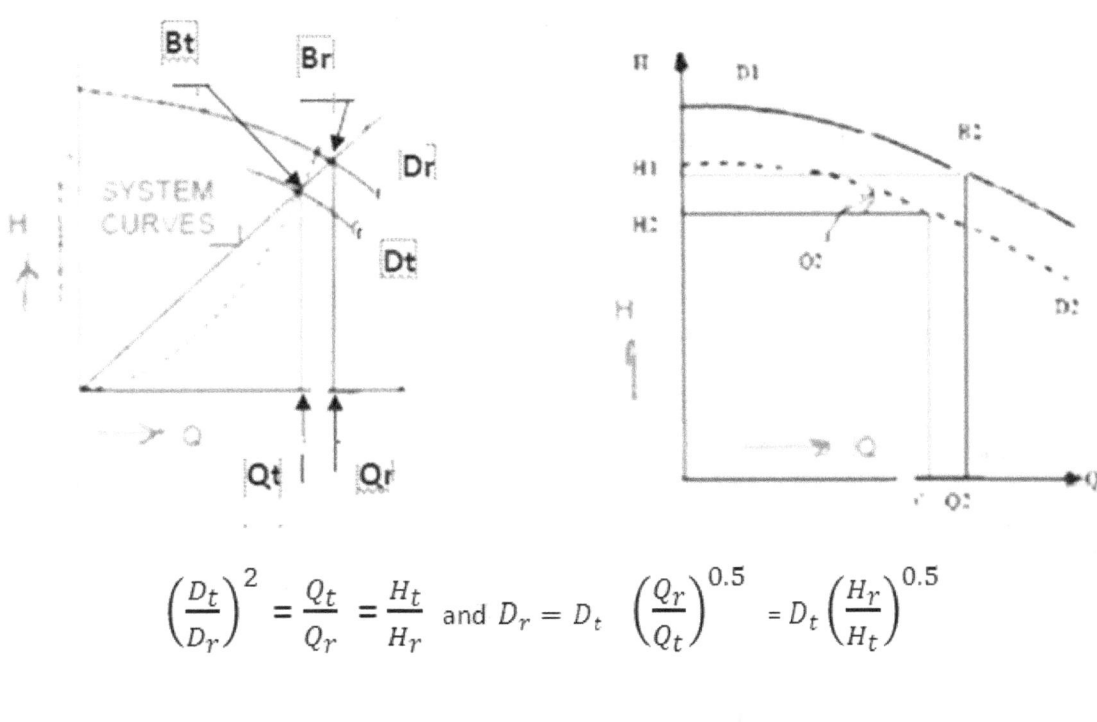

$$\left(\frac{D_t}{D_r}\right)^2 = \frac{Q_t}{Q_r} = \frac{H_t}{H_r} \text{ and } D_r = D_t \left(\frac{Q_r}{Q_t}\right)^{0.5} = D_t \left(\frac{H_r}{H_t}\right)^{0.5}$$

To know D_r, a H-Q curve D_r is drawn through the reduced operating point B_r with H_r and Q_r, (not a line as in fig.) which intersects the base H-Q curve D_t for diameter D_t at a working point B_t (with designed head H_t flow Q_t).

8.3.9 Flow Control Using a Bypass (Bypass Regulation)

The pump runs at constant speed to deliver fixed capacity but excess water to process requirement is bypassed and send back to suction sump or container. It is constructed with metering orifice and control valve to regulate the bypass flow as per demand.

- In order to vary flow, sometimes regulation is made by letting part of the bypass water back to the suction nozzle. The pump, in this case operates to the right of working point depending on the system resistance taking in account the point of cavitations behavior. Owing to recirculation losses, the overall efficiency of the pump drops considerably. This method of regulation is used rarely to control the pump flow. However, judgment is made based on overall efficiency of bypass technique and that of speed, and throttle regulation.

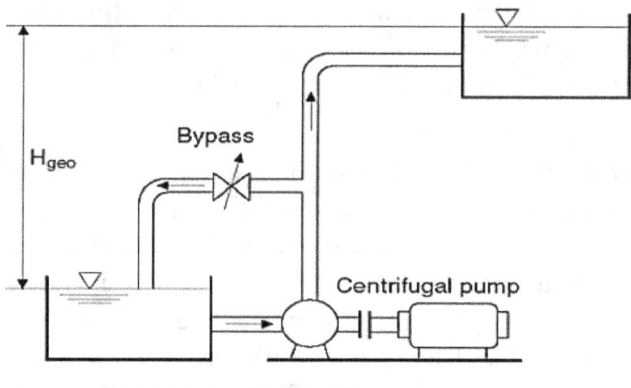

bypass control system

- The system characteristic curve will form steeper by closing a throttle valve, and it will form flatter by allowing a bypass from the discharge piping as shown in Fig. The pump operating point shift from B1 to a larger flow zone B2. The bypass flow rate is controlled and can be transferred back into the inlet tank without performing any work. Such controls are applied as the energy saving measure where the power curve falls down ($P_1 > P_2$) on increasing pump flow ($Q_1 \longrightarrow Q_2$) , especially in case of high specific speed (mixed flow, axial flow propeller) pumps.

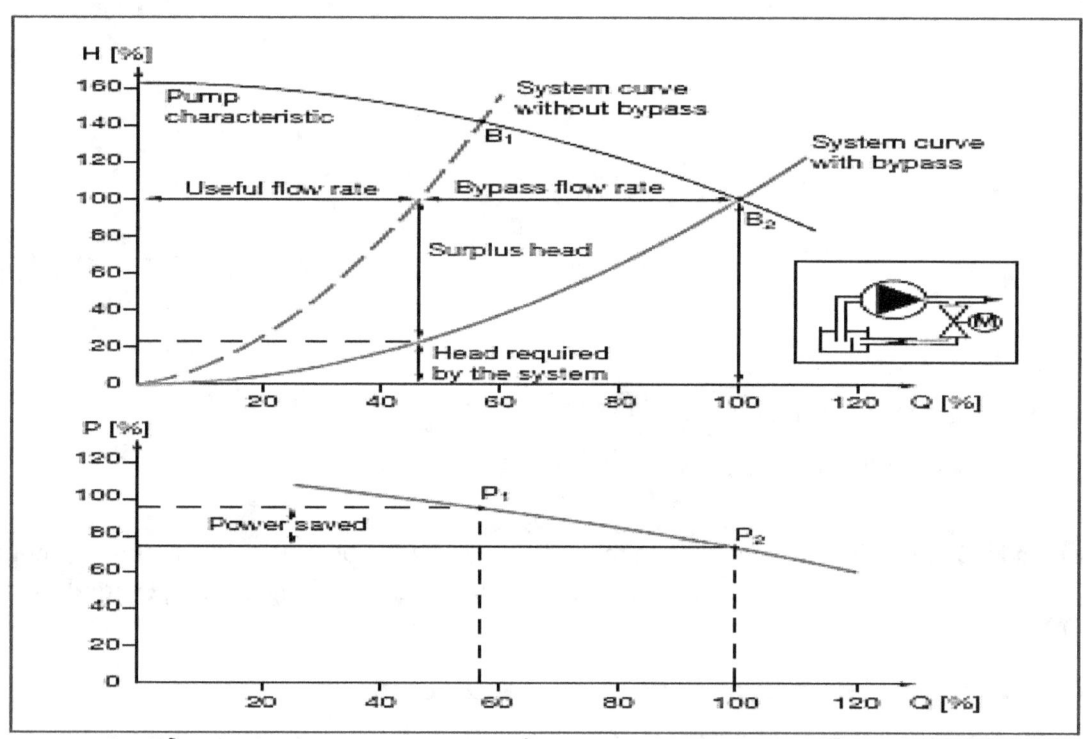

Characteristic curves and operating points of a pump with a falling power curve and flow control using a bypass.

The pump having high specific speed, Ns (axial or mixed flow) regulated by pre-swirl control or by changing the blade pitch angle is considered to be economical. The expenditure in a bypass control valve is quite costly hence, blade adjustment method is more suitable un-acceptable low flow operation of pump but in this condition, the specific speed increases & consequently the ratio of impeller outlet diameter to inlet eye diameter decreases. This ratio becomes 1.0 for a true axial flow impeller.

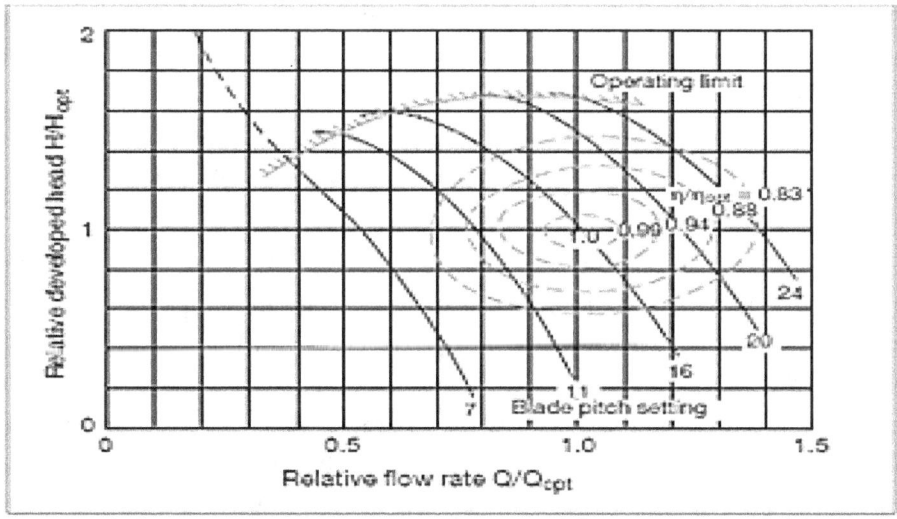

Characteristic curve set of an axial flow pump with blade pitch adjustment, $n_q \approx 200$

8.4 Handling Solid–Liquid Mixtures By Centrifugal Pumps

The centrifugal pumps by virtue of simple design, good ratio between delivery & suction eye sizes, uniform delivery and low susceptibility to trouble, are normally considered for handling mixtures of liquids and granular solids, termed as "slurries." The high flow centrifugal pumps are less expensive compared to positive displacement pumps for almost similar pumping parameters. There are varieties of solid particles like iron ore, fly ash, coal, sewage, gravel, potatoes, sugar beet, wood pulp, fish, paper stock and suspensions of lime etc. which need to be handled by centrifugal pumps. Such pumps are normally termed as slurry pumps which convey/transport slurry.

- For maximum reduction in $NPSH_{(R)}$ in handling the hot, degassed water and certain gas-free hydro-carbon, attention should be given in the following situations:

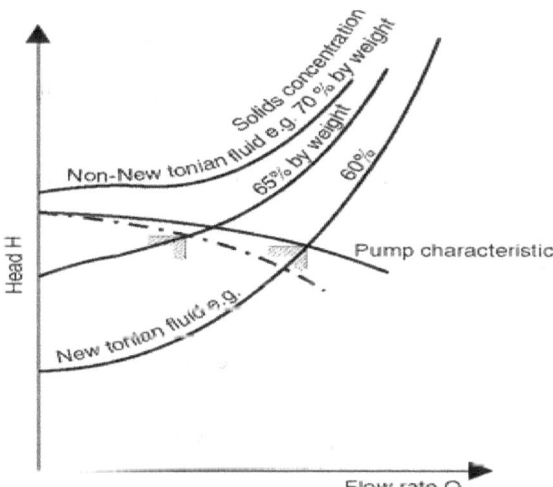

Pipeline characteristic as a function of solids concentration

√ Slurry pumps should be operated with ample (approx 10 M) $NPSH_{(A)}$.

√ Flow velocity through the impeller and volute must be kept low to minimize abrasion (increases with the third power of the velocity). The heads of slurry centrifugal pumps are kept low, presently limited to about 80 M. Admissible velocity in pipeline depends on the density, particle size, form and concentration of the solids in flow stream.

√ The impeller channels and volute cross-sections must be smooth and large enough to allow passage to the biggest solid particles expected. The back shroud blades must keep abrasive particles away from the stuffing box.

√ Flow velocity of slurry through pipe lines should not drop below critical velocity; otherwise the solids will settle into pipe from the suspension. The minimum flow velocity must be at least 0.3 m/s above the critical settling velocity

√ Slurry pumps are often operated in series in order to meet the demand of high head.

√ When the pipeline is partly blocked by settled solids, the moving solids, will encounter higher resistances during startup which need to be overcome to attain flow velocity and resuming transportation process.

√ The centrifugal pump should have a steep characteristic. If the concentration of solids is changed, the system resistance curve also alters. If the characteristic is steep, pump delivery will drop slightly and flow velocity will vary to some extent. Segregation of the solids is thus prevented.

√ In pumping and transportation of long fibers in suspension, open impellers are used to avoid clogging risk. The inlet edges are profiled and thickened so that fibers should slide off easily.

√ Slurry pumps are subjected to heavy wear. The loss of head, flow rate and efficiency should be considered while sizing the pump and drivers.

8.5 Pump Tests

The pump tests are classified as follows:

1. The **works or factory acceptance test:** It is performed at the manufacturer's test bed with quite accuracy, precise measurements and in reproducible conditions.

2. The **field test:** Measurements on a pump is carried out in the field and subjected to influence of the prevailing service conditions. Accuracy and reliability of the measured results largely depend on the instrumentation and measuring positions. If contractually agreed, the field test can be converted into acceptance test.

3. **Periodic field tests:** This teat is conducted to detect the changes in performance of pump which is largely affected by wear & tear and it is carried out at fixed schedule using reliable and calibrated instruments.

4. **Model tests:** Such tests are conducted with high measuring accuracy. The model test must be defined to substitute the acceptance test on the pump.

8.5.1 Aspects of Test Standards

• The standards are intended to determine and analyze the measured results during test on a statistical basis and ensure that the true value of the measured variable ascertain statistical reliability of at − least 95%.

• The quality classes are defined in all standards, e.g. ISO 9906

Note: Although standards such as ISO, API, HI and DIN are being harmonized,but refered individually with industrial common terminology. The overall tolerances are defined, with admissible measuring uncertainties and the construction standards.

8.5.2 Acceptance

The rules are laid down in standards to simplify understanding between manufacturer and purchaser. That are in general laid down as follows:

- Definitions of all variables needed to define the functions of a centrifugal pump and to fix the guarantees for its hydraulic performance, flow capacity and total dynamic head and pump efficiency or power.

- Definitions of the technical guarantees and their implementation.

- Recommendations for preparing and performing acceptance tests to verify the guaranteed data.

- Rules for comparing measured results with guaranteed data and conclusions should be drawn from the comparison.

- Requirements for curves and test reports.

- Description of the principal measuring techniques used for guarantee demonstrations.

8.6 System Head Curves

A pumping system may consist of pipe, valves, fittings, vessels, nozzles, open channels, weirs, meters, process equipment and other liquid-handling conduits through which flow is carried. While analyzing a particular system for the purpose of selecting a pump, the resistance to liquid flow through various components is calculated. The flow resistance with increase of flow is at the rate of approximately square of the flow through the system.

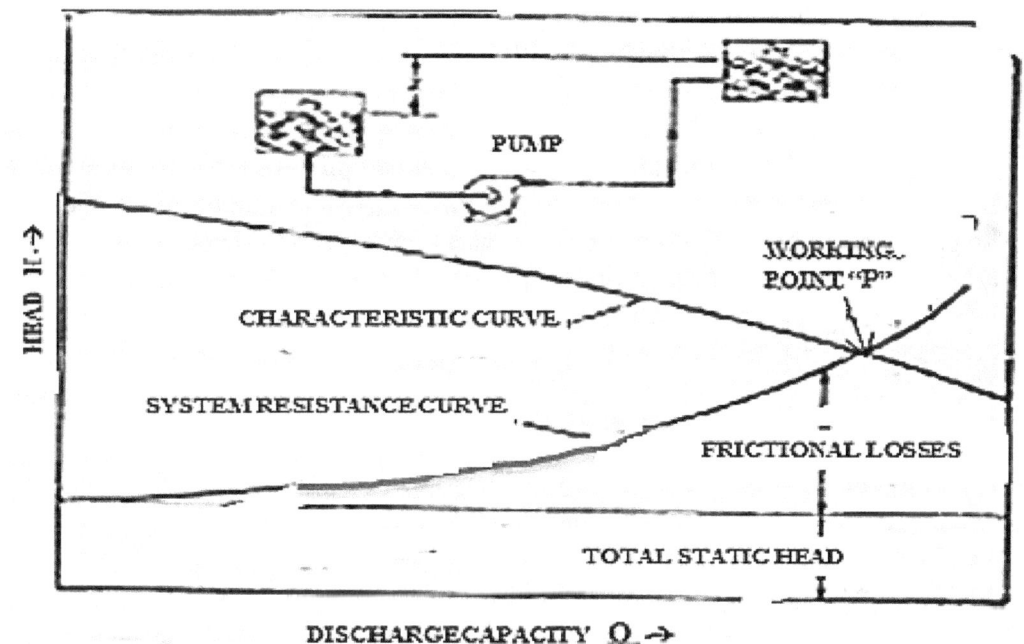

PUMPING SYSTEM WITH FIX FLOW & SPEED. DISCHARGE ABOVE SUCTION DATUM

In pumping system, energy is not only utilized for flow resistance but also overcome to raise the liquid head from suction level to a higher discharge level. When pressure at the discharge end of liquid surface is higher than the pressure at the suction (liquid surface), it will require additional energy for pumping head to overcome pressure resistance. The two heads are fixed since they do not vary with rate of flow. Fixed system head can also be negative, as in case of discharge level elevation is lower than suction elevation (shown in fig. below). Fixed system heads are also called static heads.

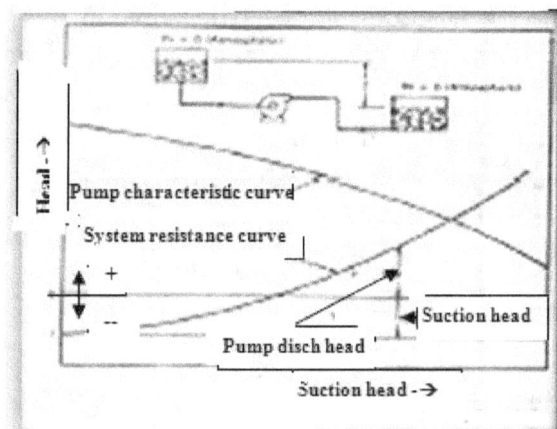

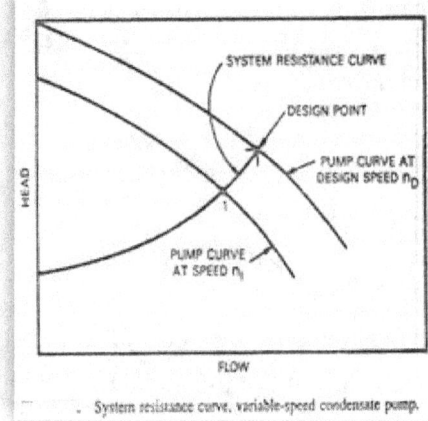

System resistance curve, variable-speed condensate pump.

A system head curve is a plot of total system resistance (variable plus fixed) for various flow rates. It is preferable to express system head in M rather than kg/cm² since centrifugal pumps are rated in M head. System head curve usually show flow in M³/sec vrs system resistance head.

8.6.1 When the System Head Is Required for Several Flows

When the pump flow is to be determined, a system head curve is constructed with following procedure–

- Define the pumping system and piping length. Calculate (or measure) the fixed, system head, which is the net change in total energy from the beginning to the end of system, due to elevation and/or pressure head difference.

- An increase in head in the direction of flow is considered a positive quantity. Calculate the variable system total head loss of pipe, valves, fittings, and equipment in the circuit for several flow rates. The fixed system head is the net change in total energy. The total head at any point is (Pd/w) + z. The pressure and liquid levels do not vary with flow. Plot variable head curve vs flow.

- The flow developed by a centrifugal pump varies with the system resistance, while the flow of a positive-displacement pump is independent of the system head. By super imposing the head-capacity characteristic curves, the flow of a pump can be determined. The system resistance curves will intersect at the flow vs head curve which is the point at which the head is equal to the required system head for the rated flow.

- When a pump is to be procured, head-capacity curve intersection with the system head curve (working point) is required to be specified. This intersection should fall in the best efficiency zone. In steady state operation, the liquid through the system gravity flow is sustained at the flow rate corresponding to zero total system head (negative static head + system resistance = zero). If a flow is required at any rate greater than that which gravity can produce, a pump is required to overcome the additional system resistance.

8.6.2 Variants in Pumping Systems

In case of fixed system resistance in a pumping system, there is just one total head for each flow rate, consequently a centrifugal pump will be operating at a constant speed and deliver just one flow. The system may operate either in controllable or un-controllable conditions e.g. changes in the valve opening at pump discharge or bypass line; thus, aging of pipes; changes in components size, length or number of pipes; changes in the process, changes in the number of operating pumps, pumping into a common header etc. are all examples of either controllable or uncontrollable system contributing the changes.

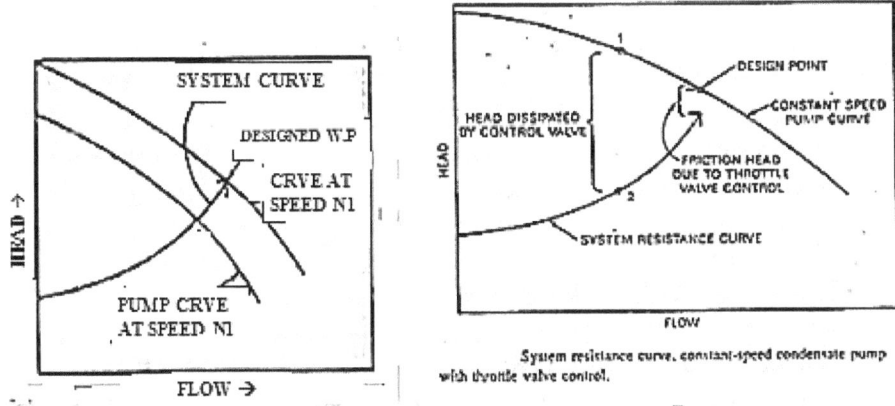

System resistance curve, constant-speed condensate pump with throttle valve control.

These changes alter the shape of the system resistance curve and in turn, affect the pump flow.

8.6.3 Priming of a Centrifugal Pump

Priming is the process of filling the suction pipe, casing of the pump and the delivery pipe up to the delivery valve with liquid to be pumped. If priming is not proper, the pump cannot deliver the liquid as the head generated by the Impeller will be in terms of meters of air which have a very low specific weight compared to water. Hence, priming is the basic step to start a centrifugal pump. Various arrangements are provided to keep the pump primed even when remains in standstill condition so that pump can be started simply by switching the power to motor.

8.6.4 Priming of a Centrifugal Pump Can Be Done by Any One of the Following Methods

- Priming with suction/vacuum pump.
- Priming with a jet pump.
- Priming with separator.
- Automatic or self priming.

8.6.5 The Flow Variation of Centrifugal Pumps Can Be Obtained By

- Changing the static head component (different water level or tank pressure).
- Changing the system resistance (e.g., by changing the setting of a throttling device, installing an orifice or a bypass line, rebuilding the piping).
- Changing/shifting the characteristic curve of pump by various methods.

8.6.6 A Pump Characteristic Curve Can Be Changed By

- Changing the speed of rotation.
- Starting or stopping pumps operated in series or parallel.
- Changing the outside diameter of pump radial impellers,
- Installing or changing the setting of installed pre-swirl control arrangement of pump in mixed flow impellers.
- Changing the blade pitch setting.

    Note: the effect of these measures for changing the characteristic curve can only be predicted for non-cavitations operation in axial flow (propeller) pumps.

8.6.7 Methods of Constructing System-Head Curves

To determine the resultant pump flow for two or more variants in pumping systems:

a. **Variable static Head:**

- In a system where a pump is taking suction from one reservoir and filling to other reservoirs at different elevations, the capacity of a centrifugal pump will decrease with an increase in the static head (i.e. filling tank at higher elevation).

- The system-head curve is constructed by plotting the system resistance head (variable+ fixed) vs. flow.

- Add the anticipated minimum and maximum static heads (difference in discharge and suction head). The resulting two curves are the total system heads for each condition.

- The flow rate of the pump is the point of intersection of the pump head-capacity (Q–H) curve with any one of the latter two system heads curves, or with any intermediate system head curve for other level conditions.

A typical head vs. flow (Q-H) curve for a varying static-head system is shown in fig.

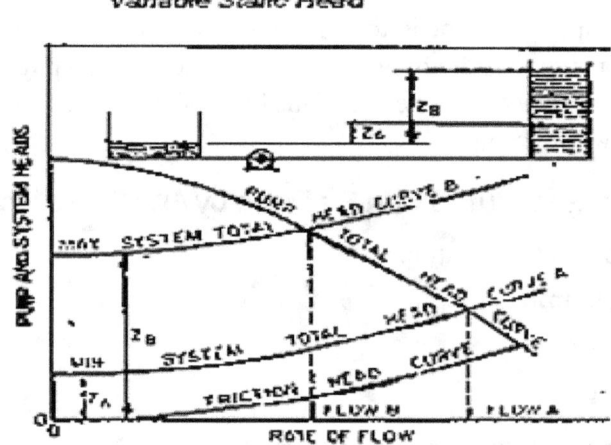

To maintain a constant flow for different static-head conditions, the pump speed can be varied to increase or decrease of the total system head. A typical variable-speed centrifugal pump, operating in a varying static head system can have a constant flow.

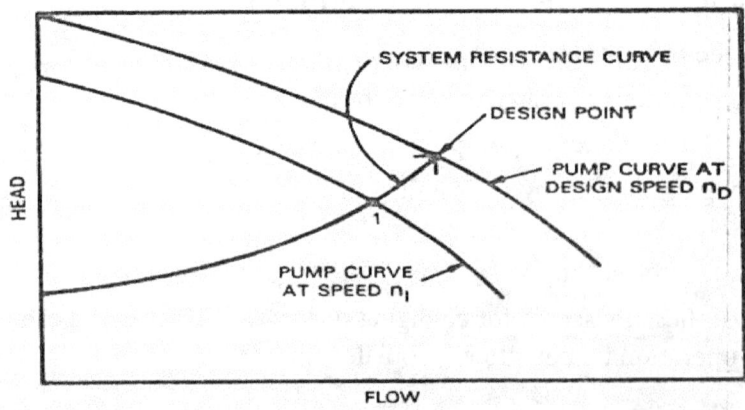

. System resistance curve, variable-speed condensate pump.

It is important to select a pump which will have the best efficiency within the system operating range, then pump should operate mostly in the variable frequency zone.

b. Variable System Resistance by Discharge Valve Operation

Alteration of valve opening position in the discharge line of a centrifugal pump changes the system head and consequently results the pump flow variation. The maximum flow is obtained with a completely opened valve when only resistance to flow is the friction from pipe and fittings. The closed valve results to operate pump at shut-off condition and produces maximum head at no flow. Any flow between maximum to shut-off can be obtained by accurate adjustment of discharge valve opening.

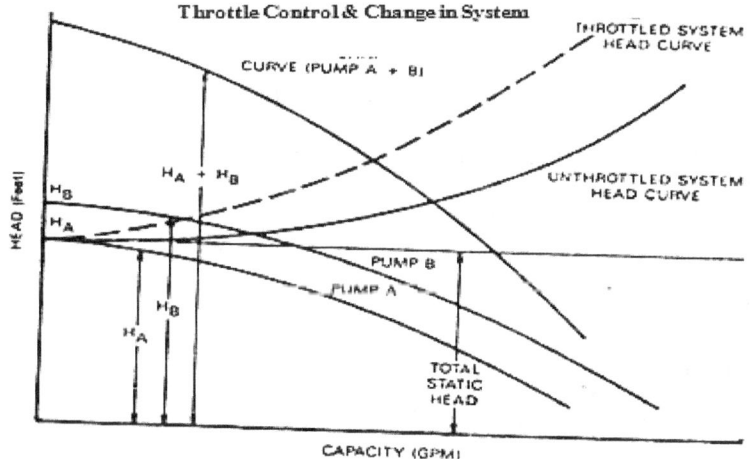

8.6.8 Flow Control by Throttling

Changing the flow rate Q by operating a throttle valve is the simplest flow control method as single adjustment for continuous flow control without any investment. But it is the most energy wasting method, since the restricted flow energy is converted to irreversible heat.

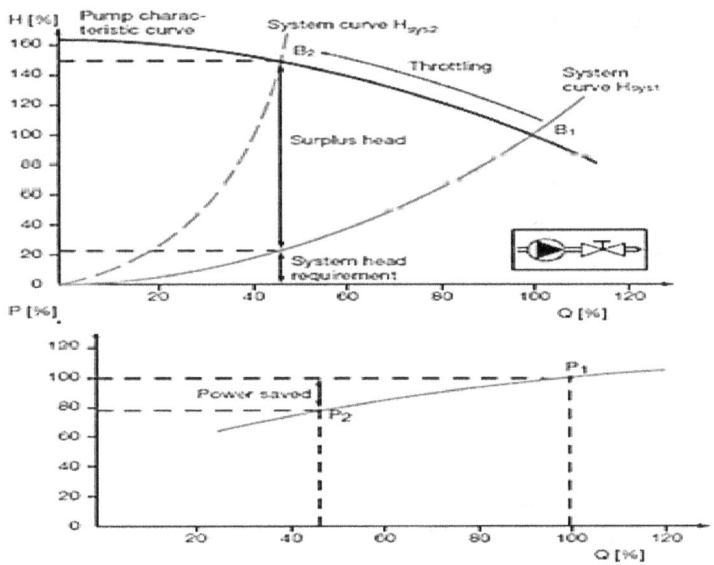

Change of the operating point and power saved by throttling a pump whose power curve has a positive slope

8.6.9 Transients in System Head

Starting of a centrifugal pump to achieve the normal flow situation, certain transient conditions are produced where required head and consequent torque is much higher compared to design value. Thus, selection of the driver of pump must be according to the starting/transient conditions rather on the basis of normal flow condition. However, pump head at shut-off is not significantly higher than normal flow. The shutoff torque in radial flow pump may be less than that at normal flow. But, high specific speed pumps (mixed and axial-flow types, Ns = approximately 5, 000) develop relatively high shut off head and shutoff torque is significantly higher than that at normal flow. Due to such behaviors of high-specific-speed pumps, special attention is paid during starting.

8.7 Starting of a Centrifugal Pump

Fluid enters centrifugal pump through suction eye and certain mass of fluid is rotated by an external source to exert a centrifugal force which throw that away from the central axis of rotation and a centrifugal force is imposed to enable liquid to transfer &raise to higher elevation.

8.7.1 The Operational Steps for Starting a Centrifugal Pump

- Close the delivery valve and prime the pump.
- Start the motor connected to the pump shaft, this causes an increase in the impeller pressure.
- Open delivery valve gradually when pump attended full speed so that the liquid starts flowing into the delivery pipe. Too much delay in opening the delivery valve will cause liquid churning into pump casing; resulting ir-reversible heat generation and rise in temperature of pump components.
- A partial vacuum is created at suction eye by centrifugal action and liquid rush from the sump to the pump due to pressure difference at the two ends of the suction pipe.
- As the impeller continues to run, more liquid is made available to the suction eye therefore impeller increases the energy of the liquid and delivers it to the reservoir through pipes.
- While stopping the pump, the delivery valve should be first closed otherwise there may be back flow of liquid from the reservoir.
- It may be noted that a uniform velocity of flow is maintained in the delivery pipe by special design of casing. As the flow proceeds from the tongue of the casing to the delivery pipe, the area of the casing increases and there will be corresponding change in the quantity of the liquid from the impeller. Thus, a uniform flow is maintained in the delivery pipe.
- The temperature of the water in the pumps rises asymptotically on obstruction of flow. To avoid overheating and possible cavitations in the pump, a certain minimum amount of liquid flow, Q_{min} (called thermal minimum flow) must pass through the pump. The temperature rise is resulted from throttling an incompressible liquid in the clearance of the axial thrust balancing device too. The total temperature rise including the balancing system must be accounted to determine the minimum flow, especially when liquid being pumped close to the vaporization pressure. Q_{min} is expressed as–

$$Q_{min} = \frac{P\,(kw)*3600}{\rho\left(\frac{kg}{M^3}\right)*C\left(\frac{kg}{kg*K}\right)*(te-ts)} * \left[\frac{m^3}{h}\right]$$

- The admissible temperature rise t_e-t_s is the differential temperature (admissible) at the suction nozzle (ts) and the temperature behind the balancing device (t_e) given by pump manufacturer. About 20°C temperature difference (t_e-t_s) can be considered safe. The minimum flow rate is maintained either by bypassing directly to the suction nozzle (if $NPSH_{(A)} >> NPSH_{(R)}$) or into the suction tank. The latter arrangement is preferred. Behavior of Centrifugal Pumps will change, if warm water is mixed into cold water tank. The hot water temperature rise at the pump impeller entry may cause a substantial loss of $NPSH_{(A)}$.

- For small pumps (up to 100 kW), the minimum flow may be calculated by above formula. For pump with power input above 1000 kW and high specific speed, the forces due to flow recirculation at the impeller entry may, depart 25 to 35% from the best efficiency point, and generate excessive vibration in the pump and piping system.

When pump is frequently subjected to minimum flow value, for extended periods of time, axial thrust force will be increased appreciably and it will be desired to use suitable higher capacity trust bearings. Single-volute pumps set up at high radial thrusts under part loads, calls for appropriate precautions such as reinforcing the shaft and bearings.

8.7.2 Starting Against a Closed Discharge Valve

When centrifugal pump is started against a closed discharge valve, the pump head (shutoff-head) will be higher than normal operating head depending on pump-specific-speed. Higher the specific speed more will be the shutoff head (% excess from operating head). When a pump is accelerated from rest to full speed against a closed valve, the head on the pump at any point of time while speeding up is equal to the square of the ratio of transient speed and proportional to full speed times the shut off head. Therefore, during starting, the head will vary from A to E. Point B,C and D represent intermediate heads at transient speeds(zero to full speed). The pump discharge valve and any intermediate piping must be designed for the maximum head at point E.

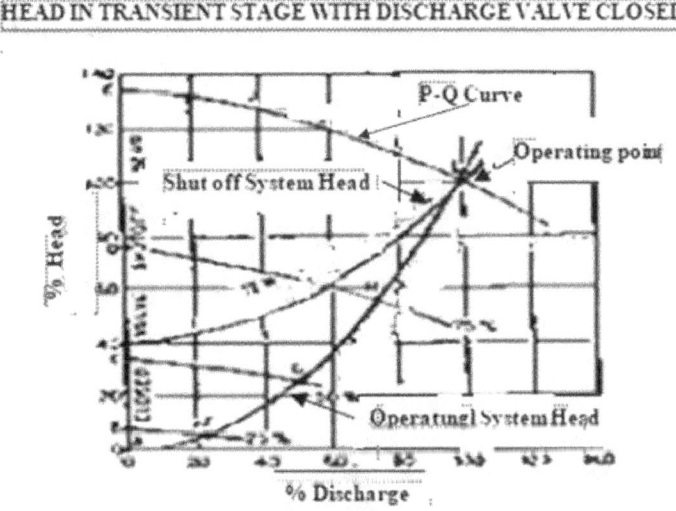

8.7.3 The Shutoff Power/Torque Increases Proportionally with Specific Speed of Impeller

When low specific speed radial impeller pump is started with closed discharge valve to prevent back flow from pressurized system, it requires less shutoff power and torque compared to that of normal flow condition. When pumps are operated in parallel and are connected to common header similar installation of discharge valve and or check valve must be used. Therefore discharge shutoff valve and a check valve, or a broken siphon installation in such system becomes essential.

8.7.4 Characteristic Curve of a Low-Specific-Speed Pump

We observe that the torque varies with pump speed when started with closed discharge valve. The torque under shut off conditions varies as the square of the ratio of speeds similar to the variation in shutoff head.

At zero speed, the pump torque is not zero due to static- friction, motor's inertia, friction in the pump bearings and stuffing boxes, cumulatively which can be greater than that of normal running friction. The horsepower input to the impeller at very low speeds, is known as dip in torque curve between 0 to 10 percent of full speed. The difference of motor and pump torque, while starting is the torque available to accelerate the pump from rest to full speed. While designing the pump shaft, not only the pump torque should be considered but also acceleration torque (excess torque) must also be taken care. Therefore, pump shaft torque follows the motor speed-torque curve to accelerate the mass inertia (WK²) of the motor's rotor.

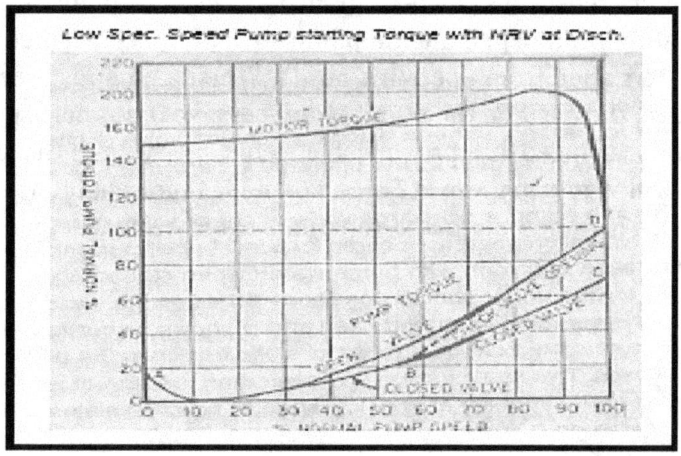

8.7.5 High-Specific-Speed Pumps

Especially propeller pump, require more than normal torque at shutoff. Hence, these pumps are started with a closed discharge valve as larger and more expensive motor is needed for starting the pump with open discharge valve. Further, the pump will also produce relatively high pressures between pump and discharge valve which should be taken care in design.

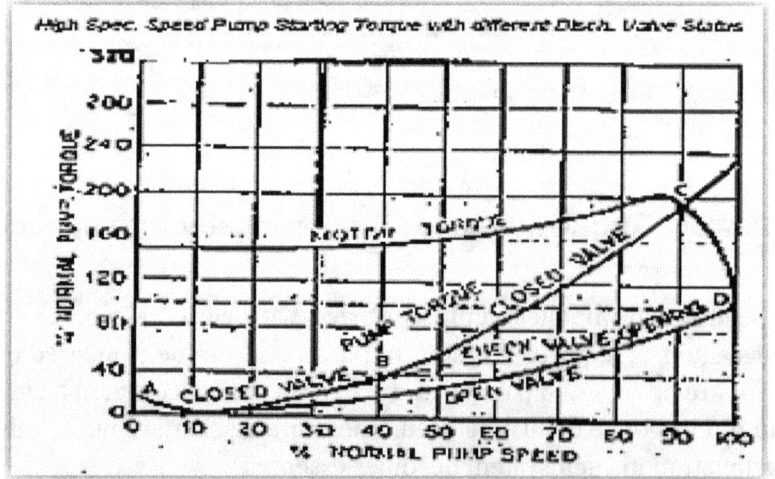

- During design, it is essential to check that motor has sufficient torque to accelerate to full-rated speed with opened discharge valve to avoid overloading of motor. Sometimes, the discharge valve opening is timed to avoid over loading of motor to some extent. To accomplish this timing it may be necessary to start opening the valve in advance of energizing the motor but it should not be more advance to open otherwise excessive reverse flow from discharge header may overload the motor.

- If the driver is a synchronous motor, it has to produce torque to overcome system- head and an additional torque for pull-in speed. A synchronous motor starts on low-torque till excitation of the field windings at pull in speed. The possible voltage drop while starting synchronous motor will lower the motor torque (varies as the square of the voltage) should be taken in account in selection of a synchronous motor for high-specific-speed pump with closed shutoff startup.

If a high specific speed pump is started against open discharge, high starting torque can be avoided by the use of a bypass valve or by an adjustable blade device in pump impeller.

8.7.6 Starting with Open Discharge Valve

If a centrifugal pump suction is connected to a reservoir and discharge to another reservoir, at same liquid elevation (ie. same head), it can be started with opened discharge valve or without check valve (NRV). The system head is essentially friction head plus the head required to accelerate the liquid in the system during starting period. Neglecting liquid inertia, the pump head would not be greater than normal at any speed during the starting period. The pump running torque may not be more than that at any speed during startup period {starting torque at any speed = normal torque * (transient speed/full speed)^2}, while the capacity varies directly with speed ratio (transient speed/full speed).

8.7.7 Starting against a Check Valve

A check valve can be used to prevent reverse flow from static head and/or head from other pump(s) in the common discharge header system. The check valve will open automatically when the head from the pump exceeds system head (including back pressure). When a centrifugal pump is started against a check valve, pump head and torque follows the shutoff torque values until a speed is reached at which shutoff head exceeds system head. When NRV opens, the pump head continues to increase. At any flow the necessary head of pump will be required to overcome the systems static head or head from the other pump(s), friction head, valve head losses and inertia of the liquid being pumped.

The speed-torque variations in starting a low-specific-speed pump against a check valve, the pump static head should overcome the system frictional head and static head resistance for establishing flow into system header. The speed-torque variations in starting a high-specific-speed pump against a check valve + static head and system friction, the use of a quick-opening check valve is required. The speed torque curve during the acceleration of the liquid in the system is drawn with assumption that the head required to accelerate the liquid and overcome inertia will be in-significant.

8.7.8 Starting a Pump, Running in Reverse

When a centrifugal pump discharges against the static head or into a common discharge header and a pump is stopped then either discharge valve or check valve should isolate from system pressure otherwise the pump will start rotating in reverse direction, in absence of non-reversing device on the pump. A pump that discharges against a static head through a siphon system without a valve will work like hydro turbine to cause a reverse rotation during priming through the siphon prior to starting.

The figure illustrates typical low and high spec. speed pumps *reverse speed torque characteristics*. When flow reverses through a pump and the driver offers very little or no torque resistance, the pump will reach higher than normal forward speed, in the reverse direction. This runaway speed will increase with specific speed and system head, speed, torque, head, and flow expressed as a percentage of the pump design head for normal forward speed. The pump running in reverse, work like hydro turbine in no load. The head on the pump will be a static head (or head from other pumps) minus head loss as a result of friction for the reverse flow rate.

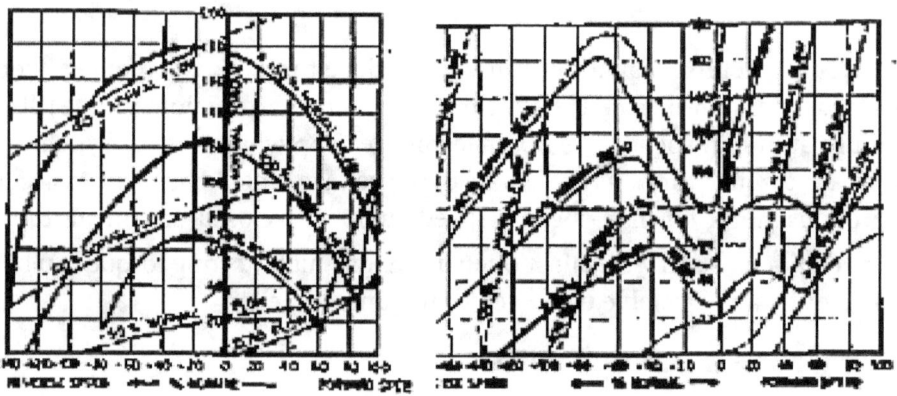

Family Curves

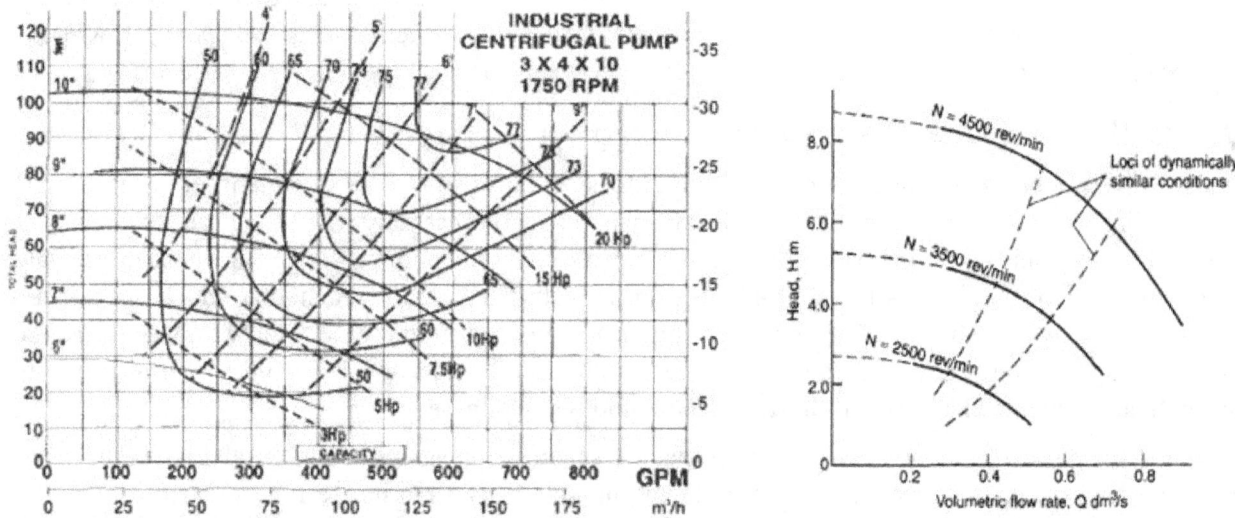

The pump while running reverse, and the electric motor apply positive torque to the pump, either motor is tripped or if not, a heavy torque will be applied on pump & motor shaft. The +ive torques applied by driver, de accelerate momentarily to stop and then may accelerate the pump to a normal speed but there will be danger of pump shaft breaking while bringing pump from reverse rotation to normal direction of rotation. It is advisable to apply some mechanical resistance to rotor eg. Wooden log, draining the suction and inching the pump motor to de accelerate the reverse speed before starting pump in normal direction.

Chapter 9

CORROSION IN PUMPING SYSTEM

DIN 50900 describes corrosion as *"the reaction of a material with its environment causing a measurable change in the material and possibly leading to corrosive damage."* Furthermore, corrosive damage is defined as *"Impairment of the function of a component or an entire system through corrosion."* This definition clearly distinguishes between the corrosion process and the corrosive damages. Corrosive damage can be defined as damage to a metallic component originating from its surface through chemical reactions of the metal by environmental constituents. In contrast to mechanical wear, corrosion is fundamentally a chemical process which can be represented by basic equation: **Me Z^+ + e^- ← → Me + Z.**

The metal "Me" enters into a chemical reaction, as a positively charged ions Me Z^+ and Z electrons (e^-) are released. Thus, metal ions remain in dissolved state in the corrosion medium or settle on the corroded surface as insoluble products. In addition to the chemical corrosion, the metal surface is stressed by surface corrosion or by some other mechanical effect and corrosion reaction is accelerated considerably.

9.1 Factors Affecting Corrosion

The corrosion reaction is considered to be thermodynamic instability of metals in relation to air, water or other oxidizing agents. All metal/compound have natural tendency to revert from refined state to the orinal state (as available in nature) under fabourable environmental conditions and huge energy is released. The energy released is approx the energy required to convert as metal from original stage of compound in nature.

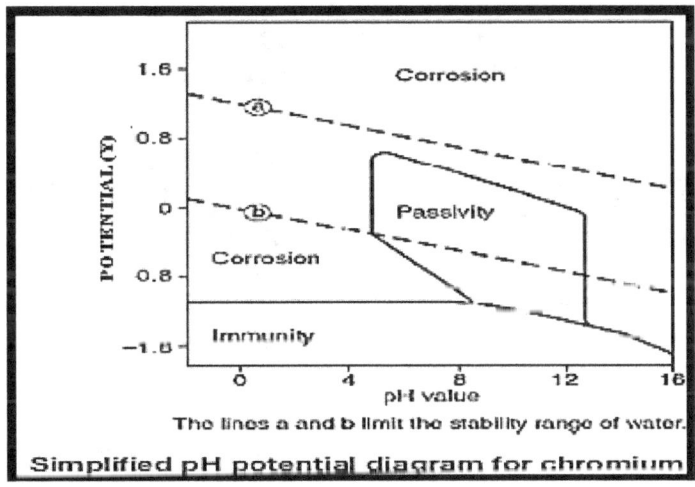

The lines a and b limit the stability range of water.

Simplified pH potential diagram for chromium

9.1.1 Concentration

For given material, the concentration of the aggressive components, present in surrounding often plays an important role in corrosion. An increase in concentration of these components affect the speed of corrosion negatively or positively e.g. the corrosion of unalloyed steels in saltless water is intensified in presence of oxygen. The similar behavior is observed with increases in hydrogen ion concentration (i.e. low pH value). The contrary behavior can be observed for CrNiMo steel in sulfuric acid, at concentrations above 50%.

The electrochemical potential replaces the motive force and a complex sum of individual resistances (e.g. reaction and diffusion resistances) replaces the resistance. "Motive force" is a thermodynamic quantity while "resistance" is directly involved with reaction kinetics.

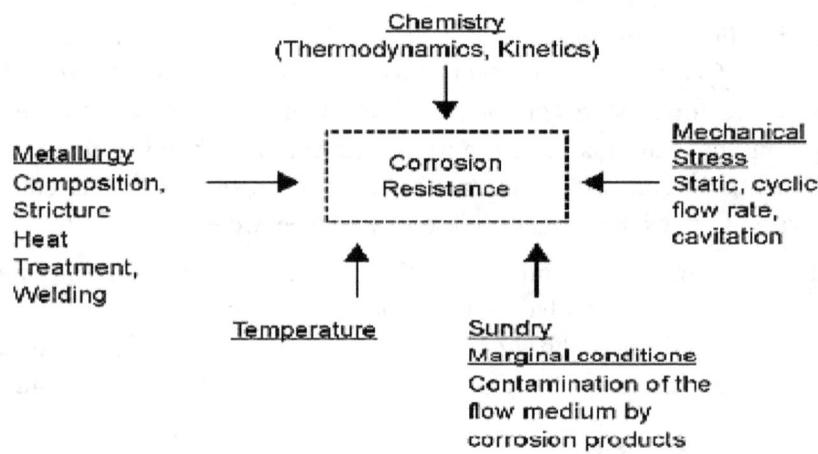

a. **Uniform surface corrosion,**

It is explained as an uniform rate of corrosion through out the metal surface (e.g. metal dissolution in acids, simple rusting processes). However, uniformity of corrosion in environmental conditions & corrodents concentration are shown in following sketches–

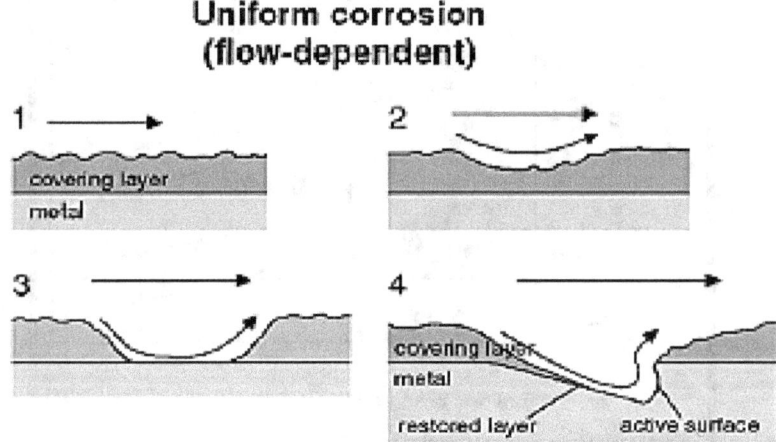

b. **Local corrosion without mechanical stressing,**

The corrosion is limited to localized area of metal surface, in narrowly defined locations (viz. pitting corrosion, intercrystalline corrosion, crevice corrosion, galvanic corrosion).

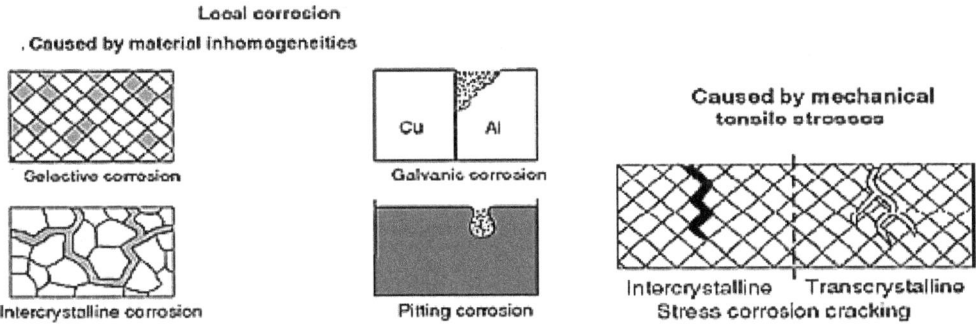

Corrosion without mechanical stressing   Local corrosion with mechanical stressing

c. **Local corrosion with mechanical stressing**

It is possible to distinguish between the mechanical stress acting on the material (stress corrosion cracking, corrosion fatigue, cavitation corrosion which cause increase of stress in corroded components) and only corrosion of meatal on its surface (eroded corrosion, chemical corrosion).

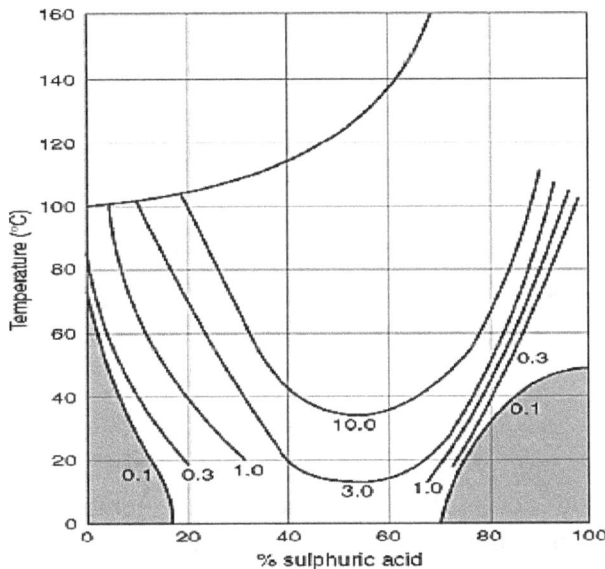

Iso-corrosion diagram for steel XSCrNiMo 1 8.10 in sulfuric acid.
mm/year metal loss

d. **Stress Corrosion Cracking (SCC) And Corrosion Fatigue**

This type of corrosion is characterized by the occurrence of cracks and/or brittle fractures by stresses even below the yield point of material. Stress corrosion cracking takes place at static tensile load. Occurrence of Stress Corrosion Crack(SCC) below 70 to 80 °C is unlikely. The use of CrNi alloys with higher Ni content (above 25 to 30%) increases the danger of SCC.

e. **Some other Reasons which Enhances the Corrosion**

• **Mass transfer**

Flow usually increases probability of corrosion. Natural and forced convection facilitate in transporting the corrosion aggressive substances to the metal surfaces and washing away the corrosion end products from place of chemical reaction. Moreover, flow-induced mechanical shear& stressing of the components consequently causes pitting and crevice corrosion in stagnant conditions.

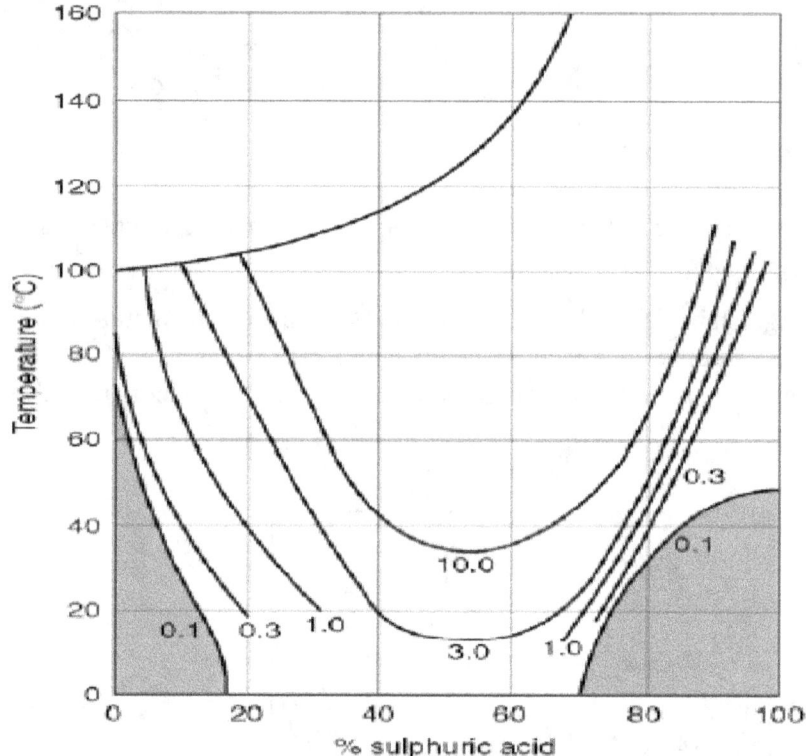

Iso-corrosion diagram for steel XSCrNiMo 1 8.10 in sulfuric acid.
mm/year metal loss

- **Temperature**

  The rate of chemical reactions considerably increases with the temperature. As a thub rule every 10 °C temperature rise enhances the reaction rate to double. Exceptions to this rule however, can be seen frequently depending on the environmental conditions.

- **Potential**

  Potential displacement from metal surface usually have a large effect on the rate of corrosion. It is established mathematically and experimentally that the rates of electrochemical reactions can change up to a factor of 105. Potential displacements will depend on the presence of redox systems, matching with other materials, external current sources, etc.

- **Time**

  Time is an important factor in corrosion processes. The corrosion reaction varies on time. This can be observed that the rates of corrosion after formation of covering layer will decrease where as that is increased on concentration of corrosion products (standstill corrosion).

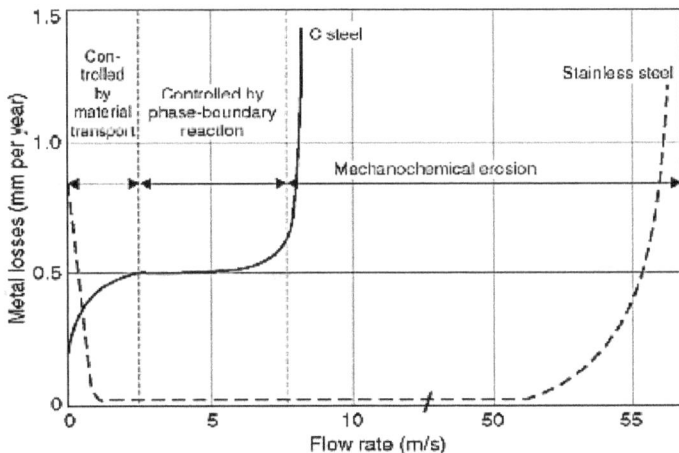

Influence of flow rate on metal corrosion

- **Pitting corrosion**

  Pitting corrosion is initiated particularly by chlorides, especially solutions in stagnant condition (e.g. during machine outages). The steel can be protected by addition of molybdenum (usually 2.5% Mo). The amount of molybdenum required to ensure pitting protection depends on chloride cocentration and temperature.

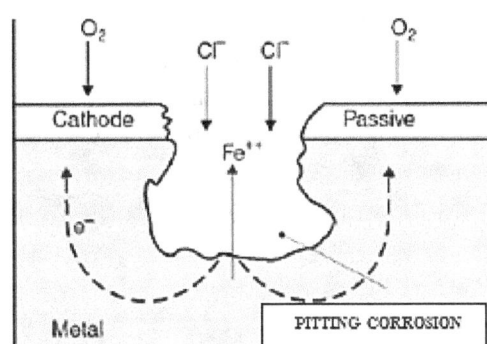

- **Choice of material**

  Corrosion resistance is not a material property like the yield point, electric conductivity etc. The "corrosion resistance" behavior not only depends on the properties of the surrounding medium, but to a great extent on other factors too.

- **Intercrystalline corrosion**

  When corrosion primarily affects the grain boundaries of a polycrystalline metal, it is known as Intercrystalline corrosion which breakdown cohesion of the crystal and consequently grains are dissoclated & disintegrated. The examples to this type of corrosion are high alloy austenitic chromium–nickel steels and ferritic chromium steels. In 550 to 750 °C temperature range, chromium carbide ($Cr_2 3 C_6$) is precipitated at the grain boundaries from these alloys.

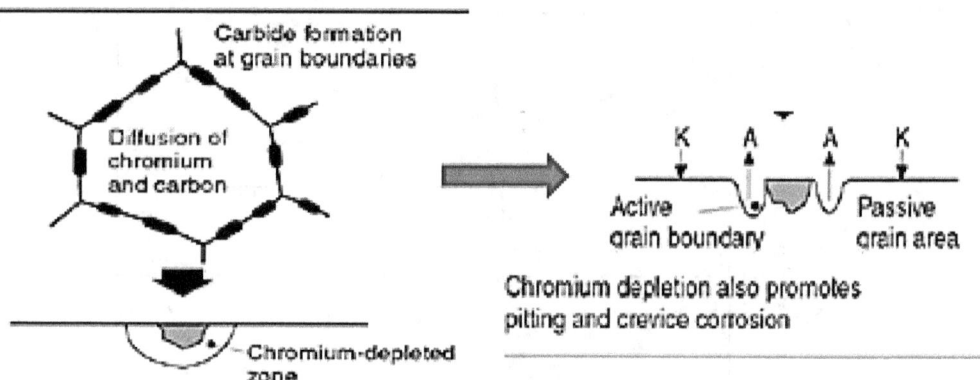

- **Selective corrosion**

  Two-phase or multi-phase alloys are suseptable to selective corrosion, where the base phase is dissolved but more noble phase remains undamaged maintaining the form of component. As for example, selective corrosion like graphitization or spongiosis takes place in cast iron, dealuminification in certain aluminum dezincification in bronzes and brass, primarly by chlorides in slightly acidic pH envoirment, reduced oxygen supply and insufficient circulation of medium.

- **Erosive corrosion**

  Erosive corrosion is the accelerated destruction of metal in fast-flowing corrosive liquid. On increase of flow rate, fresh corrosive substances meet the metal surface quickly while reaction products are carried away completely or partly by stream of flow which enhances the corrosion process.

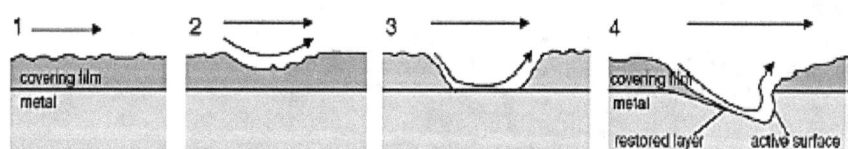

Influence of flow rate on the formation of covering film

- **Galvanic corrosion**

  If two different metals are placed in electrolitic solution (electrically conductor), galvanic corrosion will be initiated, if their individual potentials are sufficiently apart. When electronic flow is established, the base metal (anode) gets corroded. The galvanic corrosion is enhanced if concentration of salt is higher (ie. higher electrical conductivity) in solution/medium. In seawater, however, even potential differences of 100 mV can establish substantial corrosion.

- **Cavitation corrosion**

  If $NPSH_{(A)}$ of a pump approaches the vapor pressure of the liquid at operating temperature, the implosive vapor bubbles are formed which colapse by pressure and cause cavitation damages on pump interior surfaces.

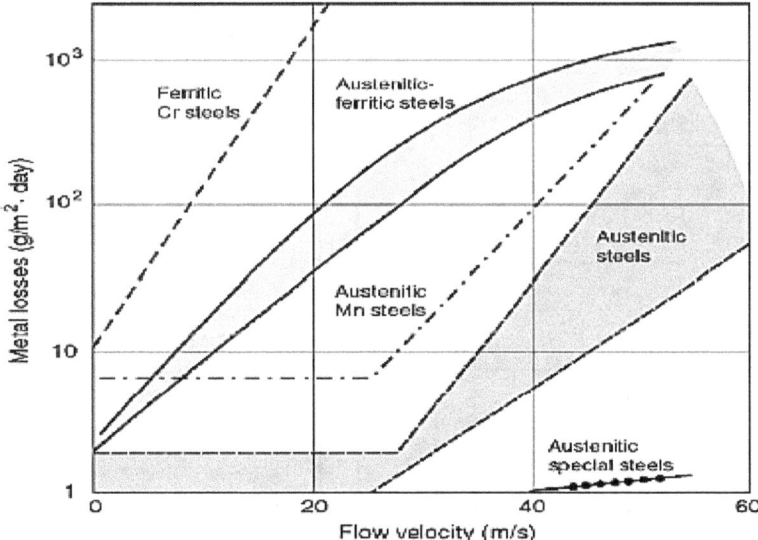

Rates of corrosion of stainless steels as function of flow vel injection water, 60°C, 23 000 mg/l dissolved salts, pH 4.5

The alternating effect of formation and colapse of bubbles in an aggressive medium increases the metal loss eg. cavitation in an Al-multicomponent bronze in seawater increases metal loss with increasing H_2S content. For austenitic CrNi alloys too the damage rate in seawater is appreciably higher than demineralized water. In slightly corrosive media, increase of hardness as cavitation resistance material does not hold good.

- **Fatigue cracking**

  It occurs by a cyclic alternating mechanical load but not specific to either materials or media. When endurance limits are no longer reached in cyclic loading without corrosion influence, fatigue cracking is not initiated.

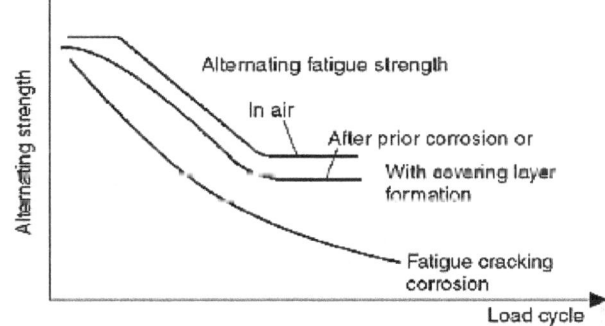

. **Alternating fatigue strength. Schematic for air corrosive medium**

9.1.2 Corrosion Problems in Hydraulic Machines

The aggressiveness of the hydraulic media and its velocity is deciding factior for choice of materials in hydraulic machines. Each types of corrosion occuring in a pump individually or combined in extremely wide spectrum of flow velocities, (0 to 100 m/s). The crevice corrosion in stagnant media and erosion & cavitation corrosion in circulating madia are genarally damages in hydraulic system.

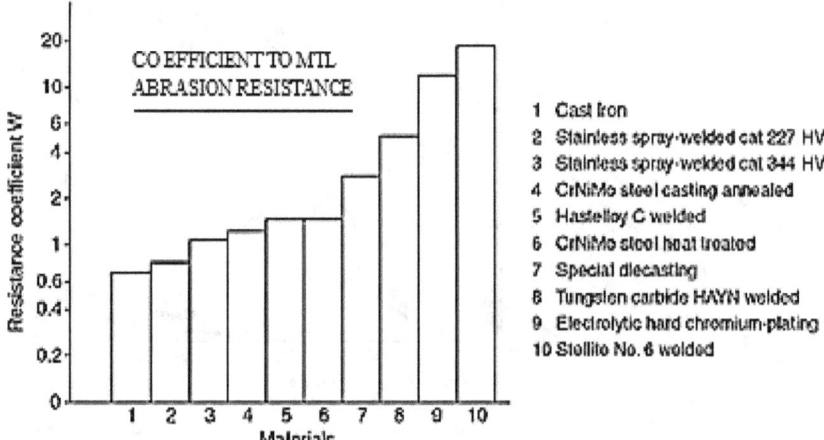

The temperature of the flow media also have a major influence on behavior of materials. Usually the rate of corrosion doubles with a temperature increase of 10 °C. Thus, in seawater temperatures over 60 °C is very much critical for CrNi steels containing molybdenum.

9.1.3 Crevice Corrosion

Crevice corrosion may be regarded as local breakdown of the passivative condition, resulted from oxygen deficiency and/or a decrease of pH due to hydrolyzing of corrosion products in the crevice by creating hurdle in emigratory diffusion. The susceptibility of stainless steels to crevice corrosion decreases with increasing stability of the passivative film, or by increasing chromium and molybdenum content (13% chromium steel < 17% chromium steel < 18/8 chromium–nickel steel < 18/8 chromium– nickel–molybdenum steel).

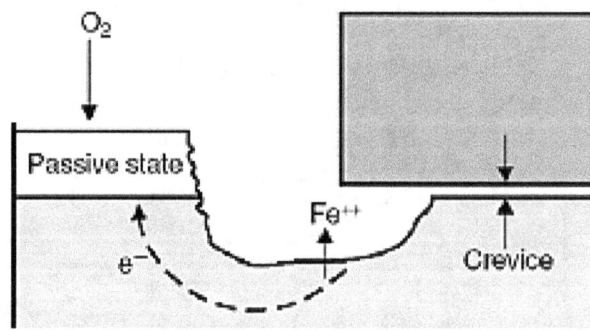

Crevice corrosion

The danger of corrosion is greater when there is higher chloride content. Higher the temperature of media or lesser chromium and molybdenum content in steel.

9.1.4 Materials Recommended for Pump Components

Behavior of some commonly used materials in different working environments with wetting factor is tabulated to illustrate suitability so that power engineers can use as ready reckoner while selecting materials for pump components.

Table   Examples of Materials Recommended for Pumps

	Material No. as per DIN (examples)	Natural waters, equil. water	Treated waters: softened, deminer-alized	Water or boiler feedwater O_2-free	Salt water Neutral	Acid	Alkaline	Sea water	Injection water	Organic liquids, e.g. oil, hydro-carbons
Cast iron, incl.	0.6025	+	−	+	%	−	+	−	−	+
Low-alloyed	0.7040									
Steel casting	1.0619	+	−	+	%	−	+	−	−	+
Incl. low-alloyed	1.7706									
Ni-Resist	GGG-NiCr202	+	+	+	+	%	+	+	−	+
Si cast iron	G-X70 Si 15	+	+	+	+	+	+	+	+	+
	1.4027									
13% Cr steel	1.4313	+	+	+	+^{1}	%	+	−	−	+
16% Cr steel	1.4405									
CrNi steel	1.4306	+	+	+	+^{1}	%	+^{1}	−	−	+
Type 18/8	1.4308									
	1.4404	+	+	+	+	%	+	+	%^{3}	+
CrNi steel	1.4408									
Type 18/8/2.5	1.4581									
	1.4460	+	+	+	+	%	+	+	+^{3}	+
Duplex steels	1.4462									
NiCrMo steel	1.4500	+	+	+	+	%	+	+	+	+
Bronze (Sn)	2.1050.01	+	+	%	+^{2}	−	−	−	−	+
Al bronze	2.0966	+	+	%	+^{2}	−	−	+^{3}	−	+

Symbols: ++ resistant

% : Conditionally unstable;   -- : Unstable;   + : SymbolicResistant

Suffixes: 1: Presence of much Chloride

2 : In absence of (NH3) or (NH4)

3 : according to (II2 S) const. & pH)

Chapter 10

MAINTENANCE OF CENTRIFUGAL PUMPS

10.1 Pump Maintenance

Safety

While taking up any maintenance job, the related safety aspects should be adhered to–

- Obtain a *permit to work* from competent authority; Permit must specify the pump(s) number and location to be taken up for maintenance.
- Steps are taken for isolating the pump from the system. Ensure that isolation is carried out as per laid down procedure.
- Ascertain the nature of liquid carried through the pipe. If line handles corrosive nature of liquid personnel protective equipment becomes obligatory. Drain the pump and connected piping before starting the work.
- Allow the pump to cool down to atmospheric or workable temperature.
- Clean the surrounding area, and make suitable access to the working place. Take care of other hot lines, if any, around the pump.
- Choose proper tools and tackles for working. Check manufacturer's instruction for special tools and other requirements.
- Allow only authorized and trained persons to work.
- Work permit should be cancelled after completion of work.

10.2 Maintenance of Centrifugal Pumps

10.2.1 Repair and Maintenance

- **Routine preventive maintenance** – Routine maintenance can be defined as the work primarily taken up to rectify the defects of normal wear and tear in a pump.
- **Major overhaul or repairs** – The major overhaul or repair is carried out to rectify the excessive wear, restoring performance of pump and its efficiency. A service record for the pump must be maintained and preserved to monitor the machine performance.
- **Condition monitoring and predictive maintenance** – Condition monitoring and predictive maintenance is a modern technique of maintenance for which a dedicated group is assigned for on line monitoring and trending the parameters, vibration & noise. From parameters and trends, balance life of machine and components are predicted. These results are passed on to maintenance group to take up the maintenance timely to avoid any breakdown. Vibration monitoring being the most powerful technique is utilized very popularly.

10.2.2 Other Maintenance Activities

- **Maintenance Schedule:** The usual intervals for routine preventive maintenance eg. Monthly, quarterly, half-yearly or annually should appear in maintenance schedule. Daily checks and running inspection should be carried out regularly for maintenance action plans.

- **Observation of Pump Operation:** Operators, on duty, should do the hourly and daily inspections and record should be maintained. Any irregularity in the operation of a pump should be reported immediately. This applies particularly to changes in the sound, abrupt changes in bearing temperatures, vibration, stuffing box leakages and deviation in operating parameters. A check of the pressure gauges and flow meter, should be on hourly basis. If recording instruments are provided, a daily check should be made to find whether there is change in capacity, pressure or power consumption.

- **Monthly Checks:** Check of temperature, lubrication and vibration of each bearing – Normally Temperature should not go beyond 72 deg C. In bigger size pumps on line temperature measuring points are provided for this purpose.

- **Quarterly checks:**
  √ Check oil & Replace every 3 months, drain the oil from sleeve type bearings. Washout the oil sump and bearings interior parts with kerosene oil

  √ Check bearing lubrication, arrangements-. Check the oil rings position, shape (circular) and movements in running condition of pump. Rings must be free from dirt, and dust. Repair or replace any defective oil rings. Refill the bearing with the specified oil.

  √ Check Grease for Specifications- Inspect grease lubricated rolling contact and sleeve bearings for healthy condition. A whitish color of the grease indicates water emulsion in grease. It is caused usually by leakage of water or other liquids passed through shaft seal. In this condition, flush all grease from the bearing and clean from warm Kerosene oil. When the bearings are cleaned and dried, fill with new grease of recommended grade.

  √ Measuring sleeve bearing clearance- Place three length of lead wire (0.5 mm) on top of the journal, place the bearing cap and tighten its holding fasteners. Now open the cap and measure the thickness of each of flattened lead wire with micrometer. If all leads are of the same thickness, the clearance is uniform, throughout the bearing. Thickness of wire gives the bearing clearance at the point at which wire was located. Record the measured clearance in service record. When clearance observed is twice the original clearance, replace the bearing.

- **Half Yearly checks:**
  √ Good, stuffing box must leak drop by drop. Check the shaft packing by observing the leakage from it, leak should be 40 to 60 drops per minute for adequate cooling of stuffing box and gland, this may vary with pump service conditions, liquid handled and the type of the packing used. If rate of leakage is higher than the recommended and does not stop on tightening of glands, or packing getting damaged frequently, check shaft sleeve for wear and corrosion and replace shaft sleeve and all packing. Don't replace one or two packs. Remove packing by extractor. Make sure that shaft is not scratched while removing worn out packing. If seal cage or lantern ring is provided, check its position. Count the number of packing in front. or in the rear side of lantern ring.

  √ Check gland follower and stuffing box housing clearances and compare with that of new installation. Check concentricity of shaft.

    To find packing thickness measure the stuffing box bore, subtract the shaft diameter, divide by 2, Stagger the packing joints while fixing new gland packing at:
    - 180 deg if the stuffing box has two packing
    - 120 deg if the stuffing box has three packing
    - 90 deg if the stuffing box has four packing

  √ **Insert packing into stuffing box and be sure that packing has reached to its position**

Be sure each ring is firmly seated before the next one is inserted. Locate the Lantern ring position; it should be in line with the centre of the cooling liquid hole. Remember that ring moves back into the box if packing are compressed. While packing is inserted, turn the shaft by hand in direction of rotation to level any high spot in the packing.

√ **Gland Adjustment**

Tighten the gland packing gently and uniformly just enough to prevent excessive leakage before starting the pump. When packing adjust itself in the stuffing box, tighten the gland follower nuts slowly and sequentially (one flat at a time) to allow trickling drop leakage.

√ **Do not back off the gland nuts**

Do not back off the gland nuts, while the pump is running without relieving pressure on packing otherwise the entire set of gland packing may dislocate in the stuffing box. If packing is too tight shaft does not move freely by hand, allow the stuffing box to cool, and then re-adjust the packing. Start the pump and check stuffing box temperature. Inching of pump can release the gland jamming with shaft.

• **Annual Inspection**

The pumps should be thoroughly inspected once a year; the bearings should be removed, cleaned and examined for flaws. The bearing housings should be fully cleaned. Antifriction bearings should be cleaned & examined for creep, clearance, damage to race and rolling elements. Immediately after inspection, antifriction bearings should be coated with oil or grease to prevent dirt or moisture getting into it.

The packing should be removed and the shaft-sleeve & shaft should be examined for wear and corrosion. The coupling should be disconnected and examined. Check alignment make correction if needed. In the case of horizontal pumps with babbit bearings, the shaft movement at both ends in vertical direction should be examined after removal of packing & coupling disconnected. The vertical movement ≥ 150 per cent of the original diametrical clearance of bearings require replacement.

If the end play is higher than recommended value by the manufacturer that need investigation and corrections. Drains, sealing water piping, cooling water piping, and other piping should be checked and flushed. If any oil cooler is installed, it should be flushed and cleaned. The pump, stuffing boxes should be re packed and the coupling reconnected. The instrument and monitoring devices need to be checked & recalibrated. If internal repairs are made, the pump should be tested after completion of the repair.

10.3 Complete Overhaul

General rules cannot easily be made to determine the frequency and regularity of complete overhauls of centrifugal pumps. The type of service for which the pump is intended, the general construction of the pump, the liquid handled, the materials used, the average operating time of the pump, and the evaluation of overhaul costs against possible power savings on overhauling will decide the frequency of complete overhaul. Some pumps on critical service may need a complete overhaul monthly, but pump used in normal applications may require overhauling every two to four years.

10.3.1 Complete Dismantling of a Centrifugal Pump

√ Centrifugal pumps should be dismantled with great care. The suction and discharge valves should be-closed and the pump casing drained. All necessary piping and parts that would interfere with the disassembly of the pump, such as bearing covers, should be removed. The upper half of pumps (split casings) should be lifted up straight after opening of dowels and nuts of the casing bolts.

Ensure that no damage to internal parts takes place while lifting the casing. The rotor should also be removed vertically, preventing injury to the impellers, wearing rings, and other parts.

√ During the dismantling, the parts removed must be marked to ensure correct re-assembly. All individual parts and important joints should be carefully examined after cleaning. If the pump has been operating satisfactorily with only a slight reduction in head and capacity due to increased leakage; a decision on reconditioning will depend on several factors–

 – Availability of spare parts.

 – Length of time the pump can be left out of service, and economic considerations and importance of getting the service from the unit without overhauls.

 – Generally, worn parts should be renewed if the pump is not to be examined until the next routine period, regardless of the performance of the unit, because when parts in new or good condition are assembled in contact with dirty or worn parts, the new parts are very likely to wear out rapidly.

10.3.2 Checks During Pump Overhauling

- Stripping the pump -Identification and marking of parts
- General Inspection of parts removed.
- Specific fault detection.
- Rectification of faults–

a. Bearing housing	b. Bearing replacement	c. Neck ring clearance
d. impeller's condition	e. Changing gland packing,	f. Check shaft for straightness
g. Keyways and keys	h. Shaft sleeve tolerances	i. Condition of nuts & bolts,
j. Condition of gland and gland housing (Clearances)		k. Assembly of pumps
l. Pump alignment		

10.3.3 Marking While Stripping off Pump

a. Clean pump, motor, bed plate etc.

b. Remove any clinger cocks, valves, gauges, tundish etc. from pump body.

c. Check if any existing marks or stampings are visible on pump, body, couplings, gland etc. It may be possible to use original stamping to avoid confusion while rebuilding of pump

d. Stamp glands. D.E. (Drive End), N.D.E. (Non Drive End) or B.E. (back end)

e. Stamp pedestal caps and bearings DE and NDE on top and bottom halves.

f. Stamp bearing housings – DE or NDE.

g. Match mark all covers which are removed including the top cover of single and twin stage pumps.

h. All parts not suitable for stamping must be marked or labeled attached I e.g. oil rings, junk rings, lantern rings, water or oil throwers, ball and roller bearings, sleeve nuts etc. and dowels

i. All bushes and neck rings must be stamped or marked from coupling end of pump.

j. While marking a multi-stage pump, chambers should be stamped from coupling end of pump i.e. suction chamber-1 chamber-2 and so on.

k. When marking impellers, the impeller whose necking enters suction chamber should be marked or stamped on the delivery side of impeller.

l. All keys should be stamped from coupling end (eg. Coupling key #.1 impeller # 2 and so on).

m. All chamber plates should be stamped from coupling end e.g. suction chamber plate No. 1 and so on.

n. All neck rings and bushes removed from chambers or plates should be stamped from coupling end of pump as used for reference.

o. Any Guide vane tips removed from chambers must be stamped for reference, numbered from suction cnd.

10.3.4 Maintenance of Specific Pump Parts

Special care is required in the re-assembly of multistage rotors having axially split casings. These casings are made from castings and, when the pump is built up, it is sometimes necessary to allow the variations in longitudinal clearances in rotor assembly. The adjustment is done in the rotor at assembly floor, in order to preserve the impellers in its correct position with respect to the volutes or diffusers.

√ While field renewal of rotors or stationary parts, all lateral distances should be compared with those on the old parts to restore lateral/axial movement for thermal expansion and running clearances.

√ The assembled rotor and stationary parts (such as casing wearing rings, stag-pieces & diffusers) should be placed in the lower half of the casing and the total radial & axial clearance are checked and conformed.

√ When thrust bearing is assembled and the shaft is in its working position, the total clearance should be suitably adjusted and the impeller is centrally located in the volutes or diffusers. The shaft lock nuts can be manipulated for final adjustments.

√ To avoid shaft distortion, all abutting joints must be squared with the shaft axis and parallel to each other. The impeller and shaft sleeve nut must not be tightened with excessive torque otherwise, the metal may get crushed. The shaft may also get bend by heavy moment due to vibration during operation. There can be possibility of rubbing and binding at the internal running joints.

√ While using locking screws, the assembly should be checked by a dial indicator, to make sure that the shaft is not bend in its bearings or at center. Check and correct if any eccentricity is observed.

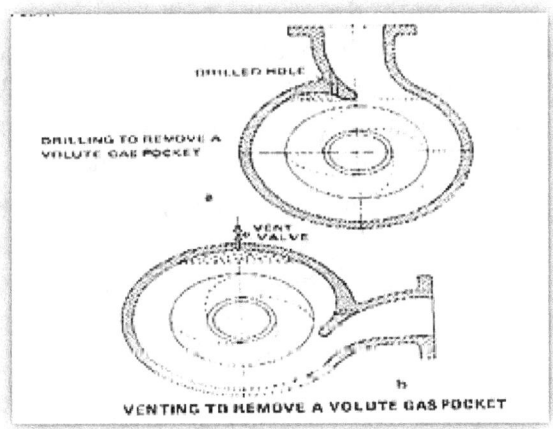

10.3.5 Pump Re Assembly

√ If the pump casing is split axially, great care must be exercised in replacing the upper casing and tightening the casing bolts.

√ If more than one row of bolts are used, the row nearest to the pump axis should be tightened up first. When all the bolts have been tightened once, re-check tightness by torque wrench to ensure the leak proof casing joint.

√ Tightness of all joints should be confirmed when the pump has been brought up to operating temperature.

10.3.6 Spare Maintenance

– The service for which a centrifugal pump is used will determine, to a great extent, the minimum spare parts to be carried in *stock*.

– In general. the minimum spares for a centrifugal pump include a set of wearing rings, a set of shaft sleeves (or a shaft if no sleeves) and a set of bearings. It is often advisable to procure a spare rotor assembly when examination indicates that the pump rotor has tendency of wear out, or it is accidentally damaged. The spare packing for the stuffing boxes and the gasket material should always be available in stock for all pumps. Two years maintenance spares parts should be purchased along with capital order of pump. Based on the contemplated method for annual/major overhaul, the spare wearing rings of original size in the assembly or undersized bored wear ring and sleeves should be available in hand.

10.3.7 Order for Spares

The pump serial number and size, stamped on the manufacturer's nameplate should always be mentioned while ordering spare and replacement parts so that the manufacturer can identify the pump & components for correct delivery. Mostly, centrifugal pumps are selected from standard design and a great number of combinations are made for each size of casing, using different impeller sizes and designs. Without an identification number (serial number), the pump manufacturer may not furnish correct spares even though size and type of the pump mentioned in order. Some manufacturers issue special formats for ordering spares which can be utilized.

10.3.8 Inspection and Service Records

√ The working schedule of the half yearly and annual inspection program should be incorporated on maintenance cards, for each pump separately. These cards should contain the pump identification, the date of the scheduled inspection, a complete record of past inspection, comments and observations of the inspecting personnel. Complete maintenance does not end up with repair or replacement of damaged parts till the written record of the conditions of the parts repaired or replaced, appearance of the wearing, method by which the repair was carried out are some important features are recorded on service card. These records will form the basis of preventive measures which should be taken to reduce the frequency of maintenance and repair cost evolved. The type of inspection records may vary with type and applications of pump.

√ It is advisable to take photographs of worn out components before they are repaired; photographs provide more accurate and authentic record of the damage. Complete records of maintenance and repair costs should be maintained for each individual pumps along with operating hours. Study of the records should reveal whether a change in materials or design had improved reliability & maintainability.

10.3.9 Major Repairs and Overhaul

Apart from the regular maintenance, the pump may require major overhaul i.e. repairs pertaining to each components in major shutdown can be taken up as follows–

- **Casing Repair**

  Sometimes, casing develops vapor lock and that can't be removed even after repeated priming. Such problem are normally faced in the volute casings pumps where gas/vapors entrapped can be removed by drilling a hole of suitable size & location. Sometimes, casings are subjected to cracks which can be repaired by gauging, pre heating (120–250 °C) and welding with suitable CI electrodes. After welding is completed it should be covered with lime powder or insulating pads for annealing.

- **Impeller Maintenance**
  - √ An impeller removed from the pump casing should be carefully examined for wear, abrasion, corrosion or damages by cavitations. General Service pumps are provided with bronze impellers which has a reasonably long life. Occasionally, these pumps operate on high suction lifts. Pumps handling water with sand use cast-iron, nickel or even chrome steel impellers depending on sand concentration & abrasiveness. While selecting impeller material character of the water plays primary role. Generally, materials that form a coating or film on impeller (protective layer), adheres firmly to the under-laid metals and do not get washed away by the water stream. However, abrasive material naturally erodes the protective film from metals surface.

    Abrasion wear is examined/tested by a sedimentation test. Some of the pumped liquid is allowed to settle in a glass container for few hours and particles are examined for abrasiveness in chemical laboratory. The abrasiveness within permissible limits of 3 MPY for ferrous & CI for CW system is acceptable. If that is detected beyond limit, change of material or application of water treatment techniques becomes necessary. For 'On Line Monitoring' of different metals, coupons are installed in water circulation stream which gives fairly good results and indications of rate of material decay.

  - √ Cavitations end result is often pitting in the impeller and casings. Cavitations symptoms can be detected by a stone breaking noise from pump during operation or performance drop by reduction in capacity and head. If impeller is rapidly pitted or eroded, it can be indication of cavitations. The corrective action for cavitations should be taken quickly. The special alloys can resist formation of cavitations pit to some extent but that become a costly proposition. In small pumps, impeller normally worm out impellers are replaced rather we go for repair.

    The rebuilding by brazing, soldering, welding etc. is feasible in bigger units. Mostly large impellers render many years of service, regardless of abrasion, erosion and wear to some extent. Unlikely sometimes, wear may take place in the impeller hub surface which is in at mounting surface in contact with shaft or at the keyway due to porosity in the impeller hub. The porosity or improper fitting of impeller permit water to creep from the higher pressure region to the fitting surface, (between the shaft and impeller hub) & material is readily attacked by aggressive water.

  - √ The, impeller cracks may develop due to excessive vibration or strains set up during the casting process and not detected at the time the impeller machining. Cracked impellers cannot be successfully repaired and should be replaced.

  - √ Balancing of impeller should be checked when the impeller is removed from the pump during overhaul. For static balancing, the impeller is placed on two parallel and leveled knife edges. If the impeller is out of balance, it will turn and come to rest with its heavy portion down. Metal must be removed from the heavy spot or same weight added at 180° position of impeller by trial and error weight till impeller rotated manually stop in any position. Drilling holes at heavy spot is preferred as it does not form eddy current.

  - √ For balancing a shrouded impeller, the best practice is to mount the impeller on a lathe and take a cut from the shroud to balance. The cut can be taken from both shrouds, depending on their actual thickness and the amount of metal to be removed. In semi open impeller, the removal of metal is not permitted rather metal removal from underneath portion of the vanes is advisable. The latter method is also used for balancing open type of impellers.

10.3.10 Shaft Maintenance

During pump overhaul the shaft should be carefully examined for any sign of wear, or irregularities, especially at all the important fitting areas, such as the hub bores, under the shaft sleeves and at the bearing seats etc. The shaft damage by rusting or pitting may be caused by the leakage of water at impeller fitting or shaft sleeves mating surfaces. It is important to check shaft surface underneath bearings fitting.

Check shaft for twisting at the keyway, due to excessive thermal stresses, corrosion at original fit looseness, impeller & subsequent key way wear/erosion. After visual inspection, the shaft should be checked for any eccentricity for correction. If the cost of the new shaft is high and proper facilities are available, a worn out shaft, should be repaired by metal spraying and re-machining. Such repairs should not be undertaken without knowing the shaft material and appropriate metal spraying methods. After the shaft has been repaired, it must be checked for distortion, eccentricity and rechecked after complete rotor assembly.

10.3.11 Pump Bearings

The bearings used on most pumps can be divided into two general classifications–

√ Sliding contact and

√ Rolling contact.

Sliding contact bearings are called sleeve bearings and rolling contact bearing are called as antifriction bearings. One should remember that bearing maintenance does not mean replacement of worn out bearings only but also proper selection, installation, storage, inspection and lubrication.

10.4 Observations During Operation

Human senses like eyes, ears and touch feelings are the most sensitive monitoring tools in monitoring the running machines. All instruments used for monitoring are simply aid to the human senses.

√ **Listening**

Place one end of a listening rod, against the bearing housing, close to the bearing as possible. Fix the ear against thumb at other end and observe sound; if all is well a soft purring sound will be heard. A damaged bearing produce a often irregular and rumbling loud noise and sound.

√ **Feel**

Check the bearing temperature either by using temperature sensitive chalk, thermometer or -often by hand skin touch. If temperature seems unusually high or change suddenly, The indication is serious. The reason may be insufficient or excess lubrication, over loading, bearing damage, insufficient clearance, pinching high friction in the seals or heat supply from an external source. Remember that immediately after re-lubrication, there will be a natural rise in the temperature, which may persist for one hours or even less.

√ **Look**

Ensure that lubricant does not escape through worn out/defective seals or insufficiently tightened plugs. Impurities generally discolor the lubricant or oil turn dark. Check seals conditions and ensure that no hot or corrosive liquids or gases penetrate into bearings. The hot spot of bearing housing can be observed by vision.

10.5 Installing Plain Journal Bearings

The bearing installation techniques vary for different type of bearings. Plain journal bearing are usually easier to install compared to antifriction bearings. However, they require great care to ensure proper alignment during installation for non self-aligning construction of bearings. The alignment is accomplished by shim placement underneath the bearing shell. If a journal bearing is of the split type, the upper half can be removed for checking the alignment after installation and short running period. A visual inspection of the bearing material will show misalignment, wiped babbit metal.

Misalignment of a plain journal bearing is usually indicated by shiny or worn spots on its inner surface. Because of its cylinder length, the inner bearing surface will not show wear over a large surface rather at localized points. Misalignment is indicated by a wear-spot on one end of the bearing surface, matching with wear-spot on the opposite corner of bearing.

Plain journal bearings should be inspected for wearing-in characteristics after a few hours of running. They should then be aligned properly, and lubricated as per recommendations. These simple procedures -can result long periods of bearing life.

10.5.1 Rolling Bearing Quality

- The bearing replacement. The bearing gaps including original clearance, fitting (interference/ clearance) and working clearance must conform to bearing manufacturer standard. As a thumb rule when clearance of installation is doubled in course of operation, replace bearing. Overall inspection of running bearings should be done to confirm that there are no crack, peeling, rust & discoloring by overheating marks. Bearing with a few flaw of peeling, streak or rusting in area less than 1 mm² on none-working faces of bearing (inner & outer races, rolling elements) can be acceptable.

- When bearing is to be replaced with different manufacture, its parameters of design (speed limit, coefficient of working ability, allowance, static load and its dimensions) are compared. The bearing can only be replaced when its entire parameters are in accordance with the original design requirements.

- Check the sizes of bearing (internal & external races) whether that conform to the design values. Generally, no clearance is allowed at the fitting surfaces of the bearing.

- Check the bearing internal & external races, rolling elements and cage for no flaws of streak, double-skin, ditch, rusting, etc. If slight rusting appears on external race working face, we can remove it carefully by honing or scrapping, but that should not affect the size of bearing.

Checking and measuring the clearance of bearing rolling element and races should be recorded and compared with designed value. Normally, clearance is measured by lead wire testing method. The lead wire of 0.5 mm to 1 mm φ is inserted between the ball and the external circle and rotating the bearing to let three consecutive balls/rollers press over the lead wire. Then, remove the wire and measure the minimum thickness of pressed lead wire with the help of 0–25 mm micrometer. The measured value is the diametrical clearance of bearing. The clearance measured by lead wire indicates approx. 0.01 mm higher value compared to the actual clearance.

10.5.2 Oil Lubrication

a. Check the oil level and replenish if necessary. Ensure that air vent of bearings are not blocked. When the oil is to be changed, old oil is drained off and the bearing sump is filled to the required level with fresh and clean oil of the same grade and make.

b. In case of oil bath lubrication of bearing, change the oil once a year provided the operating temperature does not exceed 50 °C and the oil is not contaminated.

c. The oil must be changed more frequently when operating temperatures is higher than 50 deg C,

Following guidelines should be followed

Bearing temperature deg C	Changing schedule of lubricant
≤ 50	Yearly change
50–99	Half yearly change
100–119	Quarterly change
120–129	Monthly change
≥ 130	Weekly change

10.5.3 Mounting a Rolling Element Bearing

Before pressing the bearing on shaft, several important conditions should be satisfied:

a. Cleanliness of the area is necessary to prevent contaminants getting into fitting surfaces.

b. The bearing should be supported properly and pressure is applied to the correct race of the bearing.

c. Be sure that bearing housing or shaft is free of scratches, burrs, or other irregularities, eg. ovality.

d. Apply a light coating of oil or graphite grease on shaft or housing to ensure easy installation and subsequently easy removal.

- When bearing is being pressed in place, the area in contact with bearing race should be repeatedly checked for dirt or metal chips that could wedge between the bearing race & housing or shaft & housing. On a large shaft ≥ 10 inch diameter, scrapping of metal about a mil will not cause much misalignment. However, the scrapping for a bearing having one inch diameter, abnormal misalignment will be generated.

- After the bearing has been mounted on the shaft or in the housing, check it for free movement. Make sure that your hands are clean and dry then only grasp the un-mounted bearing in your hands and rock it gently from side-to-side. Check the bearing for free movement before it is installed. Rotate the ring slowly to make sure that it turns freely without any binding or noticeable drag. If the bearing turns hard, binds or drags in a particular spot, check it for dirt obstructions. If the bearing can't be freed and there is no evidence of ovality of shaft, the bearing has to be removed and replaced with a new one.

- After the bearings have been installed, it should be tested in running of machine without load for a short period of time to ensure that all components are properly installed.

- During the test run, check the bearings for noise, high operating-temperature, and vibration. Heavy noise level indicates damage of bearing during the installation (improper mounting, misalignment, and interference of the parts).

10.5.4 Methods of Assembly of Rolling Element Bearings

- *Cold fitting of bearing* by copper rod striking and hammering through bz. Bushing on races.

- Copper hammering is the easiest assembly method of bearing, Alternatively strike on the bearing inner & outer races through hand-hammer through copper rod, The direct strike on the bearing is never permitted.

- *Bushing-hand hammer method* use bz, Cu or Al rod or strip for dispersing the hammer strike-force to inner race of bearing. The hardness of bushing should be less than that of inner race. Bush outer diameter should be slightly less than that of inner race & bush ID a little smaller than that of inner race. Besides, pay attention to prevent bushing particles falling into the bearing.

- *Hot-assembly method*: This method is used for interference fit or large bearing assembly. For details, refer rolling bearing assembly instruction by manufacturer.

- *Mechanical pressure assembly method*: This method is mainly used for fitting bearing by mechanical or hydraulic force. The bearing inner-race and the shaft slide by external force.

10.5.5 Installation of Antifriction Bearings

When roller or ball bearings are installed, misalignment should be checked after some period of operation. However, remember that even though ball or roller bearings may appear to operate satisfactorily in some misaligned conditions also but life will be shortened considerably.

- A ball or roller bearing is designed to operate with the inner and outer rings secured against the shaft and housing, respectively. However, it can be accomplished in several ways. *Generally in fittings where load is transferred from inner race, interference fit is provided between the bearing I.D. and shaft where as O.D. of housing and bearing outer race is fitted with transition fit. Contrary to this, when load is transferred from housing to outer race then fittings are provided vis a versa.*

- The bearings are usually installed using an arbor press to force them into place. If an arbor press is not available, a bz. pipe or sleeve is used. No matter how the bearing is installed, be sure that ring which is being forced in place is adequately supported while fitting.

- The force is applied to bearing or the shaft should not be too large, otherwise a bearing puller should be used. Although a bearing puller is used normally to remove a bearing, but it can also push a bearing into position as well. However, for heavy load fitting hydraulic press can be used.

If the area of contact between the bearing inner ring and the shaft is quite large, other method like oil heating or nitrogen cooling are used for mounting the bearing. A bearing with a large area of contact is a shrink fit bearing. The bearing fitting by oil heating and soaking at 200 deg C oil bath to expand the race for placing on shaft as clearance fit in proper position. Subsequent slow cooling at atmospheric temperature will produce interference fit on full cooling. Thus, bearing shrinks tightly on the shaft after cooling. Another method of accomplishing a shrink fit is to cool the shaft in dry ice or nitrogen liquid and then bearing is installed. If the interference is large, a combination of the two methods (heat and dry ice) can be applied.

10.5.6 The Noise from Bearing in Running (Idle and on Load)

- High pitch noise may be heard from interference or overlap of two noises at close frequency. A medium to low pitched noise is generated from bearing misalignment; rattles sound from poor fits or loose bearings and a rumbling sound by a poorly finished shaft or housings (out-of-roundness).

- Machine starting with excessive amount of lubricants or too high operating speed is not permitted. If the unit is running on full load, and bearing temperature indicates too-high check for high operating speed or overload.

- Scaled bearings normally have a high operating temperature during initial startup, but the operating temperature normalizes as the machine continues to run. If the operating temperature does not normalize, the machine should be shut off and allowed to cool. During the period of cooling-down, the grease redistribute itself in the bearing. On re start of machine, high bearing temperature, in general, should not appear. However, If temperature does not normalize, check the bearing thoroughly.

- Special attention should be given for locking device fitted to secure the bearing in place on shaft. Manufacturer's instructions should be followed while tightening the locking device. If no instructions available, only sufficient pressure should be used to ensure locking softly. Once the ring is locked no additional pressure should be applied. Any effort to squeeze the bearing inner-ring down may distort the inner races. The locking washer should be penned over to prevent locknut opening during operation.

10.5.7 Bearing Removal

Removal of the bearings is simply a reversal of the procedures used to install. Bearing is installed with an arbor press which applies pressure for removing the bearing in position. The bearing puller can also be used for extraction. Torch heating of bearing increase the clearance between bearing and the shaft. Extreme care should be taken to ensure that any grease or lubricant in the bearing does get fire. Ensure that shaft do not get heated up along with bearing. Pouring of water on shaft is applied continuously while torch heating to keep the shaft cool. Also make sure that the heating torch is set for a heating flame, (not a cutting flame). Remember to provide proper support for the bearing and shaft during disassembly

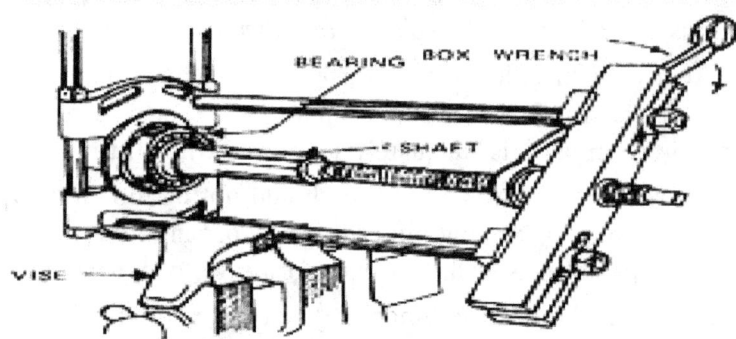

Bering Puller & Application

10.5.8 Dis-Assembly of Rolling Bearing

- Before disassembly of bearing, open the positioning lock nut using round-nut wrench or hand-hammer and special tool.

- Before disassembly of bearing, install special extraction tool on the bearing in advance and heat the bearing by pouring 110~120°C mineral oil on bearing inner race, using long-pouring-oil-nozzle – pot. To prevent the oil on the shaft, rubber sheet or asbestos ribbon is rapped on shaft. Meanwhile, let the puller be preloaded before pouring hot oil on bearing. Thus, bearing would be disassembled when its internal race is expanded & clearance is formed between shaft and inner race.

- After disassembly, inspection is carried out and by NDT Examination, if that reveals any defect is detected, rectify that as per standard maintenance procedure. Only the qualified bearing should be used.

- Follow the cleaning sequence as – Remove metal bits from shaft and shoulder by smooth file, and polish the shaft neck by oil stone, wash by agent and wipe out bearing with neat cloth.

- Measure bearing clearance by filler gauge or lead wire testing method. Stand the bearing; put the internal sleeve in a correct place, insert the filler gauge or lead wire gets through the bearing ball track. Repeating the test several times for each raw and note the minimum value radial clearance.

- Measure the diameter of journal and the bearing inner race to confirm the interference in assembly. The interference value should be in accordance with the quality standard. To ensure the accurate measurement, x journal is divided in 3 sections along the length of bearing seating. Measure two diameter at 90 deg for each section. Normally 1–1.5 mil interference is adopted in bearing assembly.

- If the interference doesn't confirm the standard, make spray-cladding, chroming, anaerobic rubber lining or installing hot-brushing on the journal portion. The interference value in hot-brushing is recommended to be 0.02–0.04 mm between inner race and journal.

- Use mineral oil heating method for installation of bearing. In this method, the bearing is hanged & immersed into heated mineral oil through firmly tied steel wire. However, it is not allowed to let bearing touch the oil container to avoid localized over-heating of outer race & losing its hardness. Bearing should not be tied through ID bore for hanging and heating in hot oil.

- Constant oil-temperature measuring should be done during heating process to limit oil temperature within 200 deg C. Soak bearing at 200 deg C for 40 minutes. The bearing need to be promptly installed on the shaft neck otherwise it can stuck midway of its travel. If during fitting, bearing start holding the shaft in the midway before reaches its right position, immediate forced strike on bearing should be made through special bronze bushing/copper-round. Install bearing in position and allow to cool down at atmospheric temperature. Move bearing to check its freeness, measure bearing clearance after installation and tighten the round-nut.

- Clean bearing by cleaning agent and smear the bearing with oil or grease to prevent rusting. Pack the bearing with clean plastic Record clearance value after hot-installation.

PRECAUTIONS

- It is absolutely undesired to use hand-hammer or hard-ironware for direct striking on bearing. The copper hammer or copper bar with wooden block can be used for striking over bearing races.
- In force fitting even external force should be applied on outer/inner race of

10.5.9 Running Inspection of Bearing

Although roller bearings are robust mechanical components which give long service, it is however, wise full to inspect frequently in course of running as well during shut down. Preferably. physical inspection can be done during a planned stoppage of the machine or when the machine is to be dismantled for some reason. Store bearings in clean and dry environment. The bearings should be readily available when needed. If drawings are available that should be studied thoroughly before maintenance work is taken up.

a. Clean the external surfaces. Note the order in which the machine components are removed and mark relative positions. Don't apply excessive force to remove any seal. Inspect the seals and other components in assembly. Follow the rule that part coming out first in dis-assembly will be go last in assembly.

b. Check impurities in lubricant by rubbing between the fingers or a thin layer spread on the back of the hand for inspection against the light.

c. The dirt or moisture can enter the bearing when defective cover and seals permit. Cover the dismantled bearings with waxed paper/plastic sheeting. Do not use cotton waste for cleaning.

d. Carry out inspection and wash the exposed bearings without dismounting. Use a paint brush dipped in white spirit and dry with a clean cotton cloth or low pressure compressed air. However, sealed or shielded bearings should not be washed. A small dentists mirror and a probe may be useful for inspection of the raceways, and rolling elements of the bearing. If the bearing is found all right, it should be re- lubricated with recommended lubricant and put in service.

10.6 Bearing Maintenance

10.6.1 The Reason of Bearing Damages

- The rolling element bearing damages can be from peeling, rusting, attrition, cracks, broken and discoloring by overheating.
- Peeling is generally referred to the drop of metal strip or particle from surface of the bearing component (e.g. inner & outer races and rolling elements). The main reason of such defects is the alternating stress when the bearing runs eccentric especially due to misalignment. Moreover, peeling could also appear from high-vibration, poor-lubrication, material de-efficiency or manufacturing defects.
- Rusting of bearing is caused by long exposure to the moist air. Bearing should be protected by application of oil/grease and well-packing.
- Rubbing of some foreign matters (dust, coal ash, rust, etc) between rolling elements and race ways. Generate scratches & localized heating during operation. The roller track sustains damage & increase the bearing clearance which may result vibration and noise.
- Overheat, discoloring indicates that working temperature of bearing reached $\geq 170°C$ and bearing steel lose its properties & get discolored. The main reasons can be lack of lubrication, failure of lubrication/or improper assembling clearance.

- Cracks in any part of the bearing e.g. inner circle, outer circle, rolling elements, support frame etc. are major incident. Such damages are mainly initiated during bearing installation, operational anomalies like attrition (rubbing), peeling and discoloring are some of the symptoms of bearing temperature rise and vibration.

- The main monitoring parameters of rolling bearing are temperature, vibration and noise. The early stage faults of rolling bearing can be detected by monitoring devices.

10.7 Bearings Lubrication

The bearings, irrespective of type, need to be lubricated. Lubricant perform three major functions to ensure good services of bearing–

- Maintains an unbroken lube-film between rotary and stationary components of bearing.

- Prevent excessive friction between rotary and stationary components to maintain the desired temperature.

- Conduct heat away from the bearing's surfaces. Remain stable under severe operating conditions e.g. temperature, humidity, vibration and noise.

10.7.1 Selection of a Lubricant

Usually selection is based on four parameters–

1. Load, 2. Speed, 3. Temperature and 4. Environment

a. Any one or all of the above conditions may be crucial in application of lubricant. This is especially true when bearings are used in outdoors or processing plant applications where they are subjected to dusty, watery, chemically polluted environment. In most cases, the manufacturer will simply specify the lubricants, with specific properties or characteristics.

b. Permanent lubricated/life time lubricated/sealed bearings require no lubrication. Most plain bearings and rolling contact bearings require direct or indirect lubrication system. An example of indirect attention would be the centralized or automatic system for of bearings. In these cases, main oil tank level is maintained.

c. Whether oil or grease should be used for lubricating bearings will depend largely on the equipment design. It is needed that enough lubricant is applied to keep the bearing surfaces oiled and slippery.

d. When oil is used, the most important factor which influences the lubrication is viscosity at the operating temperature. Slow-speed, heavy-duty bearings are lubricated with heavier oil. Bearing used in outdoors applications does not require heating device for a low pour point oil. Make sure that the bearing will run freely on cold startup. Bearings used on equipment located in warm areas will need a heavier lubricant to maintain the desired running temperature viscosity.

e. Greases are chosen in bearing lubrication for low speed and medium to heavy load applications. As a general rule, grease is used when the running speeds of a plain bearing is < 200 to 300 rpm. One of the particular advantages of grease in a bearing application is its tendency to stay put. Under normal circumstances, it won't run out of the bearing as freely as oil does. Grease also acts as a seal to restrict entry of contaminants into bearing.

f. Over lubrication of bearings is the most common cause of bearing heating. Too much lubricants will cause churning, which causes-increase in bearing temperature and reduced viscosity of oil. If the oil gets too thin, it won't carry the load intended to, and the bearing may fail. The same problem is faced when an excessive amount of grease is used.

g. The bearings are usually mounted in a housing or enclosure of some kind. Some bearings are inside the equipment and can't be accessed easily from outside. They are internally lubricated by a system designed for the purpose. As a thumb rule, if the bearing housing is too hot to rest your hand for less than 2–3 sec there is problem with the bearing/installation or lubrication system.

10.7.2 Grease Lubrication

a. Clean housing and grease nipples before injecting fresh grease, if the bearing housing is not provided with the nipple requisite lubrication should be fed during a planned stoppage of the machine.

b. Remove housing top, remove used grease and wash bearing to inspect for any foreign material, bearing defects etc. Do not use cotton waste, use clean & soft cloth to dry up housing.

c. Fill the space between the balls or rollers with specified grease, suitable for the operating conditions. The free space around the bearing should normally be filled with grease *(one third to half full)*. If the bearing is to operate at very high speed the quantity of the grease in the free space should be *just less than one third.*

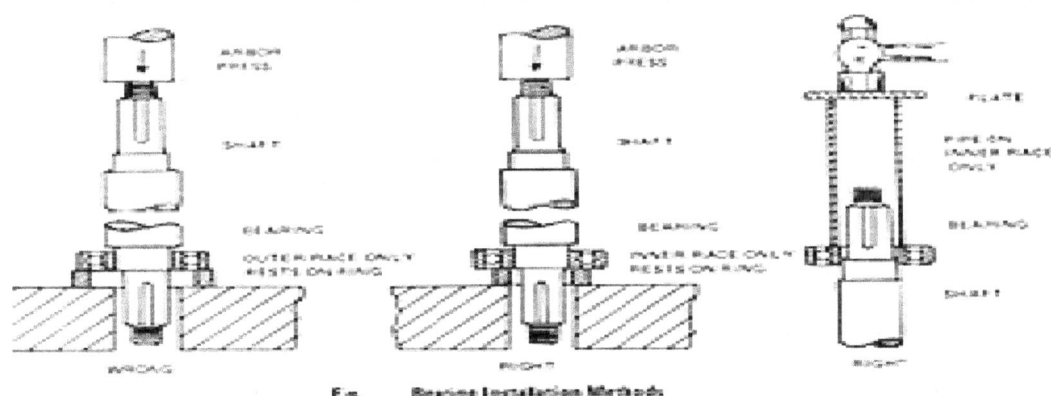

Fig.   Bearing Installation Methods

10.7.3 Application of Grease

- Lubricating grease is oils combined with thickener (generally, in the form of metallic soaps). While selecting a grease, it is necessary to consider the consistency in operation and rust-inhibiting property. Consistency is classified according to the National Lubricating Grease Institute (NLG I) scale.

- The upper temperature limit for calcium base greases is approximately 60 deg C. The calcium base greases with additive of lead soap are suitable for 'wet' bearing applications, e.g. Weir house of paper mill. Certain calcium/lead base greases provide protection against salty water.

- Sodium base greases are suitable for the temperature range −30 to +80 deg C and provide protection against corrosion. They absorb moisture and form an emulsion. However, if amount of moisture absorbed is excessive, the lubricating properties of grease will deteriorate and there is a risk of running grease out of the system.

- Lithium base greases are generally used for temperatures range of −30 deg C to +110 deg C and they are resistant to water. No rusting will take place when lithium base grease is used as it is added with rust inhibitor additive. Lithium base grease with lead soap additives provide relatively good lubrication even where free water is penetrated. A number of different types of high temperature grease are available for temperatures temperature range of 120 deg C. However, it is advisable to follow manufacturer's instructions.

10.7.4 Manual Lubricating Devices

The maintenance craftsman probably uses most of the time in manual lubrication and replacing bearings. The equipment used for manual lubrication can be as simple as the lubricator pump or hand grease gun.

Some plants use portable lubrication units consisting of pneumatic or electric driven pumps mounted on a handcart. This unit is taken around the machines for lubrication, which make saving of manpower and quality of lubrication. It also eliminates fire hazards by keeping lubricant away from the machine area.

10.7.5 Natural Oil Lubrication Systems

- Many machines are equipped with gravity or drip-feed lubricating devices. They consist of a small reservoir mounted on the bearing housing for sliding lubrication.

- Another type of natural lubrication frequently used is splash or ring lubrication. Splash lubrication uses the revolving motion of the machine parts to distribute the lubricant over journal through ring, The ring is partly submerged in the oil reservoir. It rotates and collect oil from sump, & transfers the lubricant to the upper part of the shaft. Oil is scraped from the collar or ring and then flow naturally within the bearing housing to the points of lubrication.

The large gear is partly submerged in the oil bath, contained in the reservoir or sump. As the gear revolves, it splashes oil around the housing to lubricate other gears and bearings.

- Pad and wick oilers are lubricating devices similar to the principle of ring and collar oilers. The pad oiler is simply a small reservoir fitted with a felt pad that is saturated with lubricating oil. The pad is in contact with the shaft or moving element and wick transfers the oil from the reservoir to the shaft by creeping speed. The reservoirs are filled to the marked level to ensure proper lubrication of the rotating or sliding surfaces.

Pad oilers are mounted on unit with a wick extended from the oil reservoir to the journal.

10.7.6 Pressurized Oil Lubrication

√ **Recirculation** or pressurized lubrication system has many variations in construction. It is mainly constituted of tank & pumps mounted outside the machine casing or tank located underneath/the machine. In operation, the oil is drawn from the sump, pressurized through pump and delivered to the point of lubrication. The points being lubricated include not only the bearings, but also the gears box etc. Irrespective of device in use, it is important that proper lubrication of all points of machine is provided in correct quantity, quality and at right time.

√ Lubrication is an important factor in preventive maintenance. Debris analysis, Proper selection of, lubrication system, period of lubrication & quantity of lubrication can help in enhancing the life of lubricated components.

10.7.7 Activities of Lubrication Dept.

√ Lubricants inventory and control

√ Training & developments of lubrication system

√ Continuous & steady developments

√ Follow up

√ Verities reduction or standardization of lubes and applications

√ Storage and handling system

√ Safety of man & machine.

10.7.8 Records of Lubrication

1. Lubricants consumption, trend & annual requirements
2. Schedule & compliance of lubrication
3. Daily log sheets

4. Developments in lubrication system & products
5. Test Reports of lubricants
6. Maintenance spares consumption cost & lubrication cost analysis.

Maintenance and lubrication staff must apply write quality, write quantity & timely lubrication for points of lubrication. Any anomaly should be brought to the notice immediately for timely conductive actions.

10.8 Maintenance of Gate Valve

Valve body–

√ The valve body should not have damages such as crack, sand holes, wearing out and serious corrosion etc.

√ The valve body should not have contamination inside the body. The inlet & outlet pots should be clear.

√ The screw plug at the bottom of valve body should be sealed without any leakage.

√ The camber of stem should be within 1/1000 * L, otherwise, the stem should be rectified or replaced.

√ The screw thread of valve stem should be in integrated state, without any damage crack. The wearing should be within 1/4th of original thread thickness.

√ The surface of stem should be smooth, no rust, crack, groove, erosion and other flaws. The contact surface between stem and packing should not have erosion, groove and peeling. The depth of erosion point should be within 0.25 mm, Roughness is no more than the class. 6.

√ The screw thread of connecting bolts should be integrated and pin of nut should be reliable

√ The running between stem and trapezoidal nut should be free in whole moving journey; the screw thread should be smeared with black ceruse.

√ Packing size should meet the requirements of its working temperature, pressure, and sealing area; using bigger or smaller packing as replacement is prohibited; the height of packing filling should be consistent with valve size requirement and surplus packing filling margin should remain for hot pressing.

√ The packing incision should be chopped in bevel angle of 45 degree; the incisions of each rings should be staggered 120–180 degree; the chopped length of packing should be appropriate; the connecting seam should not have gap or overlap after putting into the chamber.

√ The packing chamber and its cover should be in good integrated state i.e. no rusting, no crack & no deformation etc. The interior surface of packing chamber should be smooth without remnant. The gap between stem and packing seat and between stem and cover should be 0.2–0.3 mm but not more than 0.5 mm. The gap between the packing chamber and the cover is also 0.2–0.3 mm, but not more than 0.5 mm.

√ After tightening the packing cover, it should be flat, pressure should be even, clearance of inner hole of packing cover and stem within 0.2 mm. The cover pressed length into the packing chamber should be 1/3 of its length and surplus packing margin should be available for hot packing.

√ Lapped sealing surfaces of valve core and valve seat should have no scratch, crack and honeycomb etc. The area of sealing surface should be more than 2/3 of total. The roughness should be as per class. 10.

√ Assembled valve core should be moving free and the device for preventing its fall off should be integrated and reliable.

10.8.1 Stem, Nuts and Gearing

√ The screw thread should be in good integrated condition and the screw thread sleeve should be fixed reliably without any looseness.

√ The bearings should be in good integrated state without mottling, corrosion, peeling, rusting on bearing balls and ball paths. The gap should conform to relative quality standards. The gearing device should move free and fitting clearances of each part should be correct. The hand wheel should be integrated without any damage.

√ The fixing bolt of stop-ring should not be loose and running of stem & nut should be free. Ensure axial clearance within 0.35 mm.

10.8.2 Disassembly of Gate Valve

√ Confirm the isolating valve has been closed reliably; If the valve is electrical or pneumatic actuated power supply or compressed air should be shut off, before valve disassembling.

√ Clean valve exterior portion. Give match mark on valve body and cover before stem opening.

√ Remove the actuator and disassemble. Remove nuts of packing cover & take out the packing cover. Cean packing residuals from the chamber.

√ Dismantle the valve cover nuts, remove cover of valve, mark inlet and outlet on the valve flashboard, shovel out gasket, screw up the stem. If it is a self-sealing valve, first knock the cover with copper rod, make cover of valve fall down, knock on clip-ring inwards through four small holes with thin copper rod, make the clip-ring exit slot, then take out the clip-ring,then take out the cover of valve and screw up the stem.

√ Disassemble the stem nut and thrust bearing, the disassembled parts should be cleaned by gasoline.

√ Inspect the surfaces of valve body and cover for crack, sand hole, rust, and joint-surface smoothness and straightness. Check fatness of valve cover and radial clearance to meets the requirements.

√ Clear valve body interior, inspect sealing surface of valve seat and determine the maintenance proposal. If maintenance can 't be completed in a day, the disassembled valve should be covered.

√ Disassemble splints on the top and bottom of flashboard framework, draw out flashboards from two sides, gimbal jack and gasket, measure the total gasket thickness and record it.

10.8.3 Assembly of Gate Valve

√ Install lapped eligible 2 side flashboards to the clip-ring of stem, and fix it with upper & lower splints, add gimbal and spacer gasket inside.

√ Based on the markings made during disassembling, smear red lead powder on the surface of valve core, and put the stem and core into the seat of valve for contact sealing surface test. Inspect and confirm that core and sealing surface of valve seat is ok and sealing surface of valve core should be 5–7 mm higher than sealing surface of valve seat, otherwise adjust the spacer gasket thickness on top of gimbals until it is appropriate. For protecting fall-off, it should be sealed with no-return gasket.

√ Clean valve body, wipe the seat and the core of valve with white cloth or silk cloth cleanly, then install stem and core of valve into seat of valve and fix the cover of valve.

√ Based on the marking made before disassembling, reinstall the cover of valve, then tighten the bolts evenly. For self-sealing valve, it needs only to tighten the cover bolts slightly.

√ Fill packing in the packing chamber, packing size and material quality should meet the requirement of medium parameter.

10.8.4 Reinstall Packing, Cover and Tighten the Bolts Symmetrically

√ Reinstall pneumatic or electrical driving actuator, pay attention to tighten the bolts of connection flange, open and close valve manually to test that valve is running free & valve complete journey is smooth.

√ Make maintenance record.

√ Clear up maintenance site.

√ For self-sealing valve, when pressure gets 10 Mpa, retightens the bolts of valve cover again.

10.8.5 Maintenance of Valve Parts

√ The joint surface of valve seat can be lapped by special tools or by hand. Lapping paste 180–320 Boron carbide or silicon carbide or grinding sand or fine sand cloth, first kibble, then lap with fine paste subsequently.

√ The joint surface of core can be lapped by hand or muller. If we find deeper scratch or groove on the core of valve, we can grind it on grinding machine.

√ Cleaning of rust on packing cover, the pressing surface of packing cover should be smooth and integrated; the clearance between inner hole of packing cover and stem should conform to standard; it should not stuck when the cover is put in packing chamber.

√ Clear packing from chamber, inspect the packing seat. If that is not in integrated state, the gap between parking chamber & the stem should conform to standard and there should not be any blockage.

√ Inspect the thread of bolt and nut.

√ Clear rust from stem surface, inspect stem for bending, inspect trapezoidal thread. After clearing, molybdenum grease is smeared.

√ Inspect and confirm that turning of bolts and nuts is free, smear molybdenum grease on the screw thread. Clean and inspect stem, nut and inner bearing

10.9 Preventive Maintenance Activities

It starts with inspection (running or shut down) of equipment and plant by operatives and maintenance crues in planned manner. After inspection by visual and or instrument added, the defects for problems likely to arise in m/c leading to breakdown of machine are recognized for timely action & repair. Therefore, subsequent activities of preventive maintenance will be as follows:

a. Drawing development and records.

b. Indenting and work orders of required spare parts with records.

c. Expediting and follow up of work orders and records of material procurement – lead time, manufacturers, materials of construction and any improvement, if applied.

d. Codification of all components/spares and their proper records.

e. Storage of components with locations, quantities in hand and order with current status. Implementation of 5-S programme and review periodically.

f. Emergency, insurance and critical spares and their status records.

g. Budgeting for maintenance and inventory control.

h. Shut down planning and scheduling. PERT and CPM preparation for shut down and recordings.

i. Service records for all m/cs and components with codifications.

j. Remaining life assessment of plant and m/cs.

k. Calculation of availability and reliability of all m/cs and plant as a whole, implementation of components Life Assessment by condition monitoring in view of shut down schedules.

10.9.1 Engineering Records and Activities of Preventive Maintenance

The constituents of preventive maintenance program can be listed as follows:

1. **Equipment data:-** it is essential that maintenance personnel must have detailed data of equipment/ machins/components/suppliers name and ratings, plant inventory #, cost, last order nos, drawing/ sketch, previous life, complete list of equipment or m/c, specification of m/c or equipment, codes etc.

2. **Instruction sheets and manuals of equipment:-** manual explains correct procedure of operation and maintenance. Instruction sheet prepared for operation as well as repair & maintenance should be updated based on past experience. This sheet constitutes of information for maintenance requirement i.e. servicing, replacement, overhaul, lubrication schedule, training modules for employees engaged for operation and maintenance.

3. **Service records:-** it is important to maintain service record for each equipt/machine in chronological order or equipment code wise. This will reveal behavior of machine. We can use this record for predicting components life period of replacement, future planning for orders and procurement of spares. Service record should contain following information–

    a. Any failure and precise method of restoration.

    b. Total repair and overhauling job done

    c. Replacement of component, alteration in design/material and heat treatment or any other changes like fits & tolerances.

    d. Manpower, time, tools and tackles required or used. This record can be maintained in computer with indexing of equipment number in service, drawing number, component number etc.

4. **Delay and delay analysis:-** Delay should be recorded shift wise basis in equipment account and subject (say operation mechanical, electrical, instrumentation, outside agencies, beyond control) and it should be presented in daily, weekly, monthly and annually. Delay analysis will indicate the major delay partner where focus would be drawn, resources mobilized and corrective actions taken to eliminate the causes of delay in future.

5. **Inspection and maintenance schedules:-** As discussed earlier success of systematic maintenance lies on intensive, regular and sincere inspection of all equipment and machines by operatives, maintenance staff and sometimes external agency deployed. Inspection program should be able to satisfy four questions →

4-Questions for inspection-

a)   who to inspect?

b)   When to inspect?

c)   What to inspect?

d)   How to inspect?

6. The degree of success of inspection program is dependent on clarity of above four questions and their answer from inspection program. Based on above facts inspection format and schedules are developed for all equipment with responsibility and authorities of persons. Training to inspection personnel should be imparted to attain basic knowledge and skill of inspection. Our human senses play a major rolls in inspection, however NDT technique may be added advantage to that.

7. **NDT and condition monitoring technique** added to preventive maintenance, enhances the power and capability of inspection system in advance age of microprocessor and computer. Added techniques of condition monitoring to systematic maintenance impart more effective and powerful maintenance advantages for equipment which are inaccessible in operation for inspection. Online monitoring technique will give a picture for preventing maintenance action plan. Vibration analysis is very powerful technique for assessment of dynamic condition of equipment in process plant. The monitoring can be off line or on line based on criticality of machine. Dynamic balancing is corrective measure for unbalanced dynamic rotors for smooth operation of machine.

8. **Maintenance and overhauling schedule** are developed after due consideration of inspection reports, back logs, process requirement and development, up gradation of machine/equipment, condition monitoring based replacement of components. If particular machine has given a poor life, we should make an endeavor to improve by development of design of MOC monitoring process, fitting practices and lubrication system etc.

9. **Maintenance evaluation:-** Assessment of degree of success or failure of preventive maintenance system is evaluated by study of all preventive maintenance activities, their success & failure analysis, life of component, inspection effectiveness, engineering records, availability and reliability of plant. Cost of maintenance, capital cost, behavior of equipment/machine to the satisfaction of operatives, process requirements there should be continuous improvement of degree of maintenance system assessed as discussed above.

10.9.2 Maintenance Planning and Scheduling

Maintenance Planning and Scheduling of Major Activities Are as Follows

a. Listing of maintenance activities based on backlogs, inspection reports, Process requirement, failure analysis delay analysis, development and improvement activities.

b. Tapping of recourses: manpower, spares and components (available/to be procured), budgeting for expenses scheduling of shut down program, compliance of maintenance activities in minimum time with available recourses, economically.

10.9.3 Work Procedure of Critical Activities Is Developed Based on past Experience

a. Preparation of PERT/CPM/Bar Chart for maintenance activities.

b. Safety and statutory requirements & implementation.

c. Preparation and availability of special tools and tackles

10.9.4 Engineering Analysis

ENGINEERING ANALYSIS:- This is activity by which cause of failure by postmortem of event and activities is investigated–

a. Reason of failure

b. Failure related activities

c. Preventive measures and development of improved design and performance

d. Standard procedures of operation & maintenance for enhancing availability & reliability

ENGINEERING ANALYSIS carried out in Preventive maintenance.–

• Failure analysis-design fault/work man ship fault

• Delay analysis-equipment wise unplanned failures.

• Root cause analysis of failures

• Cost analysis-section wise/equipment wise on monthly & running average for month & year.

10.9.5 Some Common Failures Can Be Envisaged As

1. Fatigue failure/creep failure
2. Stress concentration failures
3. Failure by wear-out parts due to non availability of spares or shut down
4. Work man ship failures in assembly, manufacturing, operation or maintenance

10.9.6 Feedback & Communication for Maintenance Improvements

Regular feedback of maintenance assessments is essential to all maintenance personnel to improve their short comings. This is possible by careful and timely feedback system. Follow up for investigations, findings and periodical review.

10.9.7 Spares Planning & Control

Spares can be categorized as follows:

- Critical spares:- In absence of spares process will be paralyzed
- Insurance spares:- a critical spares which are insured & kept in stock. These are very costly critical spares.
- General spares:- (1) consumables (2) required for maintenance, generally in shut down but not critical in sense that process can be managed by repair, or stand by equipment available for operation.

10.9.8 Spare's Inventory and Its Control

Some spares which are of consumable natures are stored with min & max records (bin card). But some other spares are procured on need base. When we keep more spares in stock (inventory), the stock cost increase where as in absence of spares availability process & production cost increases. Hence, balance between inventory and production cost is maintained i.e. called as optimum inventory which is kept in hand as spares &consumables.

10.9.9 Standardization of Spares

STANDARDIZATION OF SPARES, components & machines

This is a device by which not only procurement of spares is carried out but also inventory is also controlled. This principle should be implemented from project stage and continued during operation too. Application of standardization in later sense reduces verities of similar category of spare and attempt cost reduction by import substitution by saving foreign in exchange.

10.9.10 Examination of Tools & Tackles

For using reliable tools & tackles on work, it is necessary that proper inspection & cheeks should be carried out at periodical schedule. There are some tools & tackles which are to be inspected for its healthiness, capacity & efficiency by authorized government agency. Maintenance staff should maintain a list of special tools and tackles their specification, last inspection reports, certificate of healthiness and their performance in a specified format and duly counter signed by inspector. A maintenance schedule should be drawn for such tools and tackles and compliance recorded in maintenance log sheet or computer.

10.9.11 Training for Maintenance Personnel

Training can be of following categories–

a. In-house training program, theory added.
b. Out sourced training programs
c. Specialized agencies training programs
d. Job Oriented Training programs
e. Management development training programs

Chapter 11

GENERAL ELECTRICAL SYSTEM

11.1 Motors' Trouble Shooting

TROUBLES	CAUSES	REMEDIES
Motor stall	Wrong application, over loaded motor, low voltage, open circuit, incorrect control resistance, wound rotor	Change type or size, reduce load, maintain voltage as per name plate, check fuse, over load relay, starter push button & control sequence. Replace broken resister, repair open circuit
Motor connected but does not start	One phase open, motor may be over loaded, rotor/starter coil connection defective,	Check for open phase, reduce load, open end cover & look for open bars or rings by test lamp
Motor runs and then die down	Power failure	Check for loose connection of lines, fuse and controls. Check motor terminal connections
Motor does not pick up speed	Not applied properly, supply voltage low at motor's terminal. check line voltage drop	Consult for proper application of motor, apply proper voltage by transformer tap changing or reduce load
	If wound rotor, improper control operation of secondary resistance, starting load high	Correct secondary control, check whether load is supported to carry at start
	Low pull in torque of synchronous motor, brushes mis match with rings, open primary circuit	
Too long to accelerate	Excess loading, poor circuit, defective squirrel cage rotor, applied low voltage	Reduce load, check for high resistance, replace rotor, maintain voltage by transformer tap change
Over heating of motor while running on load	Check over loading, wrong/clogged blower of air shield, ventilation problem, one phase opened	Maintain specified load, check for proper ventilation, check all load connection
	Grounded coil, unbalanced terminal voltage, shorted stator coil, faulty connections, voltage low/high, rotor rubbing with stator	Check for grounding of coil & replace/repair, faulty connection at transformer, check watt meter reading for high resistance, terminal voltage reading at motor. Replace defective bearings of motor.
Motor vibration after repair/corrections	Motor misaligned in assembly/installation, weak foundation, coupling clogging or unbalanced, defective/misaligned bearing, 3- phase motor running on single phase, excess end play	Check for misalignment in assembly or installation7 rectify, strengthen base if weak, balance rotating parts, check fr open circuit, adjust bearings, replace if defective. Check bend of rotor of motor or driven m/c..

TROUBLES	CAUSES	REMEDIES
Unbalance current in 3- phases of motor terminal	Unbalanced terminal voltage, single phased, poor rotor contacts in control, wound rotor resistance, bushes unmatched on rotor commutator strips	Check load and opened connections/contacts, and control device and rectify for defects dtected
Scraping noise	Fan rubbing air shield or striking insulations strip/papers, looseness on bed plate	Remove interferences, clean & adjust fan, tighten holding bolts, base bolts
Magnetic noise	Air gap improper, loose bearings, rotor unbalanced	Check & replace bearings, balance rotor

11.1.1 Electrical Motor Controls

- PROVEN STARTING METHODS FOR SQUIRREL-CAGE AND SYNCHRONOUS MOTORS

	Application with	
	squirrel-cage motor	synchronous motor
a Direct engaging	X	X
Starting reduced voltage		
Star-delta-starting	X	
b Starting via Impedance coil:		
— Impedance coil in supply line	X	X
— Impedance coil in star point	X	X
c Starting via starting transfomer	X	X
d Starting via block transfomer	X	X

- MOTOR STRATEGIC CONDITIONS

STARTING THE MOTOR	MOTOR PROTECTION	STOPPING THE MOTOR	MOTOR OPERATIONAL CONTROL
Disconnecting Means	Over current Protection	Coasting	Speed Control
Across the Line Starting	Overload Protection	Electrical Braking	Reversing
Reduced Voltage Starting	Other Protection (voltage, phase, etc)	Mechanical Braking	Jogging
	Environment		Sequence Control*

* An understanding of each of these areas is necessary to effectively apply motor control

11.1.2 Effective Operation of Motors

All motors must have a control device to start and stop the motor called a **"motor controller."** A motor controller is the actual device that energizes and de-energizes the circuit to the motor so that it can start and stop. Motor controllers may include some or all of the following motor control functions:

Motor control and protections are normally starting, stopping, over-current protection, overload protection, reversing, speed changing, jogging, plugging, sequence control, and pilot light indication. The controllers range from simple to complex and can provide control for one motor, group of motors, or auxiliary equipment such as brakes, clutches, solenoids, heaters, or other signals.

11.2 Motor Starter

The starting mechanism that energizes the circuit to an induction motor is called the "starter" and must supply the motor with sufficient current to provide adequate starting torque under worst case line voltage and load conditions when the motor is energized.

11.2.1 Type of Motor Starter

- Across the Lines Starting
- Reduced Voltage Starting

11.2.2 Across the Line starting

Across the Line starting connects the motor windings/terminals directly to the circuit voltage "across the line" for a "full voltage start"

- This is the simplest and least expensive method of starting motor.
- Motors connected across the line are capable of drawing full in-rush current and developing maximum starting torque to accelerate the load to speed in the shortest possible time.
- All NEMA induction motors up to 200 horsepower, and many larger ones, can withstand full voltage starts. (The electric distribution system or processing operation may not though, even if the motor will).

Across the lines starters types–

a. Manual Motor Starters
b. Magnetic Motor Starters

a. Manual motor starter

It is package consisting of a horsepower rated switch with one set of contacts for each phase and corresponding thermal overload devices to provide motor overload protection.

- The main advantage of a manual motor starter is lower cost than a magnetic motor starter with equivalent motor protection but less motor control capability.
- Manual motor starters are often used for smaller motors – typically fractional horsepower motors but the National Electrical Code allows their use up to 10 Horsepower.
- Since the switch contacts remain closed if power is removed from the circuit without operating the switch, the motor restarts when power is reapplied which can be a safety concern.
- They do not allow the use of remote control or auxiliary control equipment like a magnetic starter does.

11.2.3 Magnetic Starter

A magnetic motor starter is a package consisting of a contactor capable of opening and closing a set of contacts that energize and de-energize the circuit to the motor along with additional motor overload protection equipment. Magnetic starters are used with larger motors (required above 10 horsepower) or where greater motor control is desired.

a. The main element of the magnetic motor starter is the contactor, a set of contacts operated by an electromagnetic coil. Energizing the coil causes the contacts (A) to close allowing large currents to be initiated and interrupted by a smaller voltage control signal.

b. The control voltage need not be the same as the motor supply voltage and is often low voltage allowing start/stop controls to be located remotely from the power circuit. Closing the Start button contact energizes the contactor coil. An auxiliary contact on the contactor is wired to seal in the coil circuit. The contactor de-energizes if the control circuit is interrupted, the Stop button is operated, or if power is lost.

c. The overload contacts are arranged so an overload trip on any phase will cause the contactor to open and de-energize all phases.

11.2.4 Starting of Motors

a. **Starting of asynchronous motors**

Starting methods for asynchronous motors

Starting method	Type of equipment	Current input (mains load)	Run-up time	Heat build-up in motor during start-up	Mechanical loading	Hydraulic loading	Cost relation	Recommended motor designs	Comments
D. o. l.	Contactor (mechanical)	4–8 · I_N	Approx. 0.5–5 s	High	Very high	Very high	1	All	Mostly limited to ≤4 kW by energy supply companies
Star-delta	Contactor combination (mechanical)	1/3 of d. o. l. values	Approx. 3–10 s	High	Very high	Very high	1.5–3	All; canned motors and submersible motors subject to a major drop in speed during switchover	Usually stipulated for motors >4 kW by energy supply companies
Reduced voltage	Autotransformer, mostly 70% tapping	0.49 times the d. o. l. values	Approx. 3–10 s	High	High	High	5–15	All	No currentless phase during switchover (gradually replaced by soft starters)
Soft start	Soft starter (power electronics)	Continuously variable; typically 3 · I_N	Approx. 10–20 s	High	Low	Low	5–15	All	Run-up and run-down continuously variable via ramps for each individual load application; no hydraulic surges
Frequency inverter	Frequency inverter (power electronics)	1 · I_N	0–60 s	Low	Low	Low	Approx. 30	All	Too expensive to use solely for run-up and run-down purposes; better suited for open- or closed-loop control

b. **DOL Starting**

Complete and instant voltage is applied to motor, once it is switched on. Four to eight times normal starting torque is developed at its duty speed within short period of time. Thus DOL starting put a heavy load on electric supply system and may cause problematic voltage drop in the vicinity of low voltage grid.

c. **Star Delta Starting**

It is used to start machine used for frequent start & stop frequently started to reduce the starting current of motor. During normal running, full voltage of bus is applied to motor winding but while start up, windings are star connected to allow reduced voltage to windings. After specified time, star is switched over to delta & accelerate motor to running speed. Switch over time is approx

0.1 sec where current supply to motor is interrupted but after that full current is applied, similar to DOL. Thus, torque at starting is reduced but longer time is consumed.

d. Auto transformer

It also function to reduce the applied voltage to windings (a 70% tapping will reduce torque and current approx 49%). However, unlike Delta Star starter there is no interruption of current.

e. Soft starter

Electronic continuous voltage variation of voltage (as per dimmer principle) is applied to windings by soft starter. Thus, starting current can be selected freely within permissible operating limit of motor. It is preferred for frequent start and stop of motors or bus where current has to be limited.

f. Frequency Inverter

Usually this is provided for open or closed loop control for soft starting of motor without any other additional equipment. The output frequency and voltage of frequency are increased gradually from minimum to required value without exceeding the motor's rated current.

g. Reduced Voltage Starting of Motors

Reduced Voltage Starting connects the motor windings/terminals at lower than normal line voltage during the initial starting period to reduce the inrush current when the motor starts.

- Reduced voltage starting may be required when:
  - √ The current rush forms the motor starting which adversely affects the voltage drop on the electrical system.
  - √ Needed to reduce the mechanical "starting shock" on drive-lines and equipment when the motor starts.
- Reducing the voltage reduces the current in-rush to the motor and also reduces the starting torque available when the motor starts.
- All NEMA induction motors can will accept reduced voltage starting however it may not provide enough starting torque in some situations to drive certain specific loads. If the driven load or the power distribution system cannot accept a full voltage start, some type of reduced voltage or "soft" starting scheme must be used.
- Typical reduced voltage start include:
  - √ Solid State (Electronic) Starters
  - √ Primary Resistance Starters
  - √ Autotransformer Starters
  - √ Part Winding Starters
  - √ Wye-Delta Starters

Reduced voltage starters can only be used where low starting torque is acceptable or a means exists to remove the load from the motor or application before it is stopped.

11.2.5 Speed Control/Regulation of Three-Phase Motors

- Mechanical Devices
- Speed control of three-phase squirrel-cage motors
- Frequency change
- Voltage change

Number of poles	Synchronous speeds in min⁻¹ at		Number of stator windings
	50 Hz	60 Hz	
2/4	3000/1500	3600/1800	1
4/6	1500/1000	1800/1200	2
4/8	1500/750	1800/900	1
4/12	1500/500	1800/600	2
6/8	1000/750	1200/900	2
6/12	1000/500	1200/600	1
8/12	750/500	900/600	2
4/6/8	1500/1000/750	1800/1200/900	2
4/6/12	1500/1000/500	1800/1200/600	2
4/8/12	1500/750/500	1800/900/600	2
6/8/12	1000/750/500	1200/900/600	2
4/6/8/12	1500/1000/750/500	1800/1200/900/600	2

11.2.6 Variable-Speed Gearings/Speed Converters for Centrifugal Pumps

The variable-speed gearings mostly used on centrifugal pumps are:

- Mechanical stepless speed converters such as belt drives or friction wheel drives
- Hydraulic speed converters
- Hhydrostatic drives
- Hydrodynamic converters
- Electro-magnetic speed converters.

11.2.7 Hydrodynamic Speed Converters (Hydraulic Couplings)

a. **Advantages**

- Step less control in a range of 4:1 to 5:1 maximum. Capable of transmitting very high power outputs
- Load-free motor starting
- Gentle pump acceleration
- Damping of any torsional vibrations
- Simple torque matching by modifying the oil fill
- Assured protection against excessive heating by means of fusible cutout
- Wear-free transmission, as the power transmitting parts have no mechanical contact
- High efficiency as rated slip is very low.

b. **Disadvantages**

√ Price

√ Additional space requirement for the converter and its auxiliaries, such as cooler, etc.

Rated capacity of electric motor slightly increased to compensate for converter losses (rated slip and mechanical loss generators such as bearing and gear train driving power for oil pumps).

11.2.8 Working Principle

Put simply, the hydrodynamic speed converter consists of a pump impeller on the driving shaft and a turbine runner on the output shaft. The input torque is transmitted to the output shaft by the mass forces of a fluid (mostly oil) flowing between pump impeller and turbine runner

Power loss. $Pv = P1 - P2$

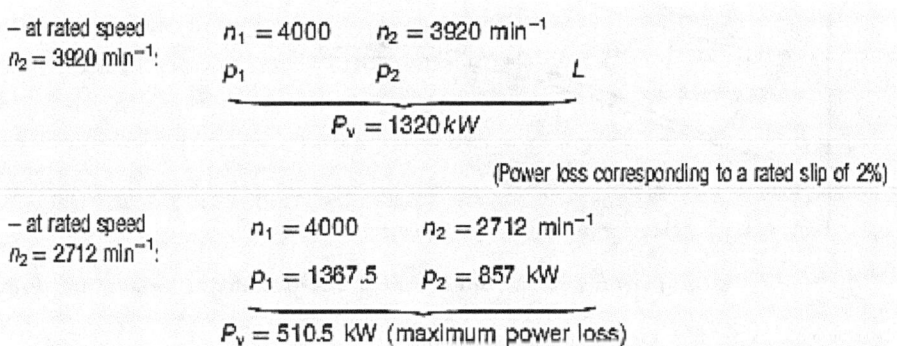

The electrical power taken up by the exciter coil in the magnetic circuit

- power loss for excitation – amounts to only a fraction of the transmitted power.
- maximum 2% for loads <25 kW
- approx. 0.5 to 1.0% for loads <500 kW. Control

11.2.9 Electromagnetic Speed Converters

Electromagnetic speed converters are generally built with a stationary excitation coil or one that turns the rotor. When the driving rotor is rotating and the driven rotor is stationary, one surface of the driving rotor is permeated consecutively by a maximum induction field (opposite a tooth) followed by a minimum induction field (opposite a gap). As the magnetic flux taken up by this surface is variable, eddy currents are generated here. Like all induced currents that follow Lenz's law, i.e. they attempt to resist the cause of their generation. The driven rotor starts to rotate. The speed differential between the two rotors is Reduced.

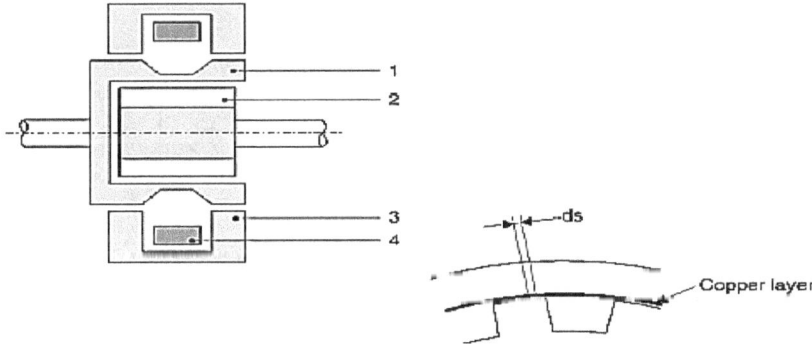

(1) Smooth driving rotor with copper layer driven at constant speed. (2) Toothed driven rotor, the speed of which is variable. (3) Stationary electromagnetic circuit excited by a coil. The magnetic field so generated flows through both rotors. (4) Coil

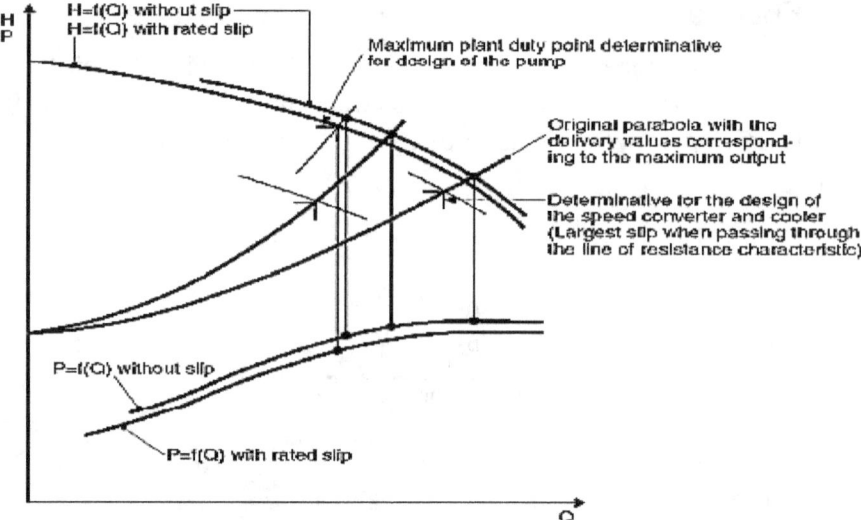

Design of a speed converter for several stated operating points

As already mentioned, speed variation is achieved by varying the exciter current. It is therefore possible to combine the speed converter with any kind of control system – preferably electronic – to form a control circuit. Pressure flow, temperature, speed and torque are among the relevant parameters for control.

Limits of application:- max. speed <3600/min; max. torque < 540 daNm.

11.2.10 Speed Control of Three-Phase Squirrel-Cage Motors

Control by means of resistances in the rotor current circuit. The control causes losses (slip losses) and is therefore usable to a limited extent only (poor efficiency).

Control by means of subsynchronous converter cascade (or static slip energy recovery system). This is particularly economical when the speed adjustment range is limited. The slip output is fed back into the three-phase grid. Hence the control is low loss, with good efficiency. Applicable for outputs of up to and more than 10 MW.

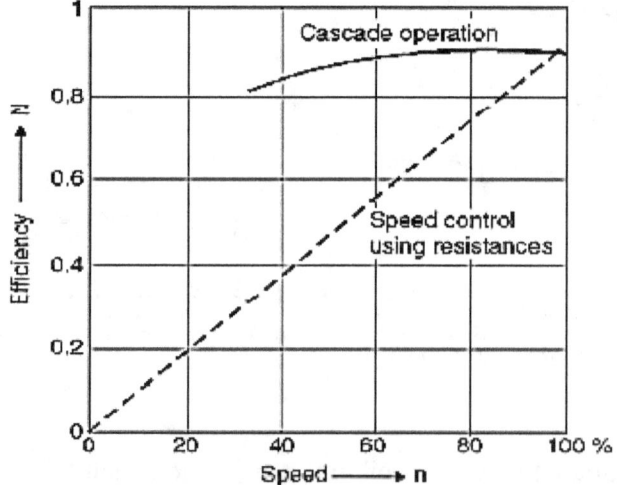

Figure · Efficiency curve of a three-phase slip-ring induction motor using resistances in the rotor current circuit and with subsynchronous converter cascade (mean output)

11.2.11 Speed Control of Three-Phase Synchronous Motors

Normally, synchronousmotors are designed to operate at fixed speed,with constant voltage and frequency.

In special cases, speed control is achieved by feeding the motor with variable frequency and voltage by means of static frequency converters. This circuitry, generally consist of controllable rectifiers, smoothing choke and inverters, is expensive, especially in the higher MW range. It can be constructed for the highest ratings and for frequencies up to about 120 Hz.

11.2.12 Some Features Relevant To The Choice Of Induction Motors For Driving Centrifugal Pumps

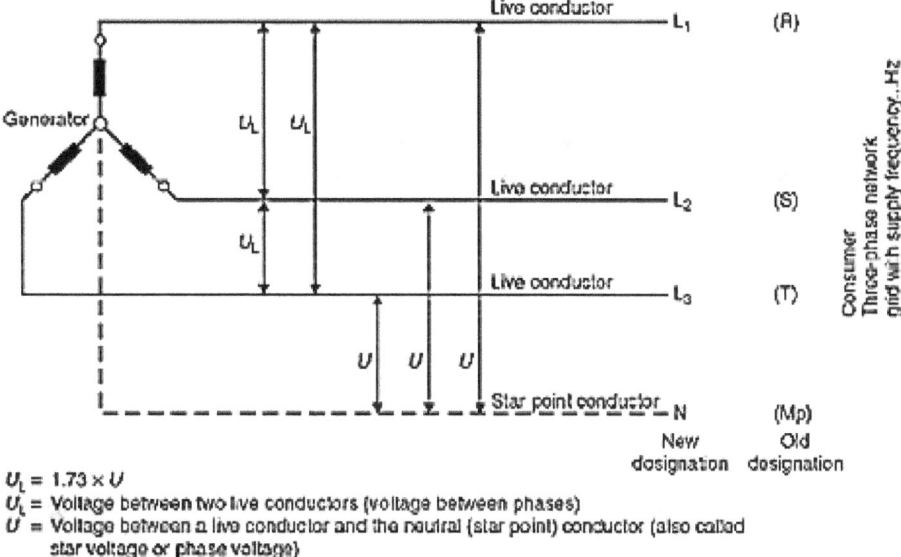

$U_L = 1.73 \times U$
U_L = Voltage between two live conductors (voltage between phases)
U = Voltage between a live conductor and the neutral (star point) conductor (also called star voltage or phase voltage)
L_1 = Live conductor (R) (also called outer conductor)
L_2 = Live conductor (S) (also called outer conductor)
L_3 = Live conductor (T) (also called outer conductor)
N = Star point conductor (Mp middle conductor or middle conductor with protection function, also known as neutral conductor)

Symbols for conductors and voltages

Frequency	Operating voltage U in V			
50 Hz	220	380	500	660
60 Hz	440			

Frequency	Operating voltage U in kV								
50 Hz	3	3.3	5	5.5	6	6.6	10	11	
60 Hz	2.3		4.16			6.6		11	13.2

11.3 Voltage Variation with Const. Grid Frequency

According to the ICE 34–1/1969 standard, fluctuations of grid voltage of up to (−) 5% . + 3% are permissible without a motor output corrections. (At boundary voltage fluctuation level, the maximum permissible temperature rise may be slightly more).

Where a motor designed to function at a standard voltage has to operate continuously at constant grid frequency with a voltage variation of (−) 5%, motor output must be de-rated accordingly. A decrease in may take place corresponding on decrease of output. Where changes in voltage is constantly (−) 5%, the motor supplier must be consulted.

11.3.1 Variations of Grid Frequency at Constant Grid Voltage

- With variations of _5% in rated frequency nominal motor output can still be achieved.
- When variations of _5% in frequency occur the motor supplier must be consulted

11.3.2 Variations of Grid Voltage and Grid Frequency Occurring Simultaneously

The magnetic conditions are then almost unchanged. Approximately the normal torque is developed by motor. Speed and output change approximately proportionately to frequency.

11.3.3 The Approximate Conversion Factors For Various Frequencies Are Given In The Table Below

The Approximate Conversion Factors For Various Frequencies Are Given In The Table Below

	(60 Hz)	(55 Hz)	(50 Hz)	(45 Hz)	(40 Hz)
Grid (supply) frequency f	1.2	1.1	1.0	0.9	0.8
Grid (supply) voltage U	1.2	1.1	1.0	0.9	0.8
Nominal output P_N	1.15–1.2	1.05–1.1	1.0	0.85–0.88	0.72–0.75
Nominal speed n_N	1.2	1.1	1.0	0.9	0.8

Basic design 50 Hz

11.3.4 Speeds of Three-Phase Motors at Frequencies of 50/60 Hz for Various Numbers of Poles

Number of poles	2	4	6	8	10	12	(14)	16	(18)
Speed n_s	min^{-1}	min^{-1}	min^{-1}	min^{-1}	min^{-1}	min^{-1}	min^{-1}	min^{-1}	min^{-1}
50 Hz grid	3000	1500	1000	750	600	500	(428)	375	(333)
60 Hz	3600	1800	1200	900	720	600	(514)	450	(400)
Number of poles	20	(22)	24	(26)	(28)	(30)	32	(34)	36
Speed n_s	min^{-1}	min^{-1}	min^{-1}	min^{-1}	min^{-1}	min^{-1}	min^{-1}	min^{-1}	min^{-1}
50 Hz grid	300	(273)	250	(231)	(214)	(200)	188	(176)	167
60 Hz grid	360	(327)	300	(277)	(257)	(240)	225	(212)	200

However, it should be noted that:

- Synchronous motors. rotor rotates synchronously with the rotating stator field (synchronous speed Ns).
- Asynchronous motors. rotor rotates asynchronously with the rotating stator field. The relative speed difference between the rotor and the rotating field of the stator (synchronous speed ns) is referred as "slip."

Motor output (kW)	1	10	100	1000
Slip at full load (guide values)	5 ÷ 8%	2 ÷ 4%	1 ÷ 2%	0.8 ÷ 1%

11.3.5 Nominal Motor Outputs

Nominal Motor Outputs (as per the manufacturer) are normally valid for continuous operation at a coolant temperature (ambient temperature) of 40 °C and at location up to 1000 m above sea level

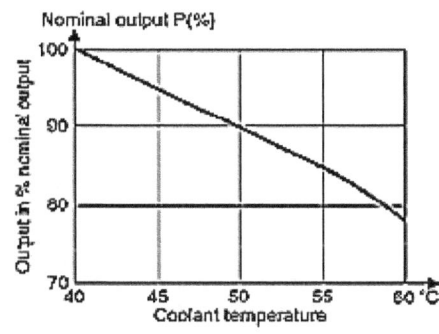

Influence of ambient temperature on output

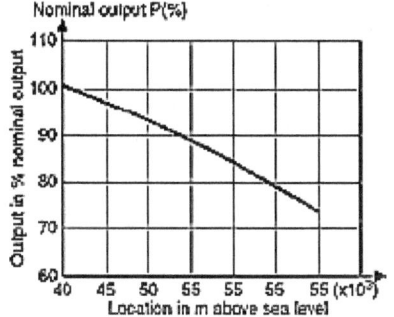

Influence of altitude on ambient temperature

11.3.6 Insulation Classification, Operating Life Of Windings – (VDE Standard 0530)

Insulation class	B	F	H
Insulated winding (heatup °C)	80	100	125

The highest safe continuous operating temperature of the individual insulating materials is derived from the coolant temperature, the boundary excess temperature value and a safety margin.

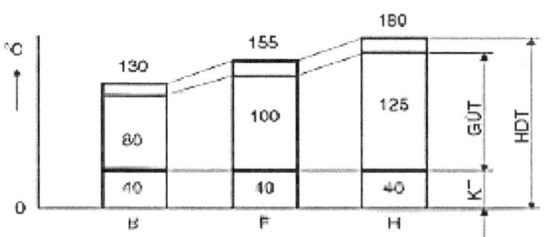

HDT = Highest safe continuous operating temperature in °C

KT = Coolant temperature in °C

GUT = Excess temperature limit (heatup) in °C

Temperature rise classes

Continuous Operating Temperature Rise vs Motor class

The mean life of motor is assumed to be 20 years. The operating life of a motor is dependent on quality of the insulation of its windings, when no account is taken for operating wear incurred in bearings, brushes and slip-rings, which are renewable with very little expenditure. Heat dissipation capability is dependent on the surface area of a motor and ventilation effectiveness. Since, the life of winding insulation decreases with increasing temperature, the boundary values for winding temperature defined in VDE 0530 must be complied with.

TYPE OF PROTECTION	CONTACT & FOREIGN BODY PROTECTION	PROTECTION AGAINST WATER
IP - 00	No specific protection against randum contact with any other component carrying voltage or in dynamic motion. No protection for penitration of solid body.	No particular protection
IP - 2		Water falling at angle, up to 15° from vertical may not effect to deteriorate
IP - 11	Protection against large area contact with internal moving or electricity carying components. Protection for solid particles ≥ 50 mm	Water falling vertically on m/c may not have much effect
IP - 12		Water falling at angle, up to 15° from vertical may not effect to deteriorate
IP - 13		Water falling at angle, up to 60° from vertical may not effect to deteriorate
IP - 21	Protection against fingure contact with internal moving or electricity carying components. Protection for solid particles ≥ 12 mm	Water falling vertically on m/c may not have deteriorative effect
IP - 22		Water falling at angle, up to 15° from vertical may not effect to deteriorate
IP - 23		Water falling at angle, up to 60° from vertical may not effect to deteriorate
IP - 44	Protection against touch contact of tools, wires etc with internal moving or electricity carying components. Protection for solid particles ≥ 12 mm	Water falling at any angle on machine does not have deteriorative effect
IP - 54	Complete protection against any component touch cintact of tools, wires etc (thickness ≥1 mm) with internal moving or electricity carying components. Protection for solid particles penitration ≥ 1 mm. Provision of air cooling system with self driven fan.	Water falling at any angle on machine does not have deteriorative effect
IP - 55		Water jet on m/c from any angle may not have deteriorative effect
IP - 56		Protection against water submersion e.g. flood/rain

- ICW37 A 81 Machine with built-in heat exchanger (cooling medium is water, inner cooling circuit – air); self-cooling.
- ICW37 A 85 As above, but cooling of internal air circuit is by means of built-in ventilation equipment, the drive of which is not dependent on the machine (independently operating fans).

The active power absorbed (P1) in kWis = $P_1 = \frac{P_N}{\eta}(kW)$

The nominal current (I_N) is given as–

$$P_s = \frac{P_N}{\eta \cdot \cos\varphi}(kVA) \quad \text{and} \quad I_N = \frac{1000 \cdot P_1}{U_N \cdot \sqrt{3} \cdot \cos\varphi}$$

$$\text{or} \quad I_N = \frac{1000 \cdot P_N}{U_N \cdot \sqrt{3} \cdot \cos\varphi \cdot \eta}(A) \quad = \quad I_N = \frac{1000 \cdot P_s}{U_N \cdot \sqrt{3}}(A)$$

where–

- P_N = nominal output (at motor shaft end) (kW)
- P_1 = active power uptake (at motor terminals) (kW)
- h = efficiency (%)
- P_s = reactive power uptake (kVA)
- U_N = nominal voltage (V)
- $\cos\varphi$ = power factor (fraction).

Chapter 12

COUPLINGS AND ALIGNMENT

ALIGNMENT in general is understood as condition when two (or more) running shafts are kept in one straight line. This does not give consideration to thermal displacement, deflection and dynamics of shaft(s) in stabilized running condition. Therefore, with due consideration of these points, we can define the alignment of two (or more) shafts as follows:

Two (or more) shafts of a machine(s) in stabilized dynamic condition, is said to be aligned when centre lines of rotors of m/c systems are co-linear or in regular smooth profile of m/c shafts configuration in such a way that any of the coupling half or part of it, transmits a force or moment within allowable limit (minimum) on that of other counter half as well other couplings in transmission line of system."

12.1 Type of Misalignments

    a.   Radial or Parallel Misalignment

    b.   Axial or Angular Misalignment

    c.   Combination of the two

12.1.1 Alignment of Pump and Driver

The following procedures outline the recommended practice for checking shaft alignment. This method is independent of the trueness of the coupling or shaft and is therefore not affected by coupling faces or eccentricity of the coupling periphery. Further this procedure is the same train of shafts.

    —   Before commencing alignment, rotate each shaft independently to check that the bearings run freely and that the shaft is true within 0.04 mm (0.0015 in.)

        The distance between the flange faces should be true within 5 mm.

FACE –PERIPHERIAL READING CAPTURING

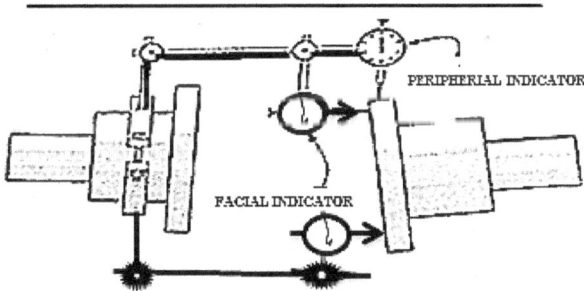

Alignment procedure in brief–

    i.   Clamp two dial indicators (DT1) at diametrically opposite points on one half coupling/the shaft behind it, with plungers of dial resting on the back/face of the other half coupling to be aligned. Mount one dial on periphery of coupling to be aligned. Rotate the coupling until the gauges are in vertical plane and set the gauges to read zero.

ii. Rotate the coupling to half revolution (180°) and record the reading on each dial. The readings should be identical though not necessarily zero, because of possible end-float of shaft into bearings which could be equally positive or negative. Refer alignment tolerances and adjust the position of bearings of shaft to be aligned with reference to other one.

iii. Rotate the coupling until the gauges are in line horizontally and reset the pointers to zero. Repeat the sequence

12.1.2 Tolerances in General

It is difficult to lay down limits of accuracy within which adjustments should be made because of the differences in the size and speed of units, but it is suggested that angular or radial alignment variations can be acceptable within limits given below:

COUPLING DIAMETER (MM)	< 1500 RPM	≥ 1500 RPM
Couplings up to 300 mm diameter	± 0.025	± 0.02
Couplings over 300 mm diameter	± 0.035	± 0.03

When the pump handles a liquid at higher temperature (ie. more than ambient temperature) or driven by a steam turbine, the expansion of the pump or turbine at operating temperature will alter the alignment. Alignment checked at ambient temperature, require suitable correction for thermal expansion so that changes in pump and driver center lines after expansion in running condition get aligned. The final alignment of the pump with driver at operating temperatures known as hot alignment should be adjusted before placing the pump in service.

For large installations, particularly with steam-turbine-driven pumps, more sophisticated alignment method is employed, using proximity probes or optical instruments. Remove machine strains caused by piping stresses while the unit is put to operation. Such procedures are followed as per manufacturer's instructions. When the unit has been accurately leveled and aligned, the hold-down bolts should be gently and evenly tightened before grouting. The alignment must be rechecked after the suction and discharge piping have been bolted to the pumps to test the effect of piping strains.

This is especially true when the pump handles hot liquids, as the pump should be disconnected after a period of operation to check the effect of the expansion of the piping, and adjustments should be made to compensate that.

12.1.3 Objective of Accurate Alignment

- To reduce axial and radial forces on supporting bearing
- To eliminate shaft failure due to fatigue, caused by vibrating cyclic forces
- To maximize Life of coupling and bearings
- To eliminate cause of shaft bend
- To reduce power consumption
- To reduce vibration level related to alignment.

12.1.4 Symptom of Misalignment

- Excessive run out of shaft or coupling or directional forces from pipings/machine foot to base plate distortion or soft footing
- Loose foundation bolts/foot bolts or damaged foundation/base plate
- Premature failure of coupling, bearing, seals, or shaft

- Excessive vibration, especially axial ≥ 1.5 * radial
- High casing temperature and bearings.
- Excessive oil or grease leakage from bearing/seals
- Coupling hot after run down test/coupling oil/grease leakage.
- Increase of shaft/coupling run out after operation
- Shaft breakage adjacent to coupling or inboard bearing

12.2 Reasons of Periodic Alignment Check

- Decay in m/c bearings due to wear
- Decay due to fatigue
- Settling of soil below foundation due to vibration or natural setting
- Foundation foot bolt loosened
- Other misc. reasons.

12.2.1 Type of Foundations Used for M/C Installations

- Rigid,
- Floating and
- Flexible

12.2.2 Tips for Good Foundation Design

- Ensure natural frequency zone of foundation, structural or soil system does not match with any running component frequency
- Provide operation and maintenance access around foundation
- Prefer to dowel foundation structures at its centre of expansion
- Minimize the height of centre line of rotation from base plate
- Protect foundation from radiation heat generated by m/c, steam or hot process piping
- Ensure concrete design mix, pouring, compacting by vibrator and water curing of concrete for 21 days from pouring date
- Check and rectify base plate distortion if any. Sandblast and paint with zinc silicate paint
- Always use matched pair of wedge blocks for leveling and lock it after final leveling
- Before final cast of foundation with non shrink cement, wash the old concrete with soda & water
- Use minimum 40–50 mm thickness of shrincom for final grout. Add DR Fixit 5–10% with shrincom for preventing molsture or oil engrace into foundation. Allow min. 72 hrs curing time before giving any load on it
- Install jack screws for alignment of m/c for horizontal and vertical movement of machine.

12.2.3 Symptoms of Piping Strain on M/C

Pipings should be free of constraint so that it can move and grow freely during operation of machine. Failures are normally from cyclic stresses rather than tensile stress. Pipings need to be supported on flexible supports/spring hangers, saddles – as a thumb rule fewer, firm, and cheaper.

- Visual look on spring hangers, spring saddle's spring compression may reveal the strained piping
- Temperature difference between piping and rotating equipment attached may occur when no fluid is moving

- Vibration increases in pump due to directional force caused by piping strain.
- Movement of piping from its support saddles in any direction may take place according to piping's force direction w.r.t. equipment flange position
- On detaching pipe from equipment flange or from piping joint flanges, pipes dislocation from its position confirm distorted piping joints
- Pipe anchored on equipment and all supports free and sliding
- Frequent misalignment of equipment is symptom of mi-alignment.
- Foundation bolts of equipment getting loose shortly or in due coarse, depending n piping distortion.

12.3 Type of Couplings and Applications

In a drive system, coupling is an important component. Coupling is device to connect two or more shaft of drive system to couple together. Couplings should perform some of the important functions as follows–

- Transmit power from one shaft to other, connected to it.
- Accepts torsional shock and absorb to its capacity provided in design
- It absorbs vibration to some extent
- Allows limited axial float of shaft connected to it in misaligned condition
- Makes assembly and disassembly of shafts easy
- Minimizes lateral load of shafts on bearings
- Coupling balanced properly does not exert any force on shafts.

12.3.1 Allowable Coupling Misalignment

It is the max limit of misalignment of the two connected shaft centre line, specified by manufacturer. However, it does not mean that while fitting the coupling, maximum recommended misalignment deviation should be utilized.

The axial thrust is also transmitted to the motor shaft, and must be accommodated by the motor bearings. Flexible couplings are employed in all cases where the drive and pump shafts are supported independently by journal and thrust bearings. These couplings must be capable of accommodating axial, radial and angular mis-alignments within specified limits.

On pumps with higher speed and power ratings such as boiler feed pumps, mainly couplings employed are full gear/half gear, flexible membrane and diaphragm types.

12.3.2 Couplings

The principal function of the shaft coupling is the transmission of torque from the driving unit (e.g. electric motor with or without gear, steam or gas turbine) to the pump shaft. There are:

- Rigid couplings
- Flexible couplings
- **RIGID COUPLINGS**

These couplings are shell or flange type. These are used primarily where no journal or thrust bearing is provided for the motor shafts (often with vertical pump. The coupling end motor bearing takes radial load of the pump shaft. Rigid couplings are the couplings which connect two shafts to function as single shaft.. Normally such couplings are fitted with fit bolts.. Coupling faces are male – female to avoid any shear force on coupling bolts.

• **FLEXIBLE COUPLINGS**

Couplings connecting two shafts, provides a flexibility to shafts in its axial and lateral movements during running. In case of misalignment of shafts, without flexibility of coupling, the probability of bearings damage, shaft bending and sever damage may not be ruled out. Flexible Couplings are designed to protect shafts of drive system by self sacrificing.

Different type of flexible couplings are shown in following sketches.

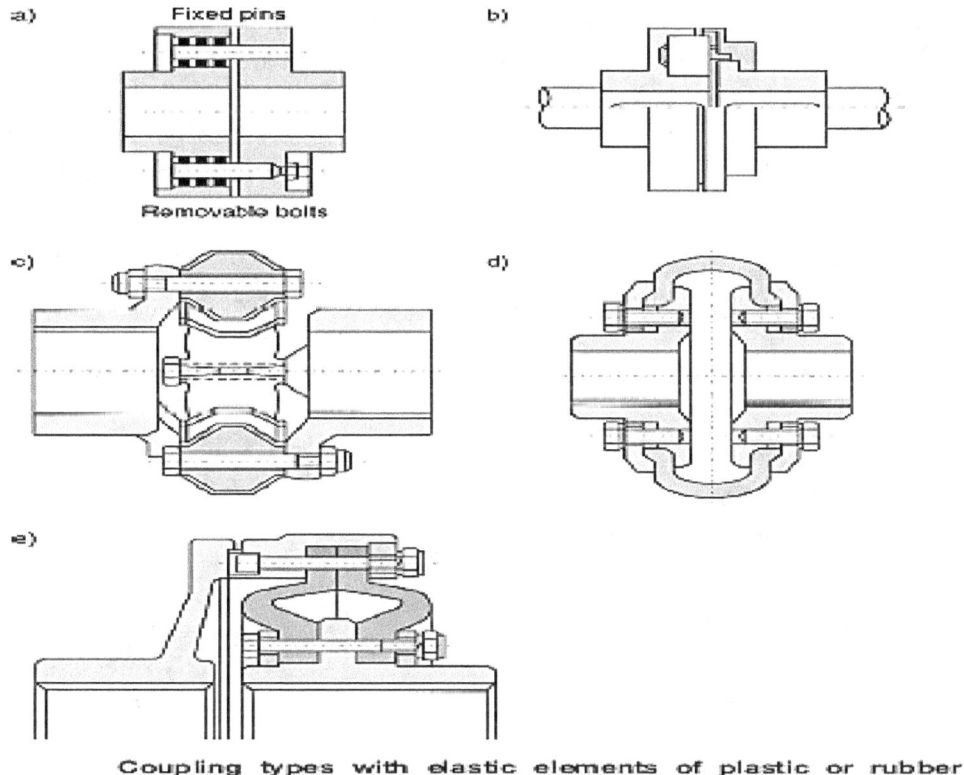

Coupling types with elastic elements of plastic or rubber

12.3.3 Different Type of Couplings and Applications

COUPLINGS	PICTURE OF COUPLINGS	APPLICATIONS
Miniature Flex Coupling		Miniature couplings are used for mini drives, instruments, electrical panel drives, limit switches etc. These couplings are meant for light duty applications

(Contd.)

COUPLINGS	PICTURE OF COUPLINGS	APPLICATIONS
Chain Type Flex coupling	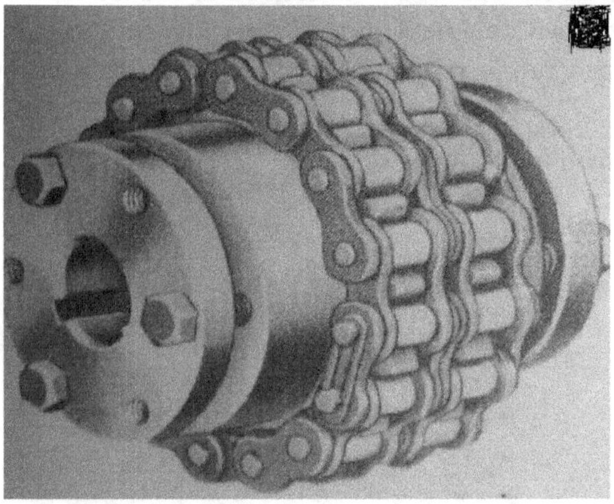	This coupling constitute of two identical sprockets, coupled with dual roller/silent chain. Sealed sprocket covers are used to protect chain and sprocket. Grease is packed into cover for lubrication. This type of couplings are able to provide large misalignment flexibility but limited torque capacity. Available capacity for roller chain = 1000 HP@1500 RPM and for silent chain = 3000 Hp@5000 RPM, maximum bore = 200 mm dia, max chain width = 6 mm
Floating Gear Coupling	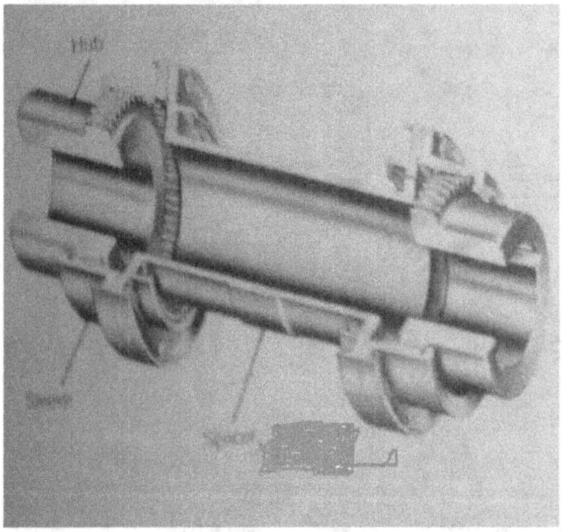	This coupling is similar to Gear coupling, divided into two half gear couplings with a small floating shaft to provide two flex points, resulting more capacity to take misalignment. Gear coupling consist of 2-hubs, 2-hub covers, fasteners. Crowned gear teeth allows misalignment. Max limits: capacity = 70,000 HP, 50,000 RPM, Shaft bore = 30", shaft-spacing = 200". Misalignment of rotating shafts directly effects wear pattern of gear teeth as shown here. Excessive misalignment will cause loading of gear at addendum portion, leading to knife edging of teeth.
Universal Coupling	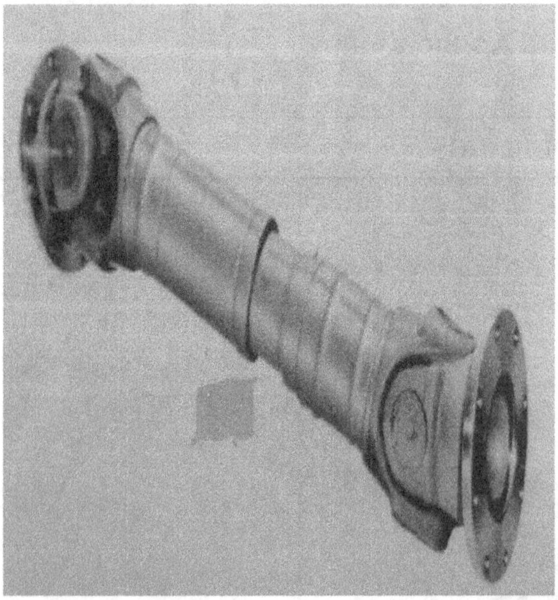	This coupling is most popularly used for connecting two shafts at angle and at a distance, or two shafts with sever angular misalignment. Two crosses used in universal coupling make two flex points which makes it suitable to be used to connect two shafts with sever angular and radial misalignments. In misalignment situation, if one universal joint is used, it will cause Cardon error but that with two universal joints it is balanced and causes sinusoidal motion in torsional direction. For correct assembly of universal shaft, both end coupler should make an equal (ie. enterance angle equals to exit angle) less than limit of universal angle.

COUPLINGS	PICTURE OF COUPLINGS	APPLICATIONS
Gear Type Coupling	 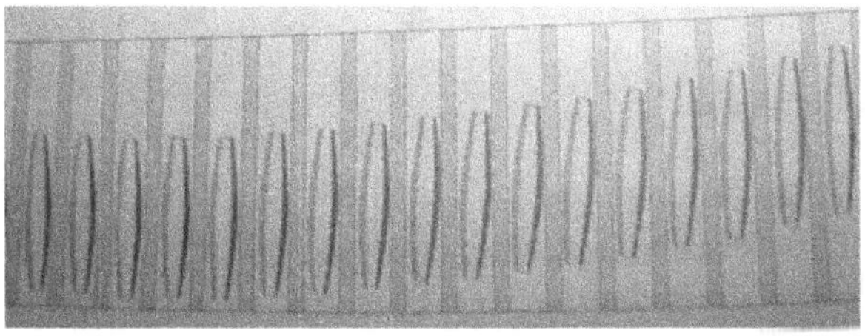	Gear coupling consist of 2-hubs, 2-hub covers, fasteners. Crowned gear teeth allows misalignment. Max limits: capacity = 70,000 HP, 50,000 RPM, Shaft bore = 30", shaft-spacing = 200". Misalignment of rotating shafts directly effects wear pattern of gear teeth as shown here. Excessive misalignment will cause loading of gear at addendum portion, leading to knife edging of teeth. Allows axial movement of shaft, capacity to work at high speed, low overhung weight, good dynamic balance, long life with proper lubrication are normal features of gear coupling.
	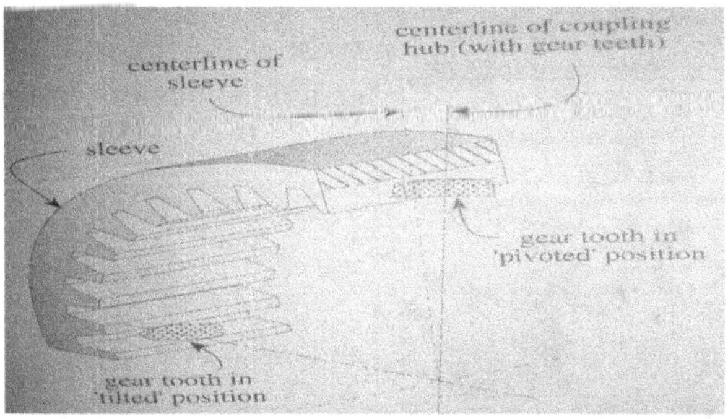	Difficult to find axial movement, reaction force in usage of turbo machineries.
Flexible Disc Type Coupling	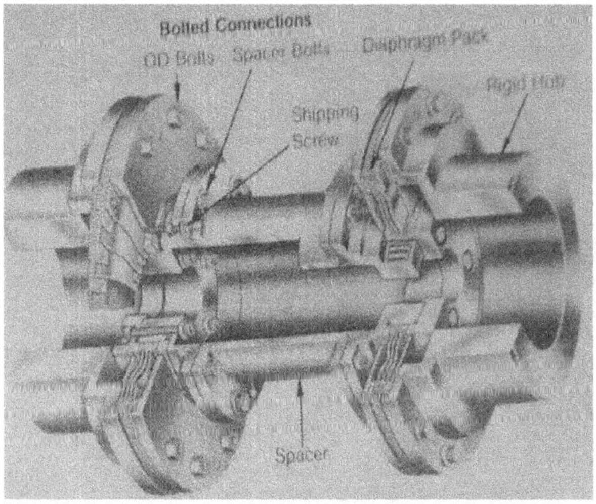 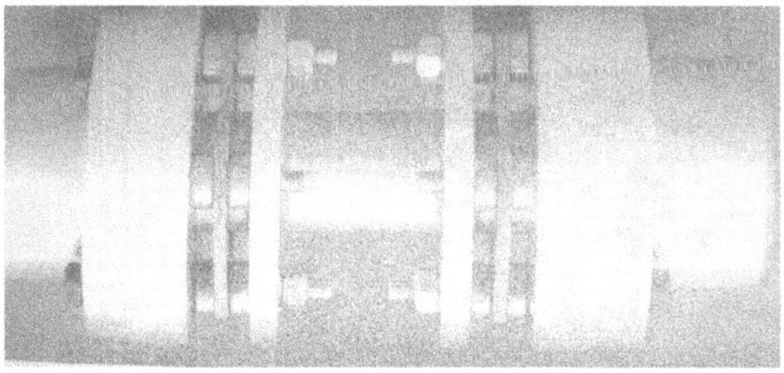	Metallic membrane or disc type couplings transmit power from one shft to other through two flexible elements (assembly of metallic shims), each bolted to the outer rims of shaft hubs and connected via a spacer tube. Misalignment and axial displacement is accomplished by flexing of metallic membranes. Max rating of such couplings are available as follows– Capacity – 30,000 HP, speed = 30,000 PM, shaft bore = 200 mm, shaft spacing = 200 mm. Now a days this type of couplings are very popularly used in TG and Turbo Blower machines with high reliability. In case of misalignment and axial shift beyond the limits, in transmission of power, elements break. In design, it should be taken care that natural frequency of element should not resonate with running RPM. It has excellent balance characteristic, and high temperature applications.

(Contd.)

COUPLINGS	PICTURE OF COUPLINGS	APPLICATIONS
Tyre Coupling		Elastomeric, a natural or synthetic nylon reinforced rubber or Teflon, medium is used to transfer torque from one shaft to other for soft torque transmission. It absorbs high starting torque or shock load. Max rating of such couplings is available as follows– Capacity = 67,000 HP@ 100 RPM, Speed = 5000 RPM, bore size = 30,' shaft spacing = 100" Elastomeric has almost no wear, acts as vibration & shock absorber, electrical insulator and soft torsion transmitting element. It accepts some axial movement of shaft. No lubrication is required. It has limitation of speed due to centrifugal force of elastomeric member and heat generated due to cycle of flexing.
Wood's Flex Coupling		
Lovejoy Flex Coupling		
KTR Rubber Disc Flex Coupling		

COUPLINGS	PICTURE OF COUPLINGS	APPLICATIONS
Pin Drive Type Coupling		Series of metal pins with springs are fitted at outer periphery (@PCD) holes. Spring pieces can swivel on pins to allow movements. Spring and pins fit sliding into hole of counter coupling hub. Max rating of coupling: Capacity = 3800 HP@100 RPM, max speed = 4000 RPM, bore = 13", shaft spacing = 13 mm. Counter coupling hub holes may be fitted with bz bushings. It can accommodate 10–12 mm axial float of shafts but with limited misalignment resistance in running.
Flexible Link Type Coupling		Flexible link coupling utilises series of crossed laced matched links with one end of each links connected to a disc mounted on driven shaft and other end to driver shaft. Links are connected such that during torque transmission, when in one link is intension that time other link is in compression. Misalignment and axial displacement are accomplished by flexing action in the series of cross links. Max rating of coupling available is- Capacity = 100 HP @ 100 RPM, Speed = 1899 RPM, Shift bore = 20," shaft spacing = 20." It has limited misalignment tolerance capacity
Metal Ribbon Coupling	 Mechanical Puller for taper fit	Popularly known as Bibby coupling. It constitute of two slotted teeth hubs, one metal ribbon & sealed cover with fasteners, Metal ribbon is made of hardened high tensile spring steel. Ribbon is fitted into slots of two hubs, mounted on two shafts to be coupled together. A gap of 4 h to 12 mm is created between faces of two hubs. Here steel ribbon is the weakest part to break on high to torque. Grease is packed into cover to reduce friction of steel ribbon relative movement with hub teeth while transmitting power/ torque fom one shaft with other. Max rating of couplings available are-capacity = 70,000 HP@ 100 RPM, max speed = 6000 RP, shaft bore = 20", shaft spacing = 12" It is easy to assemble and dismantle and torsionally soft. Due to lubricated application, it can't be used for high temperature application.

(Contd.)

COUPLINGS	PICTURE OF COUPLINGS	APPLICATIONS
Rigid Coupling		Rigid couplings applications is being made since flexible couplings were not in existence. It is used for coupling two or more shafts to work as single shaft in running. Combined behavior of rigidly coupled shaft is lie a single shaft of coupled length. This type of coupling is very extensively used in vertical drives eg. Vertical pumps. This is suitable for coupling shaft with minimum misalignment of shafts. If two shafts with mis alignment is coupled together, it will behave like a bend shaft in running.
Leaf Spring Type Coupling		This coupling employs radially spaced series of leaf springs, attached to an outer drive member and indexed into axial grooves into inner drive member. The space around each spring set is filled with oil. When spring pack is deflected, damping effect occurs in oil flow from one side of spring pack to that of other side. Max rating of this coupling s available as follows- Capacity = 15,000 HP @ 100 RPM, max speed = 3600 RPM, Bore = 12," shaft spacing = 40." I is specially designed for diesel and reciprocating engines. Its toque capacity depends on max deflection of springs 9 limited value). Combination of spring stiffness cn be installed in a coupling for different torque load design. It is soft to starting torque. It has good freedom to axial movement of shafts and limited temperature applications.

12.4 Fluid Coupling

APPLICATIONS

- To start driven m/c with low starting torque.
- It works as fully flexible cushion between driver and driven m/c as it connects the two coupled m/cs through fluid.
- Fluid coupling consist primarily a small units of centrifugal oil pump, mounted on driver shaft and hydro -turbine, mounted on driven m/c shaft. Hydraulic oil in the sump of coupling works as media for pump and turbine units.

- While starting of driving motor, oil is lifted by pump and electrical energy is converted into kinetic energy is applied on a mass of which is transferred to guide vane of hydro turbine and kinetic energy of oil start converting into mech. energy as a torque which drives the m/c. Thus in starting of m/c, slow speed acceleration of driven m/c takes place which reduces starting torque of motor.

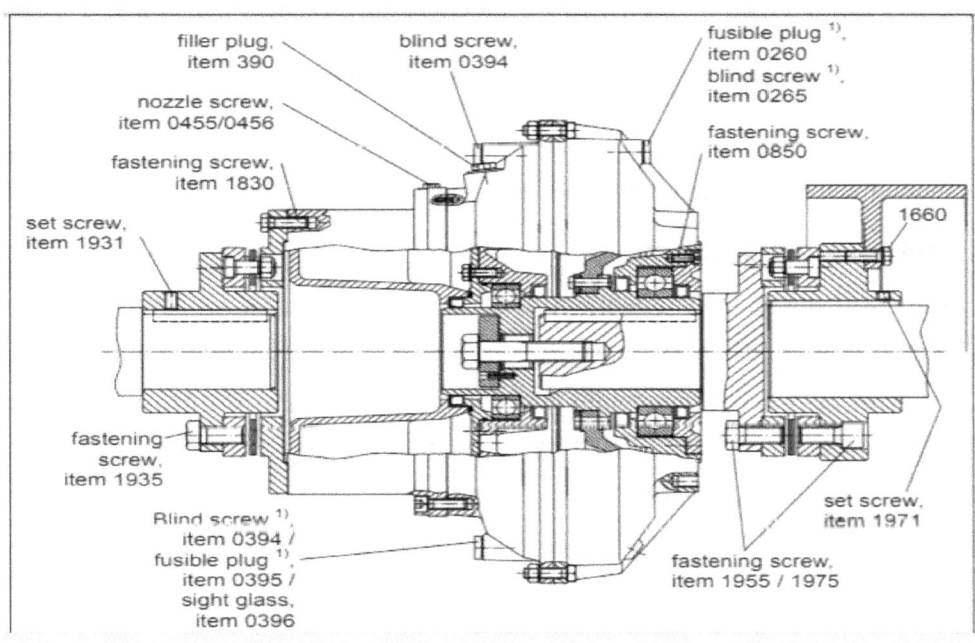

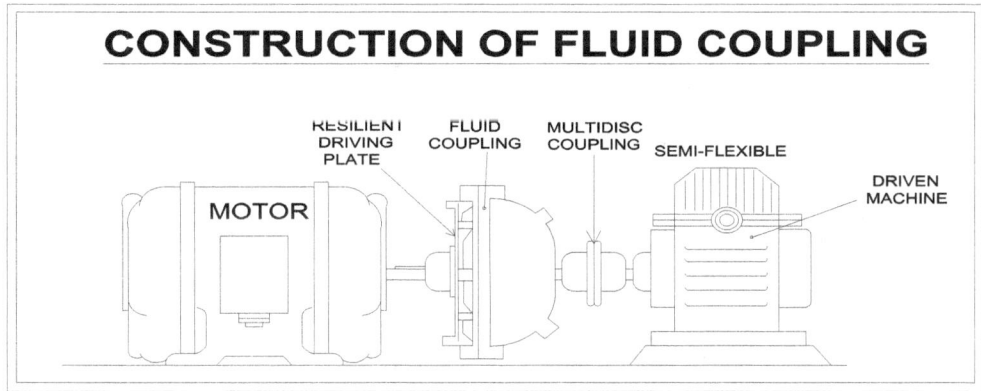

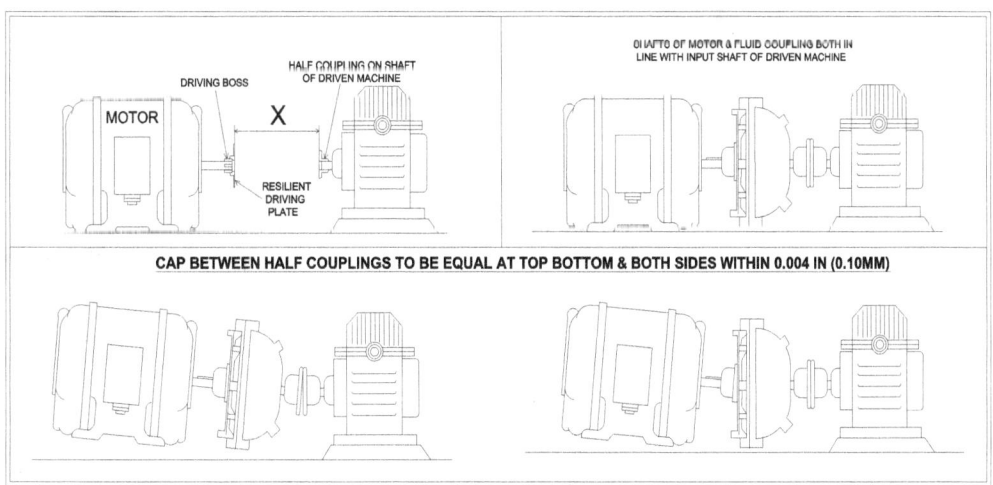

12.4.1 Fluid Coupling has Some Safety and Protection Devices

a. Fusible plug which melts when driven m/c is stalled or running jammed. This results churning of oil which will rise temp. of oil and the plug made of lead will melt & oil is drained out. In absence of oil, transmission between pump and turbine is dis-connected, resulting standstill position of driven m/c.

b. Resilience disc- resilience disc is mounted with driven m/c coupling which gives flexibility in case of misalignment and also function as weak link of system to break in case of jamming of coupling due to any reason.

c. Multi disc coupling: It is mounted on driven m/c to give flexibility and weak link to drive system to take care of any misalignment or jamming effect caused by some reasons.

12.4.2 Construction of Fluid Coupling

Input side: Impeller & casing – made of high tensile alloy.

Output side: runner shaft- made of runner of high tensile alloy.......shaft of 42 Cr Mo 4.

- Shaft carried in ball and roller bearings into casing and impeller.
- There is no mechanical connection but hydraulic connection between impeller and runner.
- A gland seal is fitted between the casing and the runner shaft.
- Impeller behaves like centrifugal pump and runner as turbine.

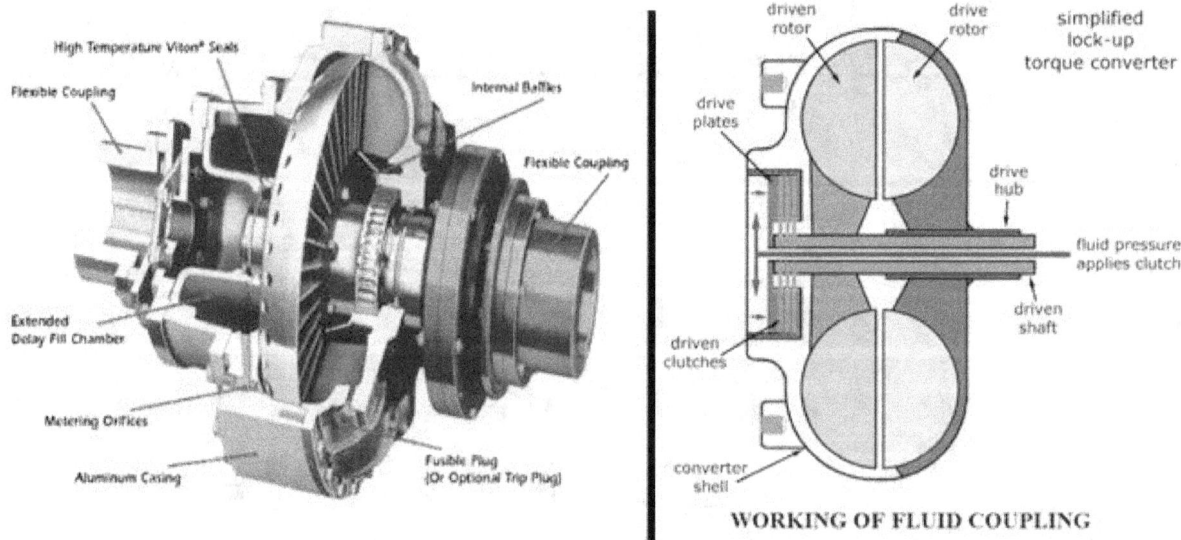

WORKING OF FLUID COUPLING

- When motor is switched on, the fluid coupling has no torque capacity. As motor accelerates the coupling torque start increasing. Thus motor starts taking load and run up to speed quickly, while the torque of the fluid coupling increases smoothly to start the m/c.

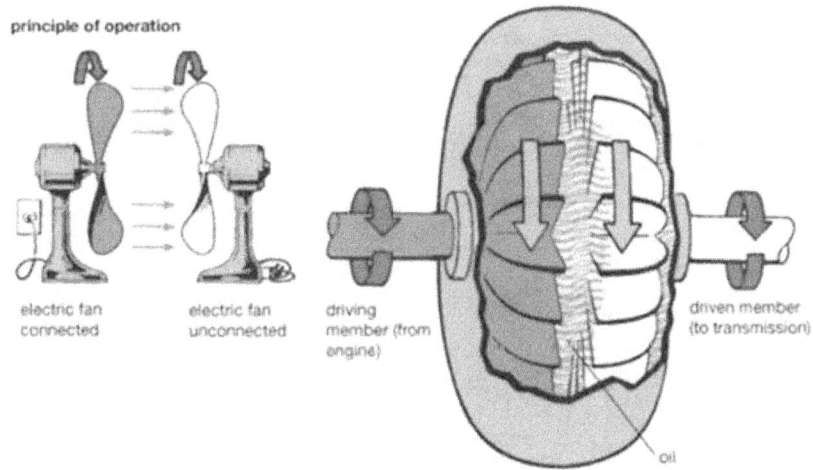

principle of operation

electric fan connected   electric fan unconnected   driving member (from engine)   driven member (to transmission)   oil

FLUID COUPLING

- Due to misalignment of coupling, life of bearings is decreased and whole installation may tend to fact.

12.4.3 Checking Points before Alignment of Fluid Coupling

a. Coupling hub of motor should be interference fit, mounted by oil heating. Key should have interference (as specified) and key top clearance of min 0.5 mm. In no case key should be loose on shaft.

b. Check run out of both coupling s mounted on driver to be within one mil. Driving shaft spigot should run true within (± 1 mil)

c. Resilience disc distortion need to be checked and if found more than ± 20 mils, replace it. For any hair crack, replace with new one.

12.4.4 Installation of Fluid Coupling

1. Mount Driving Boss and output coupling half on motor & driven machine respectively. (half coupling fit with light interference).

2. Check Trueness of Driving Boss of Half Coupling.   -------- 0.002"(0.05 mm)

3. Check Squareness of Half Coupling Face with Driven Shaft.   -------- 0.002" (0.05 mm)

4. Check Trueness of Motor Shaft   ------- 0.002" (0.05 mm)

5. If Rotor of the Motor is not located keep in Magnetic centre for Entire Process of Alignment and Checking.

6. Erection:

   a. Detach Resilience Disc from Fluid Coupling and Bolt Up with Input Coupling on Motor.

   b. Put Motor Approx Spaced from Output Coupling Half and aligned to Driven m/c.

   c. Maintain Distance between Input and Output coupling halves as specified with tolerance of +0.020"–0.000 (0.5 mm)

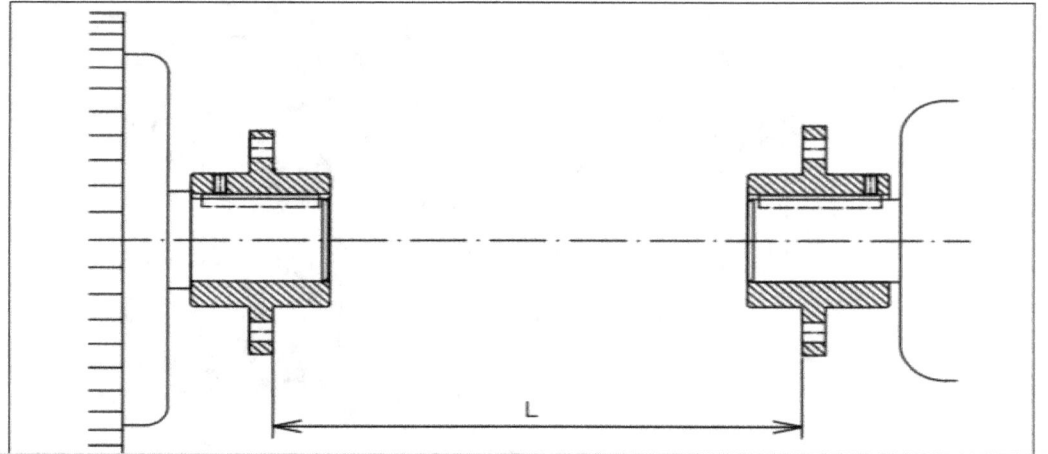

d. Remove protection coating from spigots.

e. Position Fluid Coupling in to space between motor and driven m/c. coupling halves. Centre the spigot in register in the driving boss on motor shaft.

f. Support the weight of Fluid Coupling temporarily.

g. Fix up slide disc plates and bolts.

h. Remove temporary support and other attachments to the fluid coupling.

• Color coding of Fusible plug–

Nominal response temperature	Color coding
95°C	without (tinned)
110°C for operating fluid	yellow
125°C	brown
140°C	red
160°C	green
180°C	blue

12.4.5 Alignment of Fluid Coupling

a. The fluid coupling itself is rigid unit, the shaft being fully supported in the ball and roller bearings. The fluid coupling is therefore provided with flexible mountings at input and output ends, so that it can work as cordon shaft between the motor and driven shaft.

b. In setting up the drive, the shaft of the motor, fluid coupling and driven m/c. must be brought in one line.

c. Normally the Gear Box preceding to driven m/c is assumed to be fixed in position and the driving machine is aligned with output shaft of gear box. Further, the motor is aligned with the input shaft of gear box using fluid coupling between motor & gear box.

12.4.6 Steps in Alignment

• By moving the motor up and down and sideways, bring the shaft of the fluid coupling in line with the input shaft of driven m/c. Check gaps between input shaft coupling half of driven m/c. and output coupling half of fluid coupling at 4 sides by use of taper gauge/filler gauge. It should be true within 0.004" (0.10 mm).

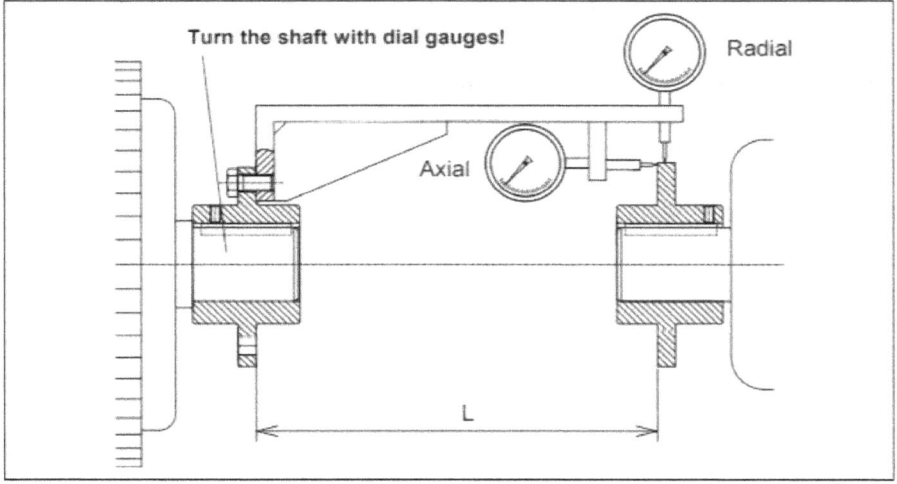

- The Shaft of Fluid Coupling now being in line with the shaft of driven m/c. within required limits, the position of free end of the motor must now be adjusted to bring the deflection of resilience driving disc to the acceptable working value. (0.004" or 0.1 mm). To measure the deflection of the disc, clamp a bar to driving Boss and attach a dial gauge with spindle of gauge on the head of one of the driving bolt. Turn the fluid coupling by hand and note dial gauge readings on one revolution. This should be true within 0.004".

Alignment tolerances in mm		
Coupling size	maximum permissible radial dial gauge deflection (radial)	maximum permissible axial dial gauge deflection (axial)
366, 422	1,0	0,3
487	2,0	0,7
562	3,2	1,0
650, 750, 866	4,6	1,5

- Disk deflection in operation may disturb the alignment of fluid coupling shaft with driven machine. shaft by a small amount. Therefore, check the gaps between the two coupling halves again and adjust the gap between motor and driven machine couplings.

12.4.7 Procedure of Alignment

Put the motor on its base frame and align roughly coupling of driver and driven machine.

a. Keep the coupling distance as specified within + 20 mils (distance is fixed for given coupling) for ease of multi disc assembly.

b. Fix the resilience discs with fluid coupling with the help of its srews and check with torque wrench for proper torque tightening.

c. Fluid coupling assembly is a rigid, with impeller & turbine rotors, supported on antifriction bearings. Therefore, flex coupling on driven shaft and resilience disc on drive end shaft is provided to give full flexibility and weak links of coupling system.

d. Now, mount the dial gauges and align driver & driven shaft within ± 2 mil by shimming at feet of driver and maintain coupling gap as specified above.

e. Install fluid coupling assembly with resilience disc mounted on driver coupling with the help of screws and tighten screws by torque wrench.

f. Now mount dial base on driver shaft and its stylus on one of the head of screws, fastening fluid coupling with resilience disc and take dial readings at four positions (90 deg apart). It will show sagging of coupling assembly downwards. Note it for correction of alignment at driver.

g. Verification of alignment–

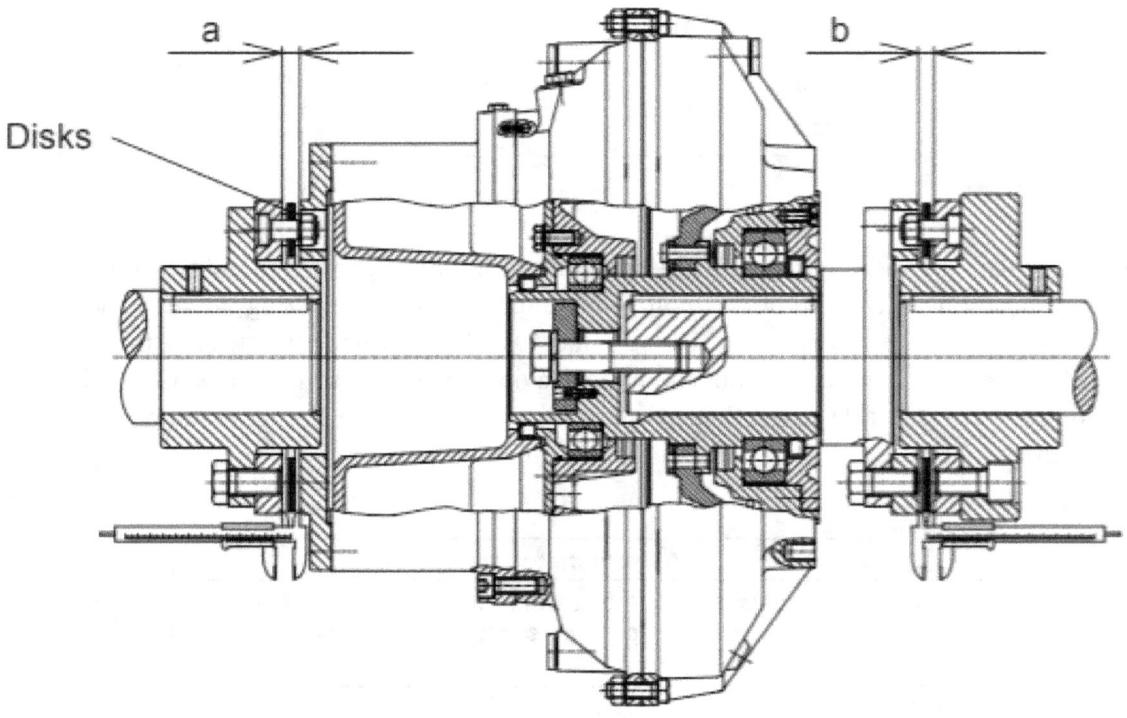

a: Distance between flanges of input side disk pack.
b: Distance between flanges of output side disk packs.
a_{min}, b_{min} : minimum value of a or b.
a_{max}, b_{max} : maximum value of a or b.
Δa: $a_{max} - a_{min}$.
Δb: $b_{max} - b_{min}$.

■ Measure a and b going around the total circumference of relevant disk pack in 45° steps, without rotating the shafts or the coupling. Compare the measured values with the following table:

Dimensions to check alignment in mm		
Coupling size	a = b	Δa = Δb
366 ˙	9,5 … 10,3	≤ 0,55
422 ˙	10,4 … 11,45	≤ 0,55
487 ˙	12,75 … 14,85	≤ 1,35
562 ˙	13,25 … 16,35	≤ 2,1

12.4.8 Filling Angle with Inclined Shaft Fluid Coupling

- Property of oil to be filled in fluid coupling–

  - Viscosity class .... ISO VG 32 to DIN 51519 *)

  - Starting viscosity............... less than 15000 mm²s⁻¹ (cSt)

  - Pourpoint .... the limit is 4°C below actual minimum ambient temperature or lower

  - Flash point .... greater than 180°C and at least 40°C above nominal response temperature of fusible plugs

  - Fire point........................... at least 50°C above max. surface temperature
    (only relevant for coup-
    lings used in hazardous
    areas (⟨Ex⟩))

  - Resistance to aging .... aging-resistant refined product

  - Compatibility with seals ..... NBR (Nitril-Butadien caoutchouc) and FPM/FKM (fluor caoutchouc)

- Proposed operating Fluid –

Manu-facturer	Designation	Pour-point in °C	Flash point in °C	Fire point in °C	Class	FE8-test satisfied
Addinol	Hydraulic Oil HLP 32	-21	195		HLP	
Agip	Agip Oso 32	-30	204		HLP	
	Agip Blasia 32	-29	215		CLP	
Aral	Degol BG 32	-27	200	250	CLP	
Avia	Avia Fluid RSL 32	-27	214	237	HLP	
	Gear RSX 32 S	-33	210	231	CLP	
BP	Energol HLP-HM 32	-30	216		HLP	
Castrol	Hyspin SP32	-28	200		HLP	Yes
	Hyspin AWS 32	-27	200		HLP	
CEPSA	HIDROSIC HLP 32	-24	204		HLP	
	EP 125	-30	206		HLP	
ExxonMobil	Nuto H32	-24	212		HLP	
	DTE 24	-27	220		HLP	
	Mobilfluid 125	-30	225		CLP/HLP	
	Mobil SHC 524	-54	234	234	HLP	
Fuchs	Renolin MR10	-30	210		HLP	
	Renolin B10	-24	205		HLP	
Klüber	Lamora HLP 32	-18	200		HLP	
Kuwait Petroleum	Q8 Haydn 32	-30	208	232	HLP	
	Q8 Holst 32	-30	208	234	HLP	
Optimol	Hydo MV 32	-38	209	234	HLP	

- In a drive where the coupling shaft is not level, evelled (i.e. inclined shaft) then the angle to which the plug hole is set for filling must be altered to compensate as shown below–

  The recommended filling angle must be modified to give the correct filling level when shaft is inclined at angles of up to 30° upwards or downwards.

For example, a recommended filling of 45° and with the shaft sloping at 15°, the corrected angle would be 58° if the shaft is inclined upwards 29° and slope is downwards.

- Oil Level Check: Turn the coupling till oil is visible in sight glass and measure the angle of centre line (radially) of sight glass. passing radially. Compare with that recommended value and adjust oil quantity if necessary.

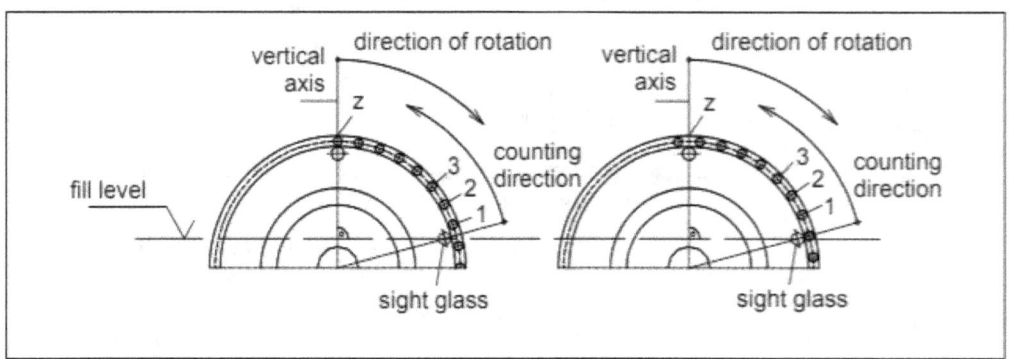

12.5 Class-I Key (Side and Top Clearance Fit)

Nominal shaft dia (inches)		Nominal key size (inches)			Nominal key seat depth (H/2) Inches	
		Width (W)	Height H			
From	To	Width (W)	Square	Rectangular	Square	Rectangular
0.31250	0.43750	0.09375	0.09375		0.04688	
0.43750	0.56250	0.12500	0.12500	0.09375	0.06250	0.04688
0.56250	0.87500	0.18750	0.18750	0.12500	0.09375	0.06250
0.87500	1.25000	0.25000	0.25000	0.18750	0.12500	0.09375
1.25000	1.37000	0.31250	0.31250	0.250000	0.15625	0.12500
1.37000	1.75000	0.37500	0.37500	0.250000	0.18750	0.12500
1.75000	2.25000	0.50000	0.50000	0.375000	0.25000	0.18750
2.25000	2.75000	0.62500	0.62500	0.43750	0.31250	0.21875
2.75000	3.25000	0.75000	0.75000	0.50000	0.37500	0.250000
3.25000	3.75000	0.87500	0.87500	0.62500	0.43750	0.375000
3.75000	4.50000	1.00000	1.00000	0.75000	0.5000	0.43750
4.50000	5.50000	1.25000	1.25000	0.87500	0.62500	0.50000
5.50000	6.50000	1.50000	1.50000	1.00000	0.75000	0.62500
6.50000	7.50000	1.75000	1.75000	1.50000	0.87500	0.75000
7.50000	9.00000	2.00000	2.00000	1.50000	1.00000	0.75000
9.00000	11.0000	2.50000	2.50000	1.75000	1.25000	0.87500

Note: Square keys are preferred for shaft sizes above line and that below line rectangular keys

12.5.1 Class-II (Key – Fit – Minimum Possible Interference, at Sides)

TYPE OF KEY	KEY WIDTH (INCHES)		SIDE FITS (MILS) NOTE: 1.0 MIL = 0.001 INCH = 0.0254 MM			KEY	FIT TOLERANCE(MILS)		FIT RANGE
	From	To	Width tolerance		Fit range		Depth Tolerance		
			Key	Key seat			Shaft key seat	Hub key seat	
Square Key	Upto	0.500	+ 0	+ 2	Class-4	+ 0	+ 0	+ 10	Class-32
			– 2	– 0	0	– 2	– 15	– 0	Class-5
	0.500	0.750	+ 0	+ 3	Class-5	+ 0	+ 0	+ 10	Class-32
			– 2	– 0	0	– 2	– 15	– 0	Class-5
	0.750	1.000	+ 0	+ 3	Class-6	+ 0	+ 0	+ 10	Class-33
			– 3	– 0	0	– 3	– 15	– 0	Class-5
	1.000	1.500	+ 0	+ 4	Class-7	+ 0	+ 0	+ 10	Class-33
			– 3	– 0	0	– 3	– 15	– 0	Class-5
	1.500	2.500	+ 0	+ 4	Class-8	+ 0	+ 0	+ 10	Class-34
			– 4	– 0	0	– 4	– 15	– 0	Class-5
	2.500	3.500	+ 0	+ 4	Class-10	+ 0	+ 0	+ 10	Class-34
			– 6	– 0	0	– 6	– 15	– 0	Class-5
Rectangular key	–	0.500	+ 0	+ 2	Class-5	+ 0	+ 0	+ 10	Class-33
			– 3	– 0	0	– 3	– 15	– 0	Class-5
	0.500	0.750	+ 0	+ 3	Class-6	+ 0	+ 0	+ 10	Class-33
			– 3	– 0	0	– 3	– 15	– 0	Class-5
	0.750	1.000	+ 0	+ 3	Class-7	+ 0	+ 0	+ 10	Class-34
			– 4	– 0	0	– 4	– 15	– 0	Class-5
	1.000	1.500	+ 0	+ 4	Class-8	+ 0	+ 0	+ 10	Class-34
			– 4	– 0	0	– 4	– 15	– 0	Class-5
	1.500	3.000	+ 0	+ 4	Class-9	+ 0	+ 0	+ 10	Class-35
			– 5	– 0	0	– 5	– 15	– 0	Class-5
	3.000	4.000	+ 0	+ 4	Class-10	+ 0	+ 0	+ 10	Class-36
			– 6	– 0	0	– 6	– 15	– 0	Class-5
	4.000	6.000	+ 0	+ 4	Class-12	+ 0	+ 0	+ 10	Class-38
			– 8	– 0	0	– 8	– 15	– 0	Class-5
	6.000	7.000	+ 0	+ 4	Class-17	+ 0	+ 0	+ 10	Class-43
			– 13	– 0	0	– 13	– 15	– 0	Class-5

12.5.2 Shrink Fitting of Coupling on Shaft

• Measure coupling bore and shaft fitting area diameter by micro meter and check interference which should conform as per tables above.

- For shrink fitting, coupling will be heated up in transformer oil to temperature as per following calculation–

$$\Delta T = i/[\alpha\,(d\text{-}0.002)]$$ where; i = interference in mils

ΔT = coupling hub temperature(heated)- amb.temp. °F

α = co efficient of expansion of coupling material

d = coupling hub bore diameter in inches

- **Taper Bore – Interference Fit:**   **HT = (12 * I)/(ST)**

  where: HT = Distance coupling hub must travel to provide an interference equal to one mil

  I = Interference fit (normally 1 mil)

  ST = Shaft taper (inch/ft)

12.5.3 Coupling Fitting on Shaft with Interference Fit

Set Screw- Key Sizes and Interference		
Shaft Dia (inch)	**Interference Fit (mils)**	**Remarks**
½ to 2	0.5 to 1.5	1.0 mil = 0.001 inch
2 to 6	2 to 5	
More than 6	0.1 to 0.35 mils/inch diameter of shaft	

12.5.4 Purpose of Interference Fit

- To prevent fretting corrosion which occurs by small amount of movement between shaft and coupling hub.
- To prevent hub from slipping on shaft when max amount of torque is experienced during startup of equipment or high running load.

12.5.5 Max Amount of Shearing Stress on Rotating Shaft "T" Is given as T = 462009 * P/N

Where: T = Torque in kg M,   P = HP (metric),   N = RPM

Allowed torsional stresses for carbon steels–

AISI – 1040	AISI – 4140	AISI – 4340
5000 psi	10000 psi	11000 psi

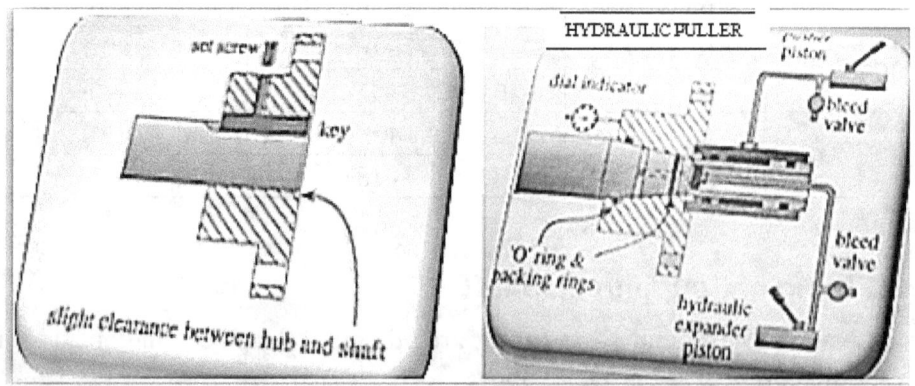

12.5.6 Interference Fit in Shaft & Hub, Hydraulically Fitted

- The surface contact of hub bore and shaft should be 65–70%

- Ensure matching of oil passages for coupling expansion through pressurized oil.

- Install o- ring seal into hub and shaft. Install dial indicator to measure movement of coupling and in turn shrinkage of hub and shaft. Knowing taper angle and movement of hub we can calculate interference-

$$\Delta y = x * \tan \theta$$

where; x – axial movement of hub (mm)

θ = Taper angle of shaft and hub bore

Δy = Interference of fitting (mm)

- Normally pressure applied hydraulically will be 120–250 bar for push up of hub on shaft

- Apply pressure slowly in gradual increase when expansion of hub is on.

- Ensure correct travel of hub, calculate actual interference obtained. hold pressure for pusher assembly and bleed off pressure ot expansion line. Release pusher piston when pressurised oil is completely drained off.

- After removing the hydraulic tackle (pusher and expander assembly) put face lock if provided.

12.6 Severity Graph of Misalignment

* Axial/Radial misalignment max dial reading out of four, say = 6 mils (0.006")

* Distance of two power transmission points (coupling faces distance), say = 4 inch

* Max **misalignment deviation** can be computed as → 0.006/4 ie. **1.5 mils per inch**

- The graph is plotted for alignment deviation against RPM of m/c indicates the severity of misalignment.

- The graph given below is divided in three zones–

  a. Excellent- If alignment deviation value against RPM of m/c falls in this zone, no action of alignment is needed.

  b. Acceptable- If alignment deviation value against RPM of m/c falls in this zone, it is acceptable but in any opportunity alignment should be done as per tolerance.

  c. Alignment Necessary- If alignment deviation value against RPM of m/c falls in this zone, m/c should be taken out of service and misalignment rectified.

12.6.1 How to Use Severity Graph for Misalignment Acceptable Limit

From the modeling graphic technique or other method virtually we observe the maximum the relative position of centre lines of rotation of the two or more shafts. Once the position ot shafts is established on graph, it is first step to determine the misalignment in terms of deviation (mils/inch) as shown in sketch. Knowing the value of deviation and operating RPM, one can find severity zone from reference graph–

  a. Excellent Zone

  b. Acceptable Zone

  c. Alignment Required Zone

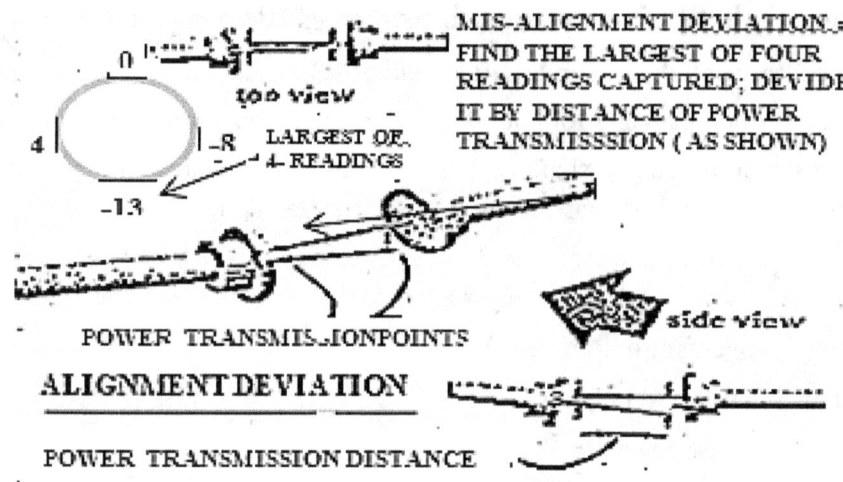

ALIGNMENT DEVIATION

POWER TRANSMISSION DISTANCE

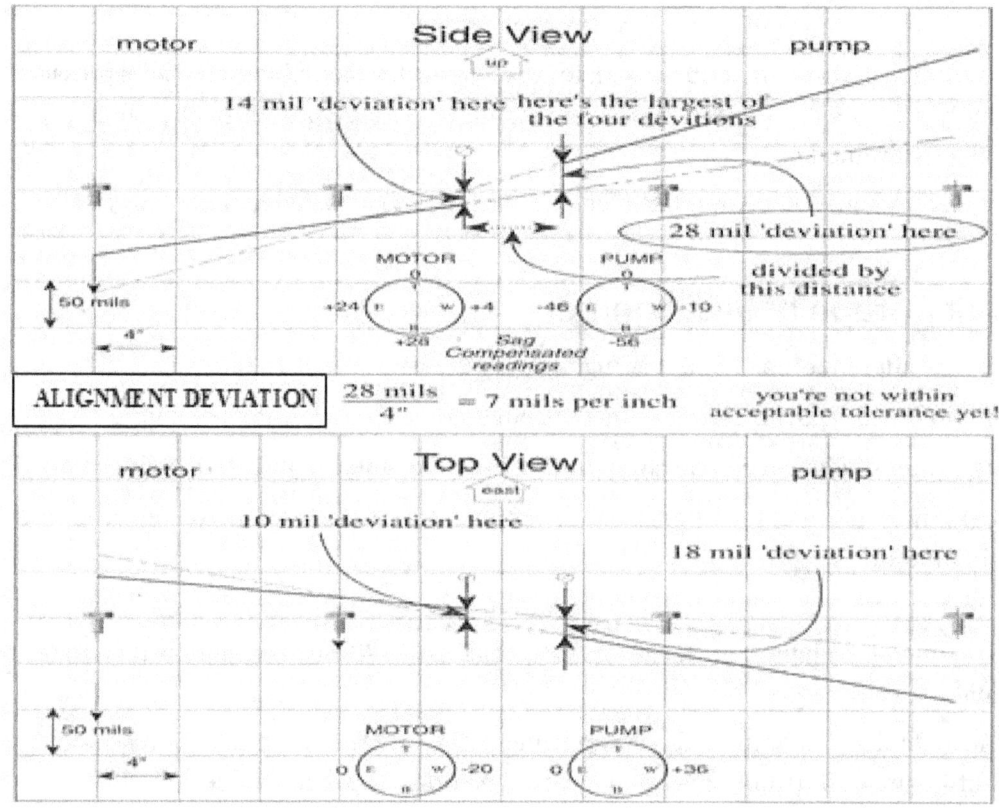

GRAPHICAL EXAMPLE FOR CALCULATION FOR ALIGNMENT DEVIATION

It is normal that some amount of mis-alignment will exist in drive system but that should be within tolerance limit as per graph. Deviation at flex point (largest of four readings divided transmission distance) plays a key role in alignment severity. The max misalignment deviation in vertical and horizontal plane of flex point is determined from elevation & top view field readings (graphical modeling). Misalignment magnitude at different points may be different but deviation (mils/inch) will be same for any point of coupled system.

    i. Check misalignment by three dial method as presented in following discussion and record the readings in specified format

    ii. Find out largest deviation of four readings radial or axial. However, for heavy rotor m/cs axial alignment is important hence largest deviation of axial readings need to be chosen

iii. Measure the transmission points distance ie minimum coupling plane faces distance of four points of measurements

iv. Calculate max deviation of misalignment as explained above

v. Note RPM of transmission couplings, under check

vi. From two values- RPM and max misalignment deviation find point of inter section and check that point falls in which zone of the three – Excellent/acceptable/Alignment necessary.

vii. Take necessary steps as per recommendations of observed zone of m/c misalignment

Note RPM of transmission couplings,

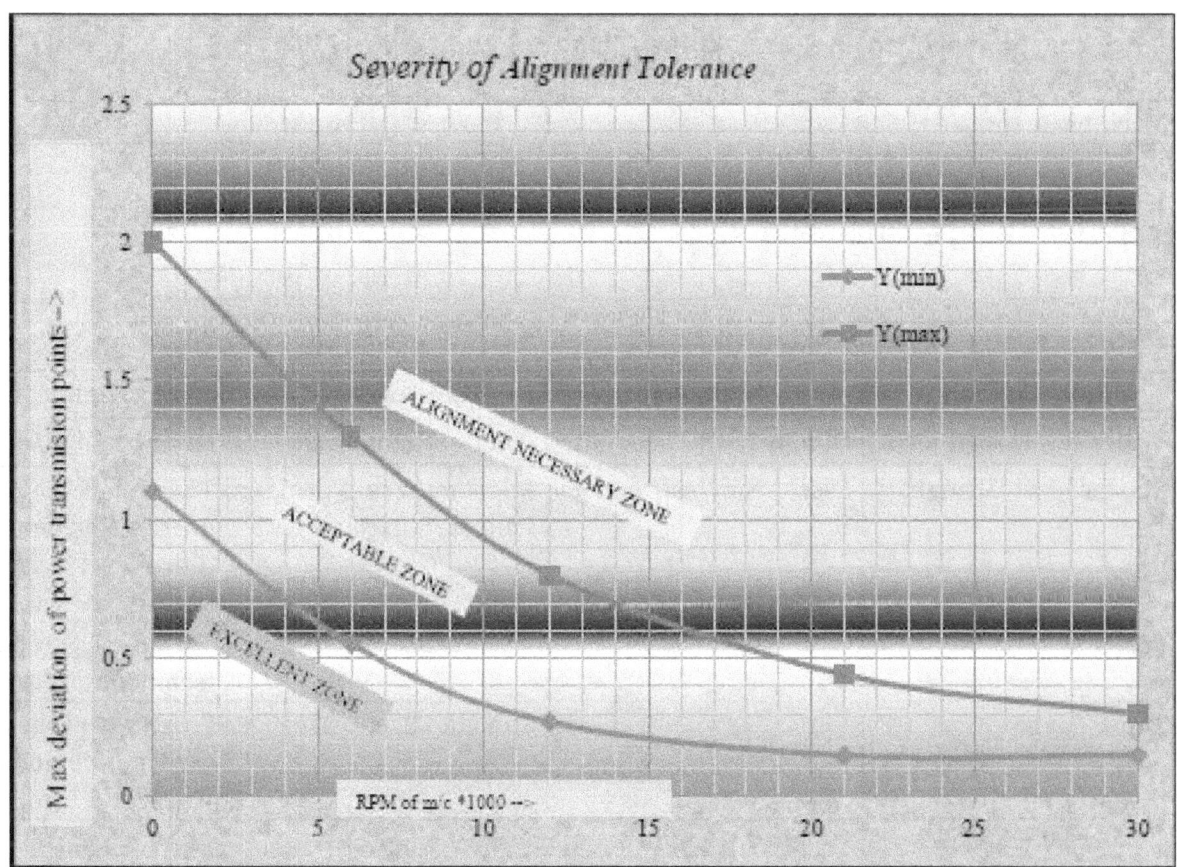

viii. From two values- RPM and max misalignment deviation Value, find point of inter section and check that point falls in which zone in severity graph?

ix. Take necessary steps as per recommendations above for observed zone of m/c misalignment.

12.6.2 Leveling of Machineries

Leveling of base frame or components of machine is process of making axis of rotation of machine perpendicular to the direction of earth gravitational force or in other words axis of rotation of machine is required to be made parallel to the tangent of earth surface at installation. Objective of leveling is to position the machine such that gravitational forces on spinning rotor are symmetrically equalized. However, absolute leveling may not be feasible due to various reasons, a tolerance values based on severity of rotating machine, accuracy of tools, error in machine frame etc. Hence, it is recomended that installations and repair guide line given by OEM/Designer should be followed. However, in absence of that follow the recommended tolerance given in following table.

12.6.3 Recommended Tolerances on Level (Horizontally Mounted Machines)

M/c Type	Recommended Levelness (mm/M)		Recommended Levelness (mils/ft)	
General Process m/cs, supported on Antifriction Bearings	0.075	0.225	10	30
General Process m/cs, supported on Sleeve Bearings(up to 500+HP)	0.375	0.115	5	15
Process m/cs, supported on Antifriction Bearings(up to 500+HP)	0.045	0.15	6	20
Process m/cs, supported on sleeve bearings (up to 500+HP)	0.150	0.600	2	8
Machine Tools	0.075	0.115	1	5

12.6.4 Levelling Instruments

LEVELING TOOLS & TACLES		ALIGNMENT TOOLS & TACLES	
Conventional	High tech	Conventional	High tech
Water level, Sprit level	Dumpy level Theodolite	Filler & Taper gauge	Laser beam m/c
Master level, Straight Edge	Laser beam instruments	Micrometers/Vermeer Caliper/Slip gauges	Theodolite, optical micrometer, Optical tilting level & jig Transit
Feeler		Dial gauges	Vermeer & Strobe technique
Taper gauge		Straight Edge, Dial gauge	Proximity probe & oscillator demodulator, Vermeer & Strobe

12.6.5 Allowable Run out in Rotor

M/C SPEED (RPM)	ALLOWABLE RUN OUT (MILS)
Up to 1800	5 mils (0.005")
1800–3600	2 mils (0.002")
Greater than 3600	1 mils (0.001")

12.6.6 Recommended Max. Radial Run Out Guide Lines

i. Measurement of run out by dial gives readings in diametrical run out and half of this will be radial run out.

ii. High spot & hills and low spot & valley creates quite confusion while taking dial readings. It should be understood that hills and valleys occurs at diametrically opposite while high spots and low spots can exist at several point and at any location of coupling hub.

12.6.7 Misalignment and Life of Equipment

The misalignment deviation at coupling of machine and effect on life of machine is presented in graphical form. Higher the deviation, lesser the life of equipment e.g. at misalignment deviation of 100 mils/inch, the initial decay will start within a month and serious damage could be possible in 20 months of time whereas if deviation is at zero range, the initiation of damage may be possible after 500 months and serious damage could be possible after 900 months. This reveals the importance of alignment and life of equipment.

12.7 Machine Casing to Base Plate Interface Problems or Soft Footing

a. This is most predominant problem in alignment of rotating machineries. When base of machinery and base frame at their interface creates uneven or even gap & do not have recommended surface contact (≥ 65–70%), such situation is called as soft footing which is more prominent in fabricated base plates due to distortion of structural in fabrication & machining. Soft footing in cast base plates is quite less. When foundation bolts of m/c and base frame are tightened, uneven/even gaps appear in alignment of m/c is and centre line of rotation of shaft appears out of desired profile

b. A directional force act on supporting bearings and rotating shaft of machine causing a sever vibration.

c. The critical clearances of machine during operation will get upset.

d. Disturbs connected machine/piping, leading to abnormal vibration behavior. If m/c is continued to operate in such condition, casing may get distorted permanently.

e. Damages critical components like bearing, seals (outer or staging), wear rings, motors etc.

Due to point contact/edge contact or toe contact of m/c feet on base frame will cause rocking or knocking (fully or partially) stress is developed in machine casing. The problems may crop up subsequently.

12.7.1 Soft Foot Checking and Correction

1. Take m/c out of service and isolate power. Open base bolts, remove all shims from base frame and, clean all shims and base frame surface thoroughly with K- oil and cloth. Re pack shims in their position and bolt up by hand tightening.

2. Check at position of all bolts where m/c foot is seating comparatively better of others and tighten that bolt with slightly more torque.

    Measure gap by feeler gauge between m/c pedestal and base plate around four corners of other bolts and record the readings. Readings will give a picture of soft footing.

3. Gaps measurements will reveal that about tapered or non uniform pads. For packing the shims in gaps, various shapes of shims can be used (say U-type, short leg partial U-type, J- type, L-type). After shim packing & tightening bolts. If bolts are getting tightened smoothly from spanner without bump torque at tightening span, it is indication of soft foot elimination. However, if on tightening feel spongy soft foot still exist and further more shim packing is needed.

4. Principle of correcting soft foot- the diagonally apposite contact pints joining axis passing through bolt holes of the corners are restricted by packing such that rocking of m/c feet corner to corner/ side to side or front to rear, in running does not occur.

12.7.2 Checking of Soft Foot Problem of M/C

It may be checked by any of the following methods–

(A) Multiple bolts – multiple dial indicators method

(B) Multiple bolts – single dial indicator method

(C) Shaft movement method

The purpose of all methods is to find the loosening effect on point of bolting (foot to base plate). On loosening base bolts one after another, movement of dial, mounted on foot (in other method on shaft) is noted and recorded. The dial movement ≥ 2 mil is considered as soft foot. Once the soft foot point (s) is detected by any method, correction is made as discussed earlier.

(A) MULTIPLE BOLTS – MULTIPLE DIAL INDICATORS METHOD

    i.   Tighten all foot bolts in position. Install one dial indicator stem, closed to bolt holes and dial base on base frame (not on m/c foot). Set dial to zero.

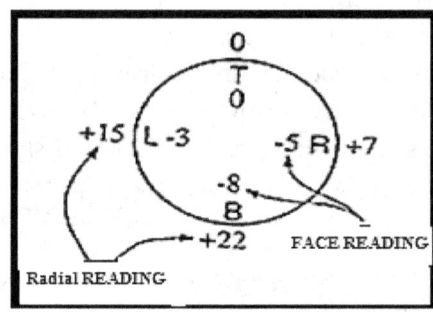

    ii.   Loosen one bolt, observe dial reading and then re-tighten that bolt. Follow this procedure in sequence one after another. However, readings less than two mils is considered as acceptable limit and need not be recorded. All bolts are numbered and soft foot readings are recorded (dial needle movement when bolt loosened). The records will reveal the warp-base-frame at point of measurement.

    iii. Pack shims into soft footing gap and tighten bots. Repeat the process if soft foot still exist.

(B) SINGLE INDICATOR METHOD (LAST CHOICE DUE TO LESSER ACCURACY)

    i.   Tighten all bolts holding m/c in position. Number the bolts in sequence and place dial indicator stem at point of one bolt hole, similar to method (a) and make dial setting at zero.

    ii.   Loosen the bolt where indicator is positioned and if 2 mils or more dial movement is observed, closed to point of bolt holes, it indicates soft footing. Measure with feeler gauge at four corners around the bolt hole and pack the gaps by shims of different shapes & sizes, as per requirements (readings of feeler measurements).

    iii. Re-tighten the bolts. Repeat the process for conformity.

    iv. Subsequently move to other bolt hole and repeat the method of soft foot identification &elimination.

         This applies to bolt holes which are indicating dial movement 2 mils or more.

    v.   Repeat the process of check and corrections once more for conformity of soft footing.

(C) SOFT FOOT MEASUREMENT ON SHAFT

    i.   Tighten all m/c foot bolts in position and mount dial indicator as shown in sketch (a). Dial base is mounted on m/c shaft firmly. Zero dial gauge-reading.

    ii.   Sequentially loosen one foot bolt at a time, observe movement of indicator needle. Observe readings loosening one bolt at a time and record readings as shown in fig (b) for all foot bolts (neglecting < 2.0 mils).

(a) Shaft Measurement Method

(b) Soft foot Checking points

iii. Measure gaps at four corners around the bolt hole with the help of feeler gauge to assess shim packing shape & size. The dial movement ≥ 2 mils only is recorded.

iv. Pack shims in foot gap positions. Packing shims of different shapes and sizes will be used as per requirement. Packing of shims and tightening is done one after another in sequential order.

v. Repeat complete process for conformity of soft foot and If at any point soft foot exist, eliminate by shim packing.

12.8 Tools and Tackles Used in Leveling & Alignment

Taper gauge, Straight edge and feeler gauge are utilized for rough alignment of couplings. Feeler gauges are most popular for any type of clearance measurements, (eg. bearing, soft foot or coupling gap clearance). Vernier calipers are used for outside or inside linear measurements. English calipers have least count of one mil where as that of metric caliper is 0.01 mm. The Vernier caliper with dial indication gives readings in digital or analogue from directly. However, this has limitation for measuring diameter at depth. This is also used for depth measurement up to its length capacity. Tape and rulers are most basic and common tool utilized for rough measurements or preliminary activities of leveling and alignment.

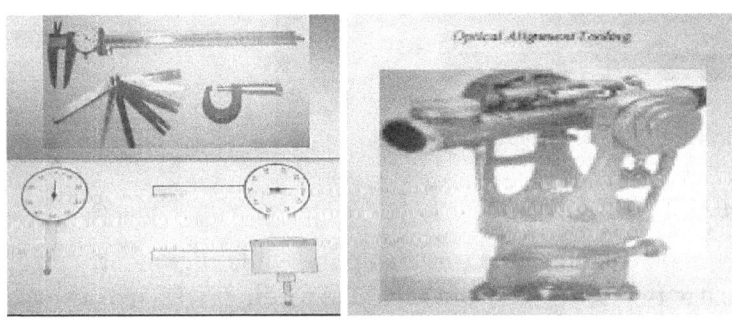

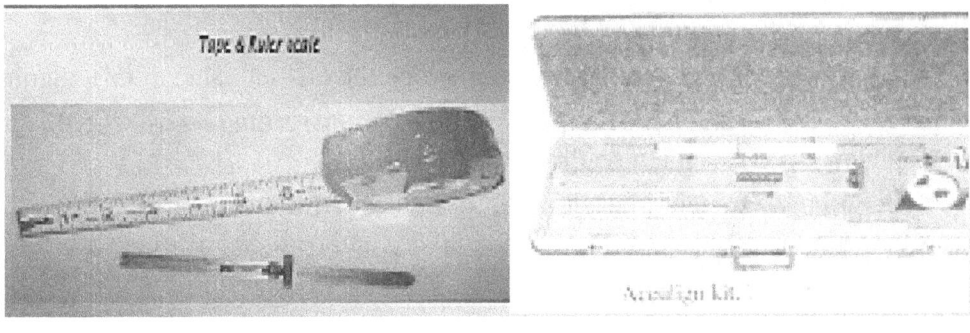

VERNIER CALIPER READING

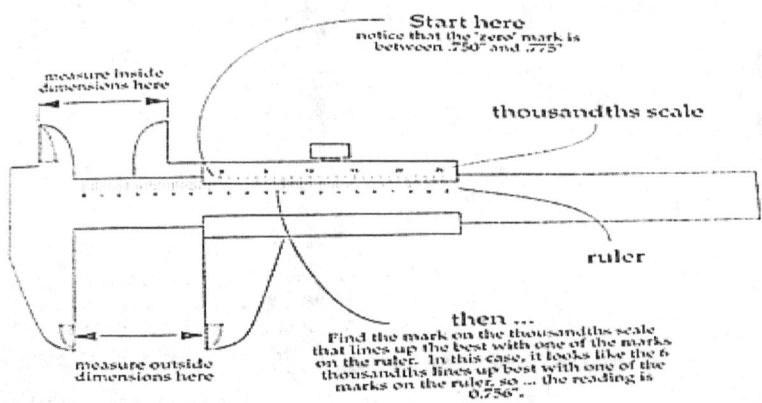

12.8.1 Dial Indicators

Basically two types dial indicators are used–

a. Bottom Plunger design – Stem moves outward and needle rotates anti clockwise

b. Back Plunger design – Stem moves inward and needle rotates in clock wise. Dial indicator checks run out or movements from selected reference plane.

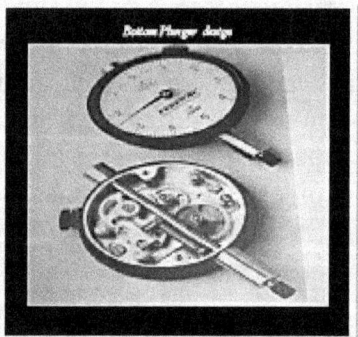

12.8.2 Golden Rules of Alignment

*   **Validity Rule**- In dial reading taken on a coupling while alignment, sum of top and bottom readings should be equal to that sum of left and right dial readings with max error of 10% of the max reading value. If not so, there can be error in taking reading e.g. looseness of gauge mounting, defective dial gauge or sagging of alignment tackle.

*   **Rim Dial Rule**: Rim dial readings are equal to twice the off- set value of mis-alignment.

*   **For plotting** side view graph for alignment, transform top reading = zero where as plotting in plan-view (top view), transform left reading = zero.

*   **Sag correction** of alignment bracket – Mount bracket with dial on a straight shaft, keeping stem of dial at zero in top position and turn shaft 180 deg and read dial reading with + ive or –ive value. For sag correction apply following rules:

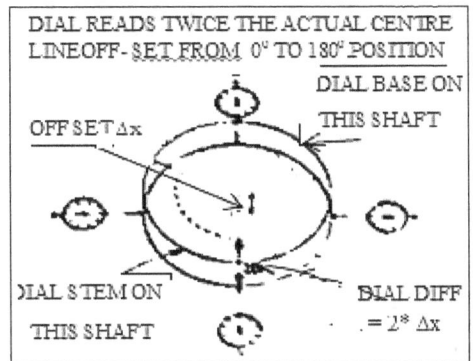

Field reading minus sag correction reading = corrected reading. Please note that all readings should be recorded with care for (+) ive or (−) ive values.

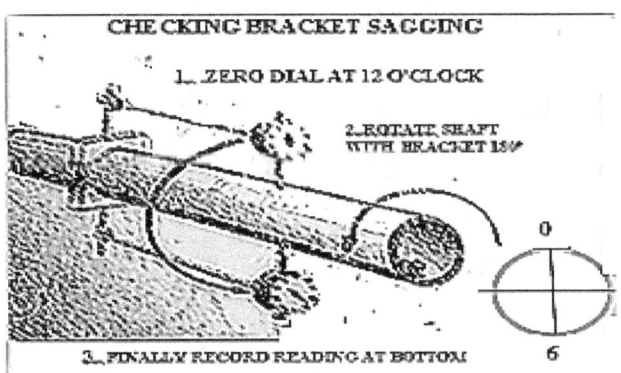

- **Reverse Dial method's graphing**- Transform both dial readings at top = zero for graphing
- **Dial bracket mounting**- Mount dial bracket base on shaft which is reference for alignment or in other words, dial stem will be mounted on shaft of m/c which is to be moved in alignment process.

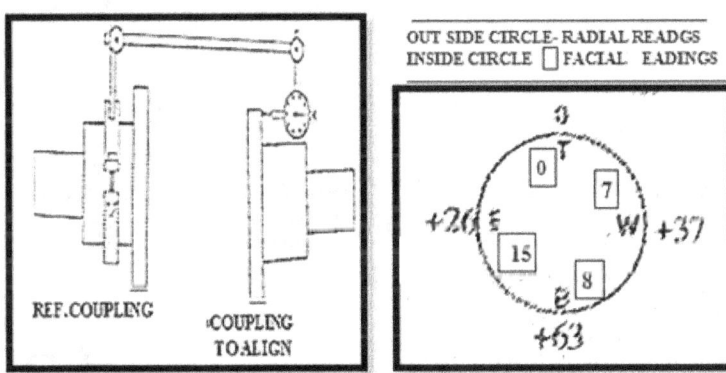

- **Face Rim method alignment**: take dial reading at 90 deg interval, rotate both shaft together to avoid eccentricity in coupling bore in machining. If it is not done so, both couplings can be aligned but shaft will remain misaligned. Further, in this method, use two dial at 180 deg for facial readings to avoid error in face reading due to shaft floating while rotation.

For alignment of horizontally mounted machines, first roughly level shafts with the help of master level and then take up rough alignment with the help of straight edge, taper gauge and or filler gauge before taking up final alignment with dial gauges.

12.8.3 Reasons For Deviation In Captured Readings

a.   Reasons for deviation from validity rule

b.   Bracket sagging

c.   Due to excess over hung

d.   Span of bar, supporting dials

e.   Stiffness of span bar

f.   Clamping of bar on shaft/coupling hub

12.8.4 Reasons of Errors in Field Readings

a.   **ERROR DUE TO BRACKET SAGGING**

Due to sag of bracket, bottom reading increases in +ive direction and top reading increases in negative direction (mostly equal amount). This error can be eliminated by modification of bracket as shown in sketch. If self compensatory bracket is not used, sag readings are taken separately and subtracted from alignment reading for correction as explained in 11.8.2.

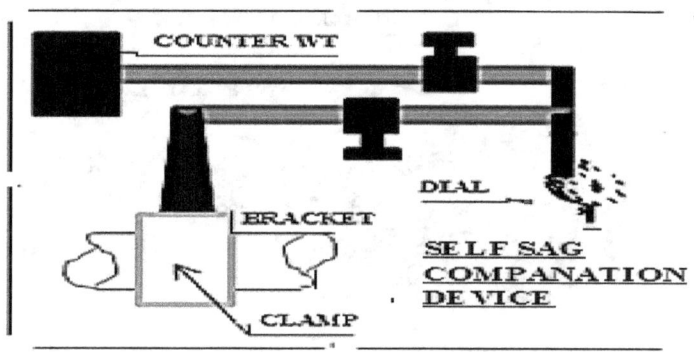

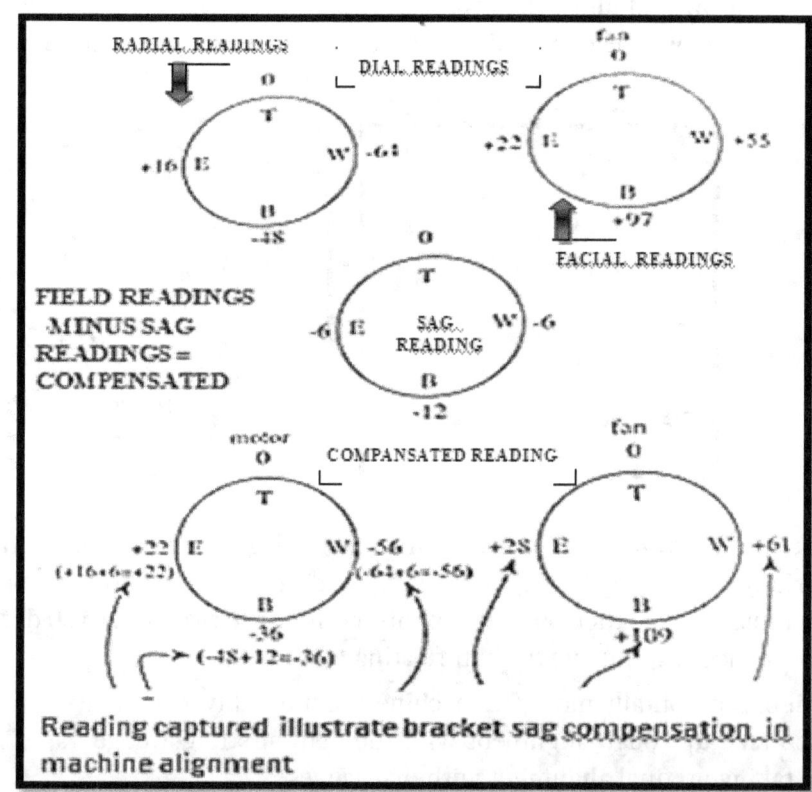

Reading captured illustrate bracket sag compensation in machine alignment

b. **Error Due to Dial Positioning Plane**

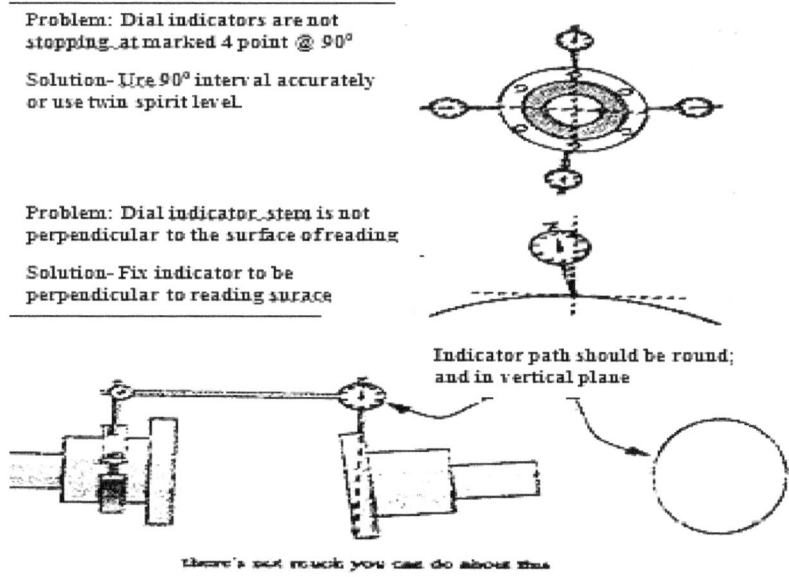

Problem: Dial indicators are not stopping at marked 4 point @ 90°

Solution- Use 90° interval accurately or use twin spirit level.

Problem: Dial indicator stem is not perpendicular to the surface of reading

Solution- Fix indicator to be perpendicular to reading surface

Indicator path should be round; and in vertical plane

there's not much you can do about this

Reasons for deviations to the validity rule.

12.8.5 Techniques to Obtain Accurate Readings

- Repeat of 360° reading and make averaging
- Check dial pressure while set at zero position
- Dial stylus should be perpendicular to rim or face surface
- Follow validity rules (indicator path round and in vertical plane)
- Dial readings should be taken while stylus at marked positions of Top, Bottom, Left, Right
- Sag correction of tackle should be made or sag compensatory tackle should be used.

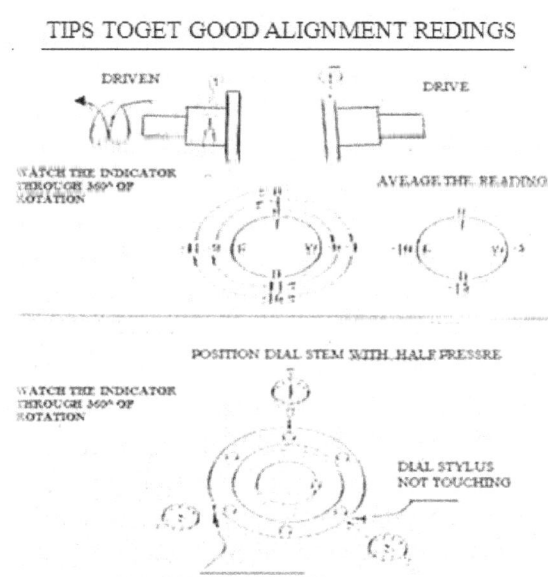

TIPS TO GET GOOD ALIGNMENT REDINGS

12.9 Techniques of Alignment for Directly Coupled, Horizontally Mounted Machines

The methods adopted for alignment of directly coupled machines can broadly divided in two parts-

a. Conventional methods

   √ Straight edge, feeler, taper gauge, slip gauge, micrometers application for alignment

   √ Face Rim (three dial indicators application) method

b. Mathematical Methods

c. Non conventional/Graphical methods

   √ Special application methods of alignment by any of the tools

   √ Reverse Indicator Method

   √ Double Radial Indicator technique

   √ Face- Rim Reverse Indicator Method

   √ Shaft to coupling Spool technique

   √ !6 or 20 point alignment technique for Rigid coupling

12.9.1 Steps in Mathematical Solution →

a. Know m/c rotor position in cold condition and stabilized running condition

b. Prepare all necessary tools & tackles for alignment

c. Level and roughly align m/c

d. Mount three dials – two nos for axial readings at 180 deg apart and one dial radial and record readings. Two axial dials are used to eliminate float effect of shafts in alignment, especially when shaft(s) mounted on sleeve bearings.

e. Transform readings into radial and axial form as per calculations shown in example

f. Calculate mathematically addition or subtraction of shims from m/c. Mathematically calculate movement of m/c to sides and in vertical plane

g. While taking readings all foot bolts of m/c should be tightened fully.

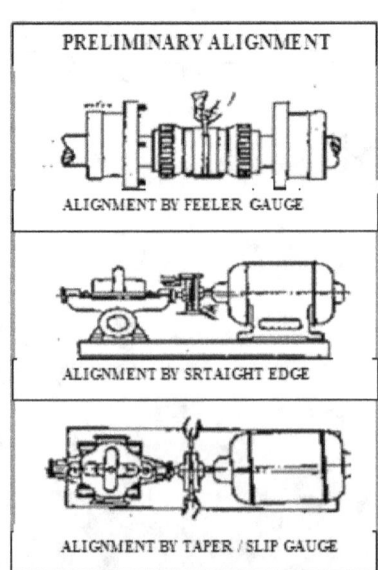

PRELIMINARY ALIGNMENT

ALIGNMENT BY FEELER GAUGE

ALIGNMENT BY SRTAIGHT EDGE

ALIGNMENT BY TAPER / SLIP GAUGE

12.9.2 Face RIM/Conventional Method of Computation

Make sag correction, of radial reading by subtracting the sag reading from vertical plane radial dial reading. The resultant value will be sag corrected reading which is utilized for calculations. The corrected dial readings are placed at left side of following sketch;, whereas at right side 4-readings are taken from Face Rim dials (axial-2 nos, and radial one) and final readings of axial as "A" and radial as "R" are computed and recorded.

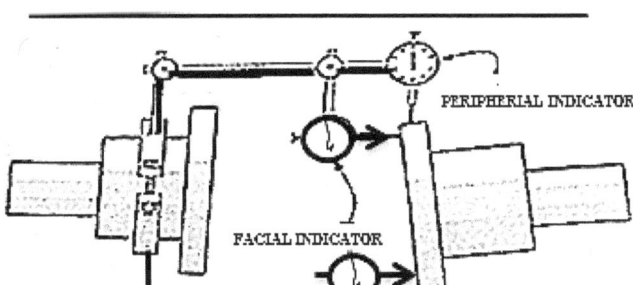

From vertical plane readings, find axial deviation. Δx and radial deviation Δy:

- $\Delta x_{(V)}$ = (Difference of top and bottom dial axial readings) = Δx
- $\Delta y_{(v)}$ = (Vertical difference dial readings, deviation/2) = Δy
- $\Delta x_{(H)}$ = (Difference of Left and right axial dial reading) = $\Delta x_{(H)}$
- $\Delta y_{(H)}$ = (Horizontal Radial dial readings difference/2) = $\Delta y_{(H)}$

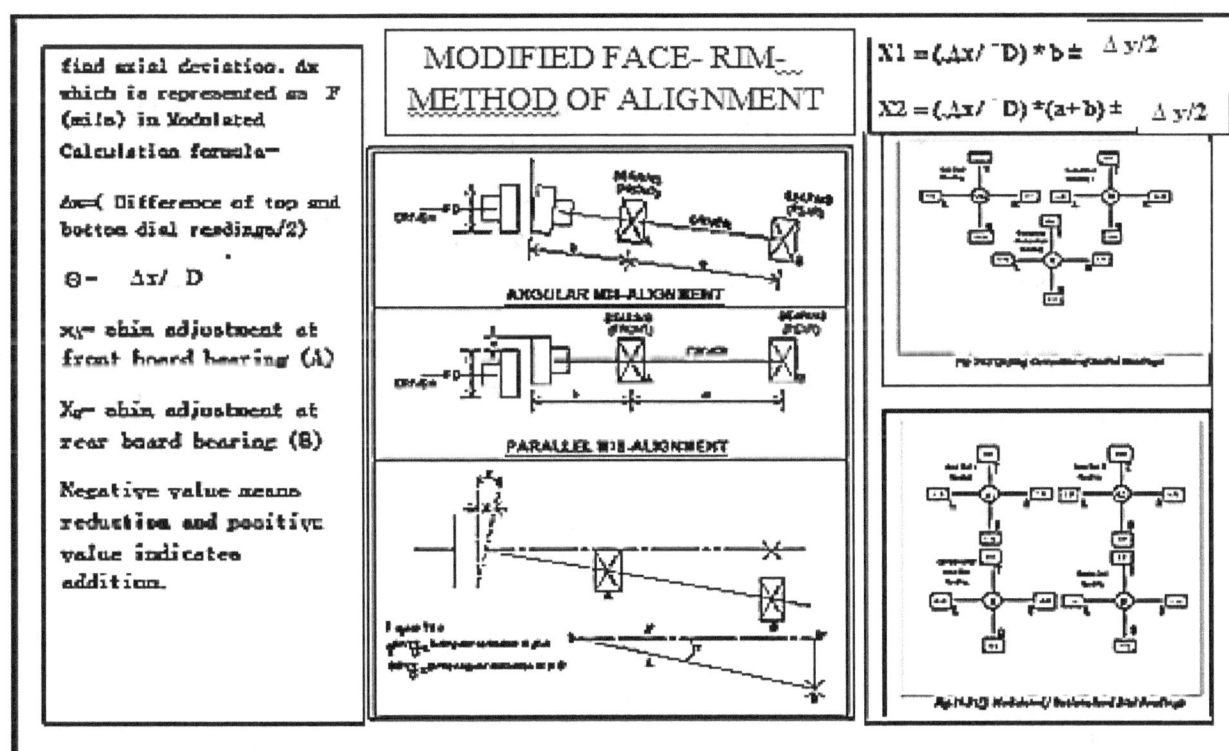

Computation of final axial reading from A1 and A2 dial readings-

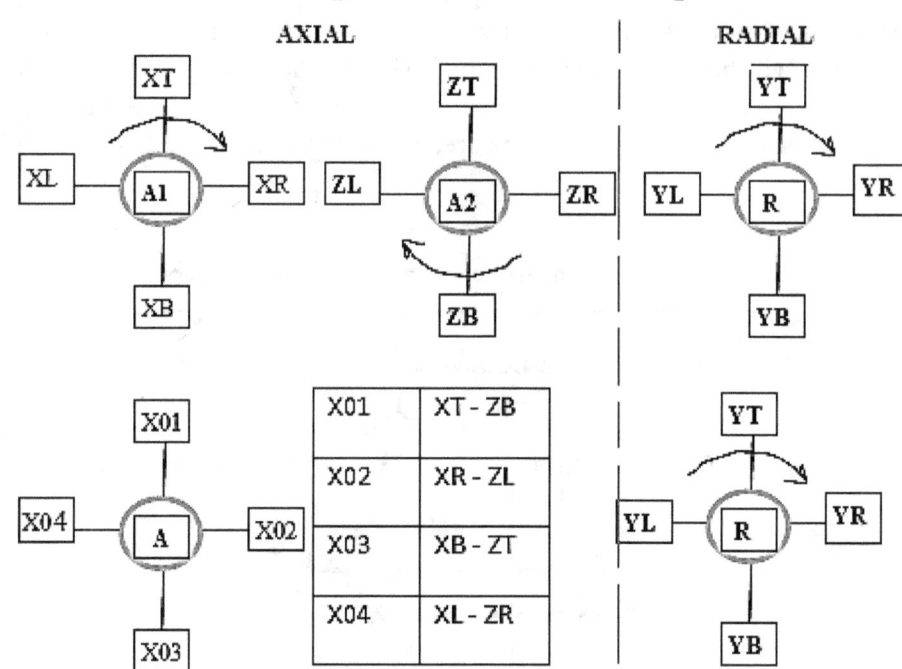

Suffix T → Top position reading;   B → Bottom position reading;
R → Right position reading;   L → Left position reading

Vertical plane axial misalignment deviation = ΔX = (X01 – X03)

Vertical plane radial misalignment deviation = Δy = (YT – YB)/2

Note: Dial A1 is the dial which reads axial reading at any point where radial dial also reads simultaneously.

$\Theta = \Delta x / D$ where D = Diameter of dial gauge stylus orbit on coupling

Shims to be adjusted for vertical plane alignment at front board pedestal (x_1) & rear board pedestal (x_2) of motor can be calculated (refer diagram & following notes)

- **Note:**
  1. When Radial Top reading is minus & bottom radial reading is + ive then Δy will be used with + ive sign in formula.
  2. When in axial reading, top is +ive & bottom minus, shim for axial alignment correction will be added (lift axis up (↑) at front, keeping rear board undisturbed. Contrary to the former case, when top reading is minus and bottom + ive, front board of motor shaft will be lowered keeping rear board undisturbed. The radial readings are not considered.
  3. Similar calculation can be applied for horizontal plane also, using left and right readings, similar to up and down positions for movement of motor in horizontal plane.

- **Assumptions:**
  1. For small arc angle, arc/radius = angle in radians
  2. Coupling gap is considered to be small
  3. Reference m/c is accurately leveled and aligned
  4. All readings are taken with foot bolts in tight condition and no soft footing exist
  5. Run out of shaft and coupling is within specified limit of accuracy

12.9.3 Shims Calculation for Alignment (Face RIM Method)

a. Shims added at point (A) coupling side bearing.

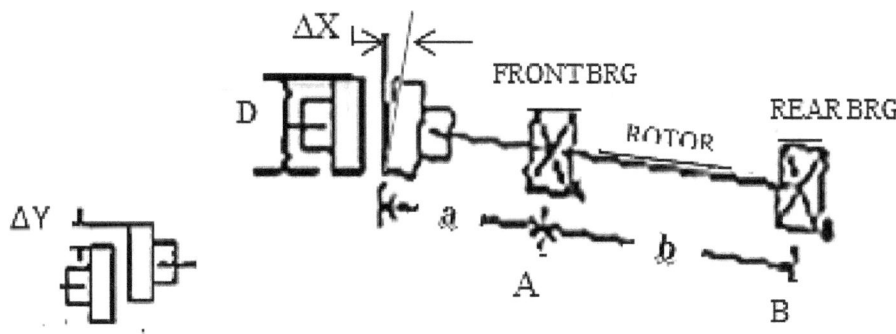

ΔX = Driver facial top & bottom readings difference   (mils/μM)

ΔY = Radial top & bottom readings difference/2   (mils/μM)

$$F(A) = \left\{\left(\frac{\Delta x}{D} * b\right) \pm \Delta y\right\}. \qquad \text{Where } F(A) \rightarrow \text{Shim adjustment at A;}$$

$$F(B) = \left[\left\{\frac{\Delta X}{D} * (a + b)\right\} \pm \Delta y\right] \qquad F(B) \rightarrow \text{Shim adjustment at B}$$

Note:-

1. If bottom and top radial difference of readings is + ive then Δy is considered with positive sign in above formulae.

2. Contrary to case- 1, the bottom and top radial difference of readings Δy is negative then Negative sign with Δy is considered in formulae above.

3. If facial readings difference of top and bottom is positive (opening at top of coupling > that of bottom then factors F(A) and F(B) are considered positive which means that shims F(A) and F(B) are to be added.

4. Contrary to case- 3, when facial readings difference of top and bottom is Negative (opening at bottom of coupling > that of top then factors F(A) and F(B) are considered Negative which means that shims F(A) and F(B) are to be removed.

5. Similar to vertical plane misalignment calculation for horizontal plane alignment can be corrected, taking left and right readings but in this case, no shim adjustment is needed, rather aligning equipment is to be shifted in horizontal plane for correcting misalignment. In this case also same formulae will apply but instead of top and bottom we have to substitute right and left readings.

12.9.4 Reverse Dial Indicator Methods of Alignment

a. **MATHEMATICAL CALCULATION METHOD**
   - A, B, C, D, E denotes as shown in sketch in inches
   - ΔX = Driver rim top & bottom reading difference (mils/μM)
   - ΔY = Driven rim top & bottom reading difference (mils/μM)

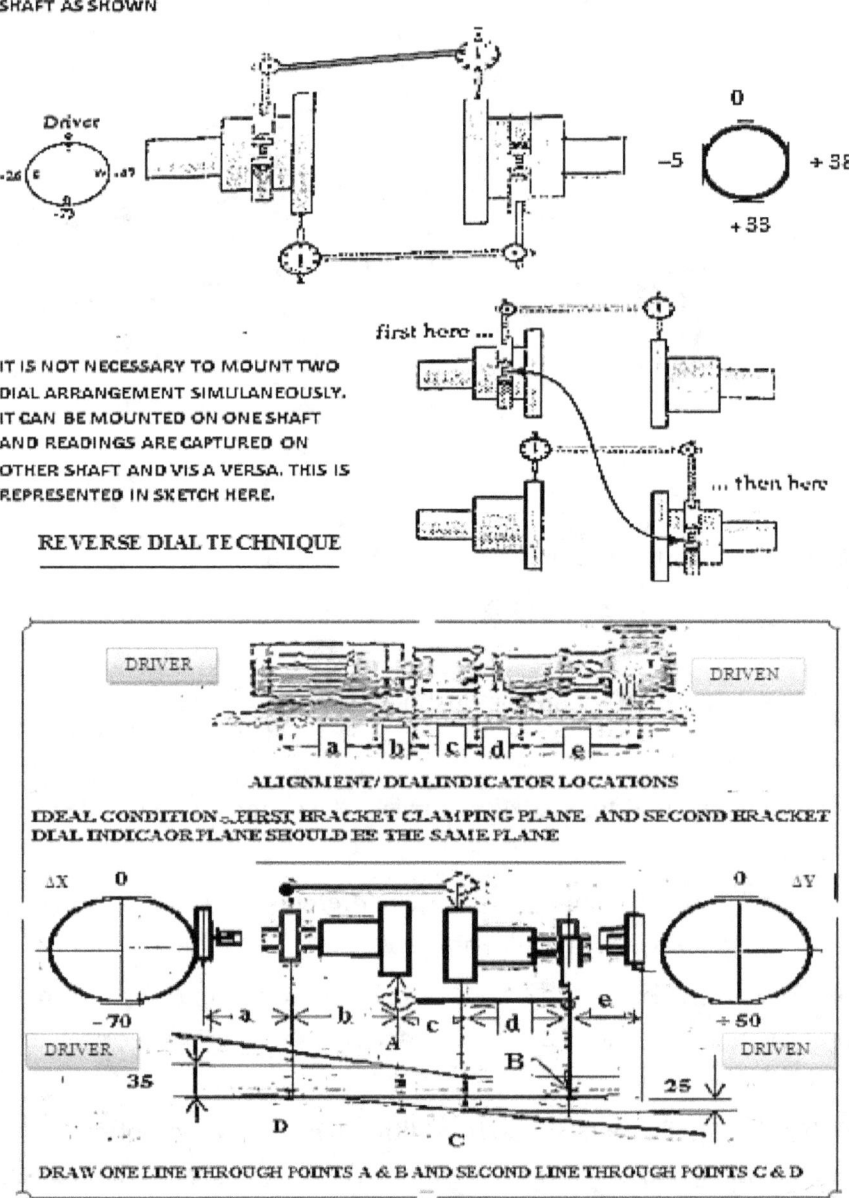

THE TRADITIONAL METHOD OF CAPTURING REVERSE INDICATOR READINGS BY
CLAMPING BRACKET ON ONE SHAFT (SPAN COVER UPTO OTHER SHAFT) AND DIAL ON
OTHER SHAFT. SIMILAR ARRANGEMENT IS MADE WITH CLAMP MOUNTED ON OTHER
SHAFT AS SHOWN

IT IS NOT NECESSARY TO MOUNT TWO
DIAL ARRANGEMENT SIMULANEOUSLY.
IT CAN BE MOUNTED ON ONE SHAFT
AND READINGS ARE CAPTURED ON
OTHER SHAFT AND VIS A VERSA. THIS IS
REPRESENTED IN SKETCH HERE.

REVERSE DIAL TECHNIQUE

ALIGNMENT/ DIAL INDICATOR LOCATIONS

IDEAL CONDITION - FIRST BRACKET CLAMPING PLANE AND SECOND BRACKET
DIAL INDICAOR PLANE SHOULD BE THE SAME PLANE

DRAW ONE LINE THROUGH POINTS A & B AND SECOND LINE THROUGH POINTS C & D

Fig. Reverse Indicator Method (Measurements & Graphical Solution)

12.9.5 Mathematical Solution in Reverse Dial Method

a. **Tabulation of measurements**

Figures used in Calculation (shown in sketch)					
Dimensional Measurements (mm)	a	b	c	d	e
Driver Rim top & bottom Readings difference	(µM)	ΔX			
Driven Rim top & bottom Readings difference	(µM)	Δy			

b. **Mathematical Calculation Chart (Reverse Dial Method)**

POSITION	DRIVER M/C FEET CORRECTION	DRIVEN M/C FEET CORRECTION
IN BOARD	$[(b + c) * (\Delta X + \Delta y)/c] - \Delta y$	$[(c + d) * (\Delta X + \Delta y)/c] - \Delta X$
OUT BOARD	$(a + b + c) * (\Delta X + \Delta y)$	$[(c + d + e) * (\Delta X + \Delta y)/c] - \Delta X$
Calculated Value will be movement on one or other m/c case for alignment		

12.9.6 Graphical Solutions in Reverse Dial

Information and dimension to be collected on Schematic diagram for Graphical solutions

• Inboard to outboard distance (ref. bolt planes)of driver and driven m/c	• In board bolting to bracket mounting planes distance for both m/cs	• Bracket mounting centre plane to indicator stem point distance on rim planes
• Dial indicators stem points plane distances	• Distance of inboard bolt plane to that of outboard	• Dial indicators readings, axial, radial or both
• Coupling diameter	• For two dial readings of axial, diameter of two indicators stem points on face (mounted at 180 deg)	• Dial to bracket and bracket to bracket distances

We limit our discussion to horizontally mounted machines or machine trains. Graphical representation is made to exaggerated misalignment part in side or top view, illustrating up or down and sides position of shafts. Once the relative position of shafts is drawn, various solutions to bring shafts in line can be depicted under restricted condition of m/c. Further, graphical modeling technique can be used for piping fittings, air gap of generator/motor, fan rotor to shroud clearance etc.

Thus, we review the following key steps in correcting the misalignment–

i. Determine the current position of the centre line of rotation of all machineries

ii. Observe any restriction on base or surrounding which will obstruct in movement of m/c to correct mis-alignment. Mark location of such obstruction on graph while plotting for misalignment correction

iii. Find out movement of all m/cs for correction so that correction can be tken up with any one m/c as per convenience and possibilities.

12.9.7 Reverse Indicator Graphical Solution Procedure

Step-1: Plot measured dimensions accurately on graph for top view and side view. Scale for distance of inboard and out board of m/cs, and readings captured. Indicate m/cs position over graph. Make sag correction for captured readings.

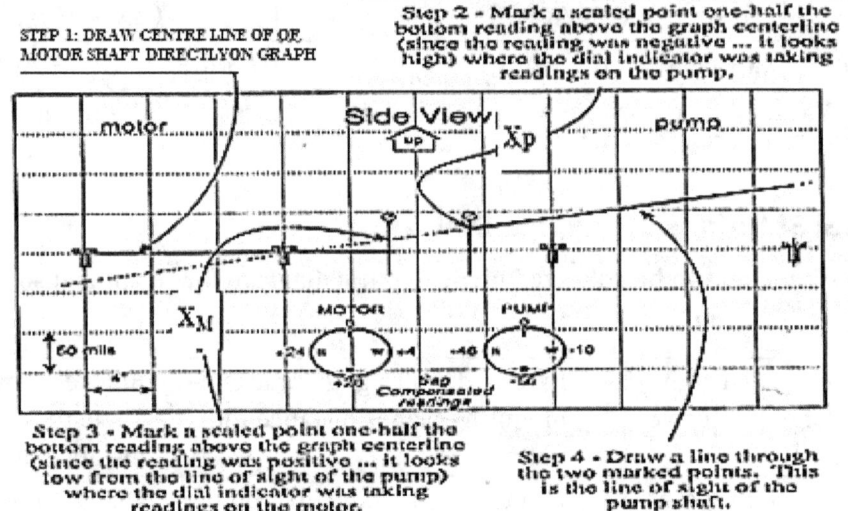

Procedure for plotting the line to point reverse indicator technique:
1. Select one of the two machinery shafts and draw one of those shafts on top of the graph centerline.

NOTE 1.: SIDE VIEW IS USED FOR TOP & BOTTOM CORRECTION AND PLAN VIEW FOR SIDES CORRECTION.
2.: IN TOP BOTTOM CORRECTION : BOTTOM NEGAGATIVE READING POINT IS POSITIONED UPWARDS FROM GRAPH CENTRE LINE

STEP 1: DRAW CENTRE LINE OF OF MOTOR SHAFT DIRECTLYON GRAPH

Step 2 - Mark a scaled point one-half the bottom reading above the graph centerline (since the reading was negative ... it looks high) where the dial indicator was taking readings on the pump.

Step 3 - Mark a scaled point one-half the bottom reading above the graph centerline (since the reading was positive ... it looks low from the line of sight of the pump) where the dial indicator was taking readings on the motor.

Step 4 - Draw a line through the two marked points. This is the line of sight of the pump shaft.

Reverse indicator line to point modeling technique example where the motor shaft is placed directly on top of the graph centerline.

Step-2: Select appropriate up and down scale factor, start with the shaft that has larger of the two bottom readings. See that entire shafts fit into graph paper. Dial readings of both m/cs are recorded on graph at the respective side of shaft of that m/c. Dial reading is considered corresponding to the shaft on which indicator stem point is placed. For plotting shafts point, two planes of dial stem points on both shafts are selected on graph paper. Based on dial reading (corrected for sag) for Elevation/side view graphing, use top and bottom readings of both shafts with reference to horizontal axis. Up location of shaft will be plotted up (↑) in graph and vis-a-versa (i.e. negative bottom reading will be plotted above center line and contrary to this for +ive bottom value point will be below the center line of graph). Thus, one point for both shaft centers correspondingly to dial position is marked on graph. Now mark the in board and out board bolt positions. on x axis and correspondingly projected on shafts locations. Rim reading is twice the off set of shaft. Hence, while plotting the shaft position, under line, not the dial reading.

As shown in graph, plot motors shaft center line on graph center line. Position dial point Xp and X_M on dial point vertical plane. Join Xp and X_M points and extend full length of pump shaft. Extend pump sight center line as dotted line to cover motor shaft length. Project vertical line from motor –base- bolt -points on the motor shaft center line (shown dotted) and measure vertical distance from motor shaft center line (sight position) to graph center line. These measurements are shim corrections for motor base points.

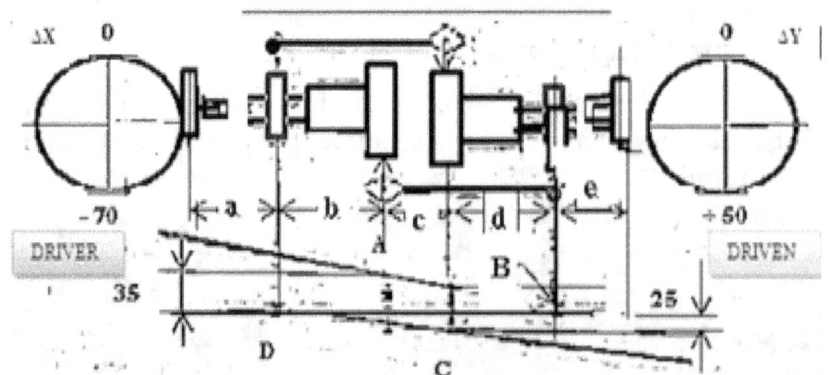

Step -3: From the plotted graph of both shafts, use shim correction readings to move and align motor with other shaft. The shims will be removed or added is decided by position of motor shaft if in graph it is below the graph center line, shims will be added). Accordingly alignment should be completed.

12.9.8 Face–Rim Reverse Dial Graphing Procedure

Step-1: Similar to reverse indicator graphing, outside circle radial readings are marked whereas inside circle readings are face readings. In side circle face readings & half of radial readings at bottom are defined for graphing.

Step-2: Take half of peripheral (radial) reading & full of face reading and plot coupling position of pump shaft with reference point (horizontal central graph axis and coupling face vertical line intersection). The coupling face point of inter section). The positive rim reading (at bottom) is positioned bottom of axis and positive face reading (at bottom) is positioned left to vertical face of coupling. Overlay the pivot "T" centre on point A and slide "T" to let just cut the horizontal graph axis line. Now extend this line towards motor (shown in dotted line in the graph).

Step-3: Measure shaft movement of m/c (Say motor) at supporting points with respect to reference position at x-axis and make arrangement for shim removal or addition to align m/cs together. If thermal expansion compensation is to be provided, shim adjustment can be made accordingly.

D = Coupling dia at which face dial stylus locus is formed (say 10" in example here)

Δx = Difference of top and bottom facial readings (say – 12 mils in example here)

Δy = Difference of top and bottom facial readings (say +30 mils in example here)

X_{M1} = Motor In board shim addition/subtraction (mils) say addition in this example

X_{M2} = Motor Out board shim addition/subtraction (mils) say addition in this example

Xp_1 = Pump In board shim addition/subtraction (mils) say addition in this example

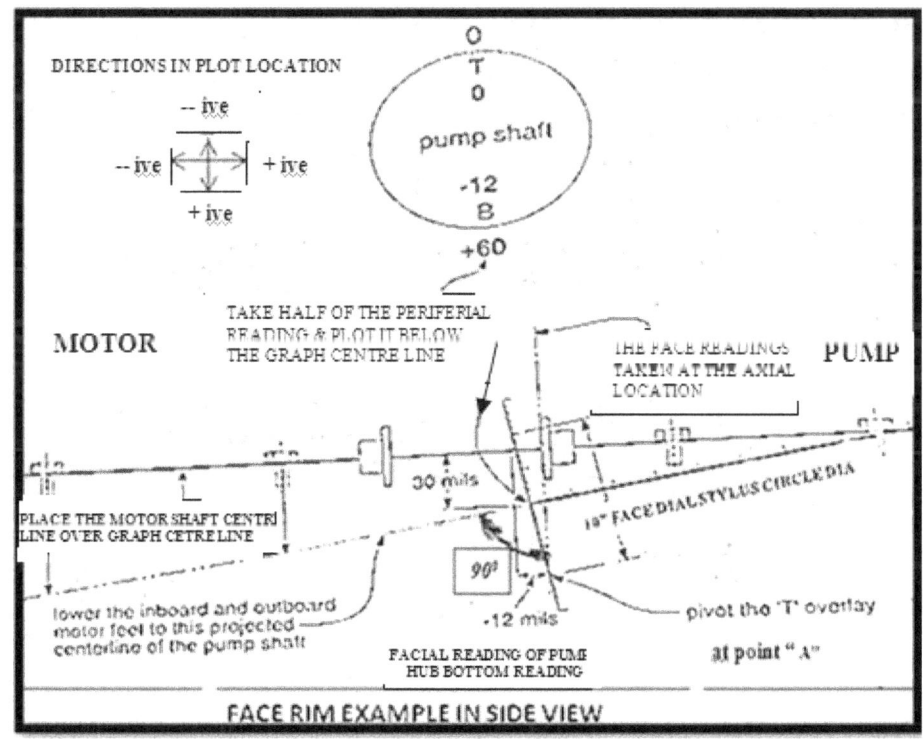

- **SAG CORRECTION IN REVERSE DIAL READINGS–**

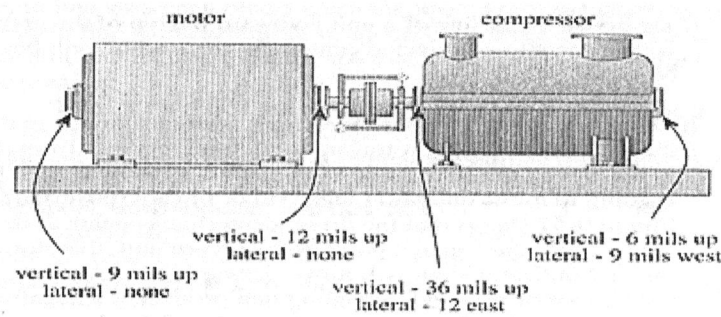

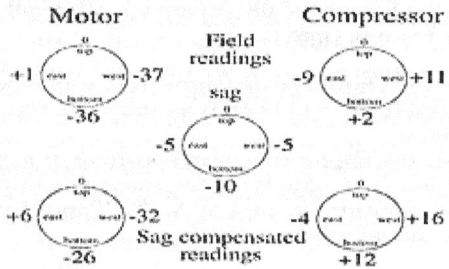

SAG CORRECTION IN REVERSE DIAL TECHNIQUE OF ALIGNMENT

12.9.9 Alignment by Mathematical Calculation Methods

The mathematical relationship is developed between machinery dimensions and dial readings captured and shim adjustment required for aligning shafts in Face Rim or reverse indicator method. From given equations, calculation for shim can be done for driven or driver machine to align in in-situ condition. One can decide driven m/c as reference and driver can be adjusted to align or vis a versa.

a. **MATHEMATICAL CALCULATION IN FACE RIM METHOD:**

H = Coupling Diameter (inches);　　F = Rationalized axial reading (mils)

R = Radial reading (mils);　　C = bracket to dial stylus distance (inches)

A, B, C, D, E are in inches as shown in sketch (Note: All sizes taken in inches)

Y = Rim dial reading difference from top to bottom or right to left in sides

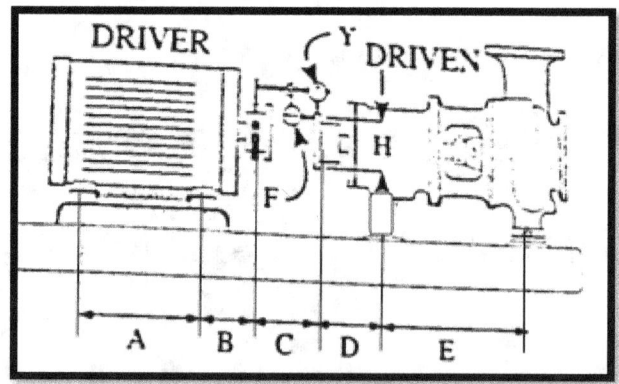

i. **MOTOR FEET SHIM ADJUSTMENT** (If motor- to shift) (from vert. plane readings, axial deviation. Δx is represented as F (mils))–

    √   **Inboard feet of driver** shim adjustment = $[F * (B + C)/(H^2 + F^2)^{1/2}] - Y$

    √   **Out board feet of driver** shim adjustment = $[F * (A+B+C)/(H^2 +F^2)^{1/2}] - Y$

  ii. **Or CORRECTION AT DRIVEN M/C-**

    √   **Inboard feet of driven** shim adjustment = $[F * (D)/(H^2 +F^2)^{1/2}]+Y$

    √   **Out board of driven** shim adjustment = $[F * (D+ E)/(H^2 +F^2)^{1/2}]+Y$

  b. **MATHEMATICAL CALCULATION IN REVERSE DIAL INDICATOR**

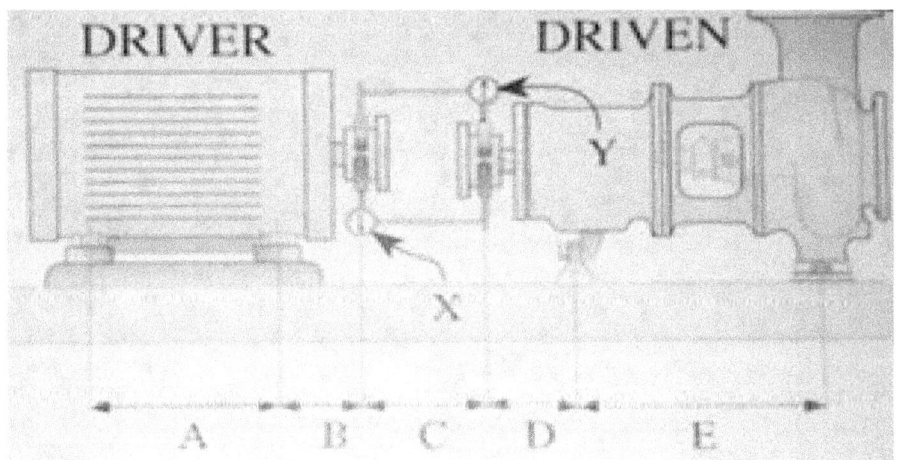

    H = Coupling Diameter (inches);

    X = driver Rim readings difference top to bottom (mils)

    Y = driven Rim readings difference top to bottom (mils);

    C = bracket to dial stylus distance (inches)

    A, B, C, D, E are in inches as shown in sketch (Note: All sizes taken in inches)

  i. **CORRECTION AT DRIVER M/C**

    √   **Inboard feet of driver** shim adjustment = $\{(B + C) (X + Y)/C\} - Y$

    √   **Out board feet of driver** shim adjustment = $\{(A + B + C) (X + Y)/C\} - Y$

  Or

  ii. **CORRECTION AT DRIVEN M/C**

    √   **Inboard feet of driven** shim adjustment = $\{(C + D) (X + Y)/C\} - X$

    √   **Out board of driven** shim adjustment = $\{(C+D+E) (X + Y)/C\} - X$

12.9.10 Face – Rim Graphing Method with Axial Readings (Elevation & Top View)

The basic measurement principle of this method is to find angle between shaft centre line and that of coupling spool/jack shaft. There is only one flex point for each shaft.

  a. **DOUBLE RADIAL & AXIAL GRAPHICS METHOD**

    √   Prepare modeling diagram and label all required dimensions specified for graphing

    √   Mar x- axis over central horizontal line of graph)

    √   Mark motor shaft, coupling and bracket position on x-axis

    √   Draw vertical line on x- axis at position of pump coupling face

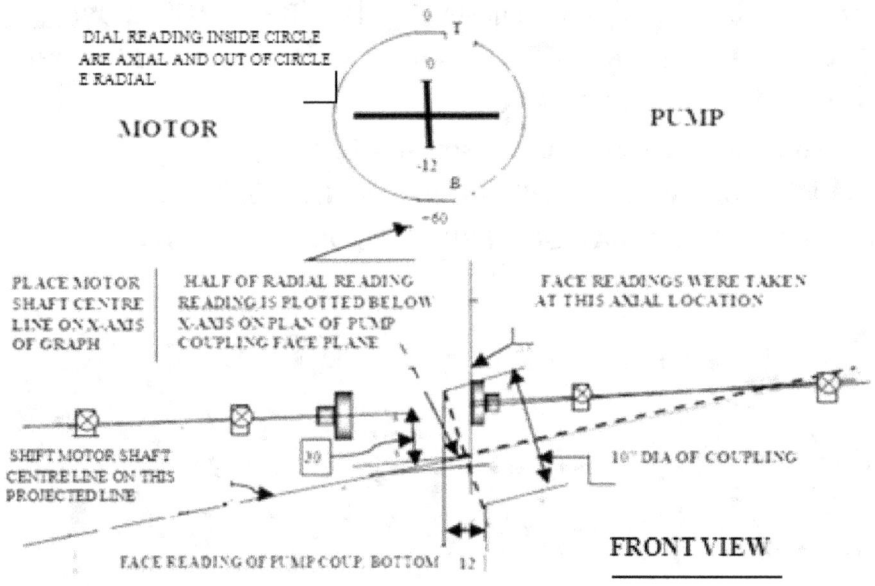

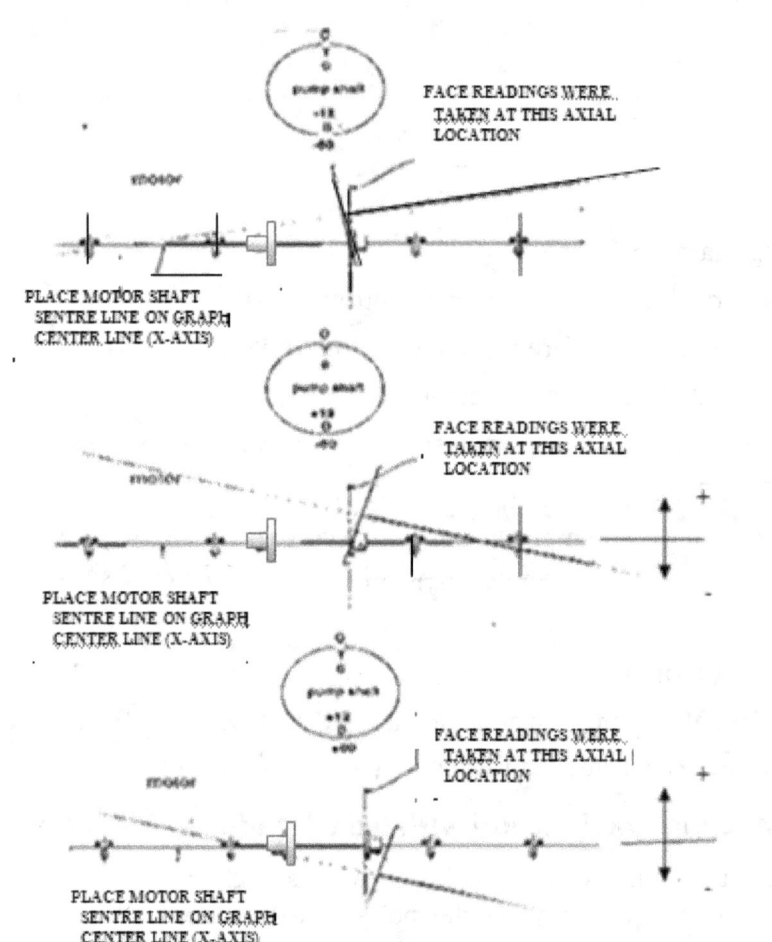

√   Draw Elevation view first followed by top view. Mark coupling halves position on x-axis.

√   Plot both shafts centre line, bolt positions, coupling position, dial capturing positions. If bottom or side readings are negative, place a point half of radial reading at top of centre line (x-axis) and for + ive reading half of radial reading at bottom of x-axis. Record dial indicator readings on graph.

√   Start at inter section of graph x-axis and dial capturing reading points.

√ Alignment bracket was attached to the motor shaft and indicator positioned at two different locations along the length of fan shaft, capturing the circumferential/radial readings. The bottom reading at motor end indicator say reads (−20) mils that means point is 20/2 = 10 mils higher to x-axis or so called centre line. Along the plane of indicator point on motor shaft, measure as per scale 10 mils upwards (say point p_m). Similarly, plot reading of fan shaft indicator's half reading (bottom reading say + 12/2 = +6) along the plane of indicator point on fan shaft, 6 mils downwards from axis line, say point p_f. Join point p_m and p_f and extend the line towards rear board. This will represent the fan shaft centre line.

√ Now read shim adjustment reading from front view by measurement of distance between shaft line of pump and x-axis (centre line of graph) at point of front and rear feet fasteners. These readings indicate amount of shim. adjustment (removal or addition) to align along motor shaft say x-axis.

√ Similarly from plan view, horizontal position shifting will be known.

12.9.11 Shaft to Coupling Spool Reverse Indicator Method Alignment Correction

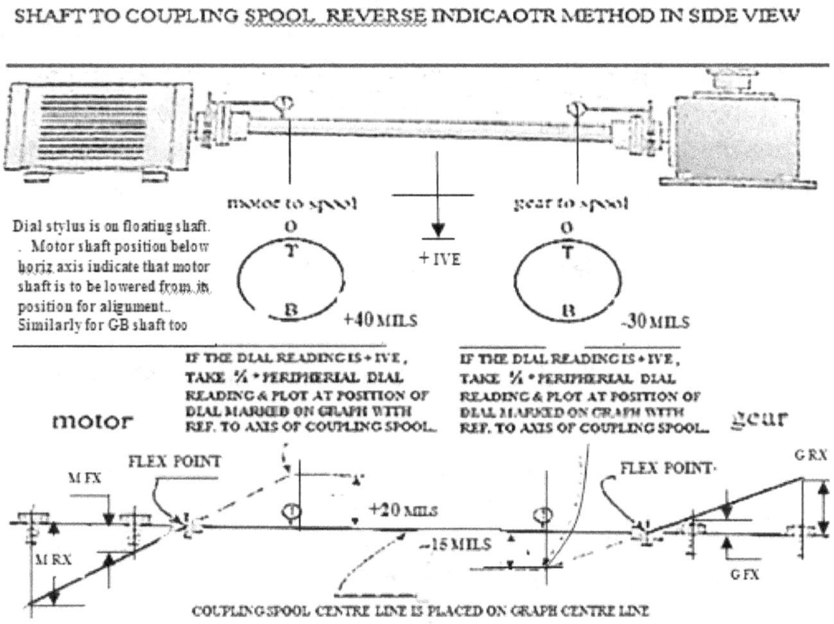

SHAFT TO COUPLING SPOOL REVERSE INDICAOTR METHOD IN SIDE VIEW

√ Positive reading means lower than horizontal axis and negative reading means above horizontal axis.

√ After graphically shafts position is drawn (as shown above), the shaft movement dimensions at feet fastening positions (MFX, MRX, GFX and GRX) are measured from graph.

√ Motor front foot is lowered = MFX; motor rear foot is lowered = MRX – by removing shims MFX and MRX respectively.

√ Gear box front foot is lifted = GFX; Gear box rear foot is lifted = GRX – by adding shims GFX and GRX respectively.

√ Thus, floating shaft will be aligned with motor and gear box.

12.9.12 Movement Restrictions in Alignment and Their Remedial Action

a. **Bolt bound restriction-** It is found some cases when one m/c is to be moved in a position but it can't be shifted side way to full amount of movement due to holes restriction. It is advisable to–

      i.   Turn bolt shank up to root diameter of bolt or make uniform strength bolt

     ii.   Use under sized bolt if it permits designed stress value

    iii.   Use HT material of bolt with under sized diameter to accommodate movements

    iv.   Make movement of both m/c. If pipings restriction occurs, modify with vertical and horizontal shift of m/c or re locate pipes by cutting and welding.

  b. **Restriction in down movement-** It may be case that m/c to be shifted vertically down does not have any shim. This may be resolved as–

      i.   Level and align both m/c freshly by adding shims in both m/c

     ii.   Make graphical position of m/cs and pivot the other foot of m/c (not to be lowered) and accordingly alignment is completed.

  c. **Upward movement restriction–**

It is possible that m/c to be moved upward is restricted by some object which can't be shifted or removed. It will be necessary to go for any of the following solutions–

– Level and align both m/c freshly at level which does not fall in restriction zone. for this make graphical position of m/cs and pivot the other foot of m/c (not to be lowered) and accordingly alignment is completed

– In graph plotting, pivot the foot of m/c falling under restriction and finding the shims addition or subtraction for movement of one m/c or both. Choose the solution to meet the requirement. Combination of pivoting of both m/cs can be tried in graphical solution of shafts rotational centre lines.

Note: whichever foot bolt is made pivot point that should be machined shank with some side clearance with its hole.

12.10 Compensation in Cold Alignment for Running Position Changes

Practically all machines undergo to some extent of alignment changes in running. In order to run shaft co linearly, it is necessary to know the amount and direction of changes which will be compensated during cold alignment of m/c. The change in position causes misalignment. If deviation of alignment as per severity chart is within acceptable limit, it can be tolerated and cold alignment correction may not be necessary. Thus, if correction is made, it will give better performance of m/c.

12.10.1 Machines Which Change Its Position More Prominently

THE M/CS WHICH CHANGE ITS POSITION MORE PROMINENTLY during operation can be listed as follows:

  a. Rotating machineries > 200 HP, running at or above 1000 RPM

  b. M/c that undergoes changes in casing temperature in operation eg. Electric motors, Generators, Steam turbines, Gas turbines, IC Engines.

  c. M/c attached with Speed Changers (gear boxes)

  d. M/cs used for pumping and compressing fluids, gases which undergoes change of temperature of fluid ≥50 deg C from intake to discharge e.g. centrifugal and reciprocating compressors, centrifugal pumps, HVAC air moving equipment, furnace fans.

12.10.2 Cause of M/C Movements in Running

  a. The most common reason movement is thermal expansion from change of temperature (ΔT) in machine due to compression of fluid/liquid, frictional forces in m/c, flow of hot liquid/gases, weather effect on equipment, heating or cooling of pedestals, change in operating condition of m/c, restriction by natural forces like centrifugal force, gravitational forces, wind forces.

b. Startup kick of m/c also change position for a short while which may be for some minutes or hours but settle closed to its original position. This is a transient phenomenon of m/c.

c. Setting down of m/c foundation due to soil setting.

d. Wear of bearings, coupling elements, loosening of foundation bolts/fasteners, burs or foreign matter while fitting components during overhauling of m/c.

12.10.3 Measurements from off Line to Running Condition of M/C

a. Movement of line of rotation of centre line of m/c compared to its base.

b. Movement of line of rotation of centre line of m/c compared to remote reference or observation line.

c. Movement of line of rotation of centre line of m/c compared to other m/c casing or foundation.

d. Movement of line of rotation of centre line of one shaft compared to other shaft of m/c.

This type of measurement is very much time consuming. The survey of Offline to running movement is specified by manufacturer for very critical m/cs.

CO EFFICIENT OF THERMAL EXPANSION FOR SOME MATERIALS USED IN INDUSTRIES					
Material	Inch/inch °F	mm/mm °C	Material	Inch/inch °F	mm/mm °C
Al – alloys	$12.5 * 10^{-6}$	22.5	Invar	$0.68 * 10^{-6}$	$1.224 * 10^{-6}$
Brass (70% Cu 30%Zn)	$11 * 10^{-6}$	19.8	Ni steel	$7.3 * 10^{-6}$	$13.14 * 10^{-6}$
Carbon steel (AISI-1040)	$6.3 * 10^{-6}$	11.34	Stainless Steel	$9.8 * 10^{-6}$	$17.64 * 10^{-6}$
Grey Cast Iron	$5.9 * 10^{-6}$	10.62	Vulcanized Rubber	$45 * 10^{-6}$	$81 * 10^{-6}$
Concrete	$7.2 * 10^{-6}$	12.96	Nylon	$55 * 10^{-6}$	$99 * 10^{-6}$

12.10.4 Expansion Movement Calculation

The movement of shaft center line in vertical and/or horizontal planes of rotation due to thermal expansion are computed from the following data, collected from m/c–

i. Base frame to shaft supporting pedestal or any component linear measurement in mm. Temperature of pedestal in running and stabilized condition of m/c

ii. Material of construction of pedestal and its linear expansion, given in table above

iii. Poision's ratio of pedestal MOC, to calculate lateral movement of supporting pedestal or frame

Computation of Linear expansion of pedestal, $\Delta L = \alpha * L (t - t_a)$

Where–

ΔL = Movement of line of rotation vertically up or down

T = Temperature deg C of pedestal in stabilized condition of m/c

t_a = Atmospheric Temperature deg C

α = Coefficient of thermal expansion for MOC of pedestal

For computation of lateral movement – multiply with poision's ratio "m" in above formula; use lateral length as "L"

The infra-red radiation, emitted from an object can be observed by thermography equipment. Infrared captured by this equipment is converted into electrical signal and further though CRT photographic thermal image of the object is recorded. From thermo graphic image temperature gradient can be observed by down loading image to computer and point to point temperature can be graphed. This can be also visualized from color change from point to point. Average temperature value can be computed from thermo graphic picture for calculation of thermal movement of axis of rotation.

12.10.5 Measurement of M/C Movement in Running Condition

a. **By Proximity Probes**

Install a machined rectangular target block closed to centre line horizontal plane of m/c, if flat surfaces are not available on m/c. erect a water cooled pipe, supported on base plate, closed to machined block position and install three proximity probes perpendicular to surfaces in x, y and z direction, facing on block with specified gap. Readings at cold condition and hot condition will be recorded. From these readings movement of m/c centre line of rotation from cold to hot conditions can be computed. Readings of all probes are recorded by proxy meter in digital or analogue form. Thus by this method difference of hot and cold position readings will indicate the change in position of a m/c from cold to hot conditions.

In this method, customized block and probe holder is required to be designed m/c to m/c. If m/c is changing its position from cold to hot in high magnitude, probes will be bottom out and will not be possible to use this device.

b. **By Vermeer strobe System** (Direct shaft to shaft):

It utilizes the face to face measurement principle, adopted to measure off line to running (OL2R) machineries movement. Two small Vermeer scales are firmly attached across each flexing point in a coupling. A set of 4 readings (at 12, 3, 6,and 9 o'clock) in cold condition and similarly in hot condition when m/c is stabilized. From variable rate strobe Vermeer can be freezed at different positions (12, 3, 6, and 9 o'clock) for taking readings. Check reading from validity rule. While freezing at different four positions, accuracy of positions should be maintained. For checking bottom position (6 o'clock) reading, mirror can be used, still if not possible to read, calculate from validity rule [(top + bottom readings] = (left + right reading)]

- This method is fairly accurate, simple and cheaper.
- It can measure change in position in vertical as well as horizontal plane of axis of rotation.
- It can be utilized as on line continuous measurement device.
- Fixing of Vernier may cause unsafe act, if loosened by motion of shaft/coupling.
- It can be used in spool piece coupling system only where two flex points are working in coupling assembly.
- Plotting of graph is similar to Face to Face Method but cold alignment readings plotted against x-axis will work as reference position for hot or stabilized situation. Hot position readings in plotted similar to the cold position readings and change of position of axis of rotation, in stabilized condition with reference to cold position is measured to find deviation.

c. **Optical Remote Reference Line**

The optical tooling is most powerful measurement device for finding change in position of rotating axis from cold condition to running (OL2 Running) condition. This is discussed in alignment tooling. The position of rotating shaft centre line position in offline and stabilized running conditions is measured by installing optical meter, leveling accurately. Readings are observed in cold condition and plotted against x-axis. This plot of position is taken as reference line for change in position of line of rotation of axis when running m/c position is plotted similarly as that of off line readings. Difference of off line and running condition readings in graph indicates the change in position of axis of rotation of shaft from off line to running condition.

12.10.6 Accuracy of Readings Will Depend on Following Points

- While observing reading, scale should be held firmly on stable & cleaned platform of m/c and perpendicular to axis/horizon. It will be better to use magnetic base for holding scale. Scale should be erected closed to bearings.
- Readings will not be accurate in high temperature environment.
- Use peg test to calibrate the instrument for perfect leveling.
- Proper adjustment of focusing knobs to get clear image as well as eye is unstrained.
- Tilt optical instrument slowly so that its level is not disturbed.
- Readings should be taken in stable temperature environment, preferably in night Keeping above points in practice, an accurate readings can be assured. This is very practical for vertical movement readings.
- This instrument can't give an accurate reading if m/c has high vibration.

12.10.7 Optical Alignment Tooling

This is perhaps the most versatile tool available for wide variety of applications such as leveling of foundations, bearing pedestals for squareness of m/c tools or machineries drive trains, measuring off line to running m/c movements, checking of roll parallelism in paper mills or steel plants. For accurate readings, following precautions are to be taken while operating the tool–

a. Use magnetic base or stable platform for holding scale. Ensure scale verticity by hand held square level gauge.
b. Ensure that scale target is in vertical position. Use bubble clamp to hold target in position. Move slowly tilting jig so that level is not disturbed.
c. For out-door equipment, take reading in night so that thermal instability do not disturb the stand legs position.
d. Capture two set of readings.

When m/c is vibrating, focusing on target scale becomes difficult, otherwise tool is simple to operate like Theodolite or dumpy level with accuracy of 1–2 arc sec when precisely leveled.

12.10.8 Laser Beam Technique of Alignment

It uses following major components–

- Laser beam emitting device
- Laser detector device
- Prism refraction, reflector or beam splinter device

Plug-in, back zeroing, laser / target mounts holding
FixtureLaser AB components.

12.10.9 Important Features of Laser Beam Technique

a. No fear of bracket sag. Its accuracy of ±3 micron is quite accurate
b. Couplings disconnection may or may not require

    c.   Relatively Laser instruments are expensive

    d.   Range of measurement is limited by detector size (10 × 10 mm mostly used)

    e.   Incapable of reading run out condition or bend shaft

    f.   Both shafts have to be rotated while capturing alignment readings

    g.   Difficult to capture readings in bright sun light

The laser emitter is mounted on one shaft and detector on other shaft. Deviation in emitting and detecting points is misalignment at that location of shaft. Observations are made for 4 set of readings at 12, 3, 6, and 9 o'clock positions for alignment calculations. When "Off Line To" (OL2) running position deviation is to be observed, a second set of readings should be taken after stabilizing the m/c, stopping and immediately taking the readings same way as that done in off line position.

Difference of Off Line to Running readings will indicate the change of position of line of rotation of shaft from off line to running condition.

12.10.10 Alignment of V-Belt Drive

Though misalignment of V-belt drive is not so critical but it poses different type of problems. In general, v-belt alignment is accomplished with straight edge, or string but objective is to use more accurate method of alignment for running smooth and long life of v- belts. Steps of alignment are as follows-

    a.   Check radial and axial run out of both pulleys and rectify within 10 mils

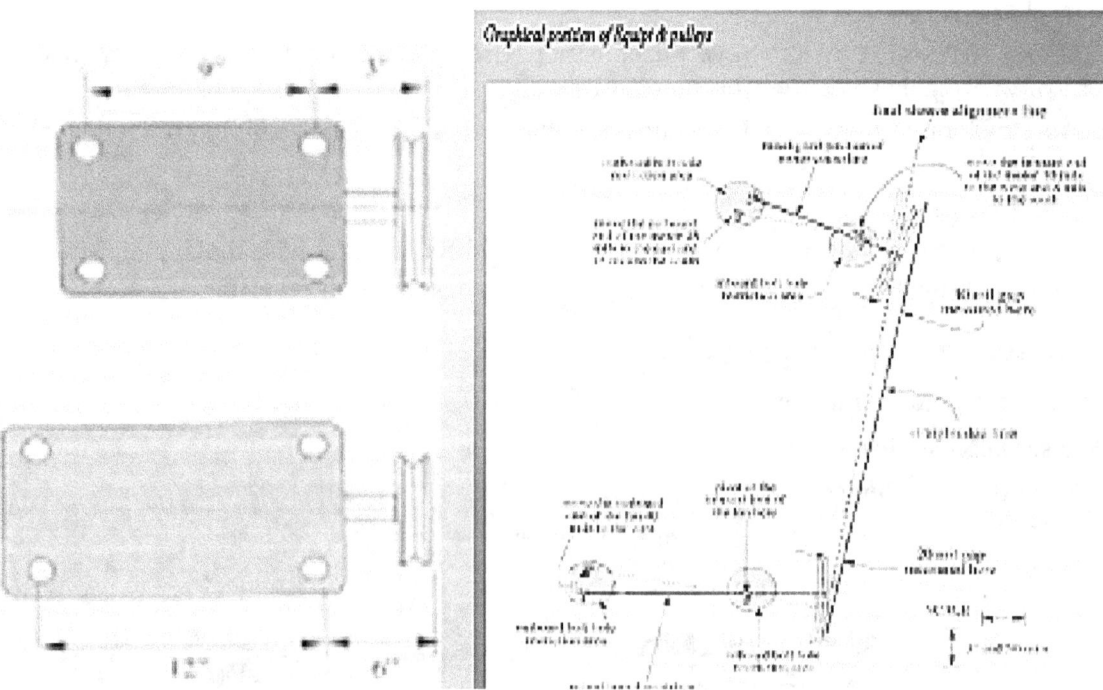

    b.   Align pulleys with the help of chord roughly and tighten bolts to keep m/cs in position

    c.   Check for soft footing for both m/cs- driver & driven and correct that within 2 mils

    d.   Check wear of pulleys and belt from gauges of pulley and belt

    e.   Take all dimensions, as shown in sketch for plotting graph to a scale on graph paper.

    f.   Draw reference lines for centre line of two m/cs. From actual dimensions plot of m/cs as shown here

    g.   Draw new position of motor shaft axis, keeping its pulley centre position as pivot point. Check rom graph movement required for motor and do it in field accordingly.

Table .. Loss coefficients ζ_1 for various types of valves and fittings (referred to the velocity of flow in the line connection nominal diameter DN)

Type of valve/fitting		De-sign	Loss coefficient ζ for DN =																			Comment
			15	20	25	32	40	50	65	80	100	125	150	200	250	300	350	400	500	600	800	1000	
Shut-off valves																							
Slide-disc valves (d₀=DN)	min	1	0.1																0.15	0.12	0.11	0.1	Including DN of diameter)
	max		0.45		0.4		0.45	0.4	0.25	0.25	0.22	0.21	0.19	0.18	0.16	0.17		0.18	0.16	0.14	
Round-body gate valve (d₀=DN)	min	2						0.25	0.24	0.25	0.22	0.25	0.25	0.25	0.16	0.17	0.16	0.15	0.13	0.11	0.11	
	max		0.45		0.5		0.32	0.31	0.30	0.30	0.39	0.36	0.35	0.23	0.23	0.22	0.20	0.19	0.18	0.15	0.14	
Ball and plug valves (d₀=DN)	min	3	0.10	0.10		0.09	0.08	0.08	0.07	0.07	0.06	0.05	0.05	0.04	0.05	0.05	0.03	0.02					Partly <DN ζ=0.4 to 1.1
	max		0.16														0.15					
Butterfly valve PN 2.5-10	min	4						0.59	0.38	0.26	0.20	0.14	0.12	0.09	0.06	0.33	0.33	0.33	0.33	0.30	0.04		* also for PN 40
	max						0.91	1.00	1.00	0.70	0.42	0.34	0.59	0.42	0.40					0.38	
PN 16-25	min						2.04	1.60	1.55	1.30	1.10	0.84	0.75	0.56	0.48	0.41	0.33	0.71	0.47	0.41	
	max						1.50*	2.30*	2.10*	1.93*	1.79*	1.59*	1.30	1.10	0.90	0.74	0.42*		
Globe valve, forged	min	5	10		6.0		6.0																($\approx$ 2 m $\approx$ 5 cm to be achieved for operating value
	max		6.0		6.8		6.8																
Globe valve, cast	min	6	10		6.0		6.0	3.7	5.0	3.2	2.2	1.5	2.0								
	max		6.0		6.8		6.8	5.0	6.4	5.2	4.6	2.7	4.0								
Compact valve	min	7	0.3	0.4	0.6	0.6	1.0	1.1	1.9	1.6	1.5		1.1									
	max		0.1	0.9	1.9								3.5									
Angle valve	min	8	2.0			3.1	3.4	3.6	4.1	4.4	4.7	5.0	5.3	5.7	6.0	6.1		7.0			
	max		3.1														6.6					
Y-valve	min	9	1.5		1.5	1.6	2.0	1.9	1.7	1.6	1.5						1.5					
	max		2.4		2.2	2.1											2.6					
Straight-body valve	min	10	0.6															0.6				
	max		1.6					0.5										1.6				
Diaphragm valves	min	11	0.8					0.9				0.8						1.2				
	max		2.7									2.7									
Non-return valves																							
Non-return valves, straight-seat	min	12	10		1.5	1.6	1.5	3.2	3.7	5.0	7.3	4.3	3.0	1.0	4.3	4.5							Axially repeated holes from DN 125
	max		6.0					4.2	5.0	6.4	8.2	4.6	4.0		4.6							
Non-return valve, axial	min	13	3.2	1.4	1.5	1.6	1.5	1.9	1.7	1.6	1.5		1.5									
	max		1.6	2.4	2.3	2.1	2.0	3.0						2.0								
Non-return valve, disc-seat	min	14	2.5					1.0	0.9	0.8	0.7	0.6	0.5	0.4	0.4	0.4	0.4					
	max		1.0					3.0							1.0	1.0						
Foot valves	min	15	0.5															17.0	[6.5]	[5.5]	[4.5]	[4.0]	1:1 is open
	max		1.0					0.4									1.0					
Swing check valve	min	16														0.4	0.3		0.3	0.3		Swing check valves with continuous travel length)
	max															1.0					3.0	
Hydrostops v=4 m/s, v=3 m/s, v=2 m/s		17						0.9			3.0		3.0	2.5	2.5	1.2	1.2						
								1.8			4.0		4.5	4.0	1.8	1.8	2.4						
								5.0			6.0		8.0	7.5	6.5	4.0	7.0						
Filters		18					2.8											2.8					
Strainers		19					1.0											1.0					In clean medium

Table .. *Loss coefficients ζ for various types of valves and fittings referred to the velocity of flow in the line connection nominal diameter DN)*

Type of valve/fitting		De-sign	Loss coefficient ζ for DN =																			Comments
			15	20	25	32	40	50	65	80	100	125	150	200	250	300	400	500	600	800	1000	
Side-disc valves (d₁=DN)	min	1	0.1	→																	0.1	For d₁ < DN cf. footnote ⁵)
	max		0.45	0.4	0.35	0.5	0.5	0.45	0.4	0.35	0.3	0.3	0.25	0.2	0.17	0.16	0.15	0.15	0.13	0.11	0.2	
Round-body gate valve d₁=DN	min	2						0.15	0.24	0.15	0.17	0.1	0.15	0.16	0.17	0.16	0.13	0.13	0.13	0.11	0.11	For d₁ < DN cf. footnote ⁵)
	max			0.15				0.32	0.31	0.30	0.39	0.26	0.35	0.31	0.23	0.30	0.19	0.18	0.16	0.15	0.14	
Ball and plug valves d₁=DN	min	3	0.10	0.10	0.09	0.09	0.08	0.08	0.07	0.07	0.06	0.05	0.05	0.04	0.05	0.05	0.02	0.02			0.04	For d₁ < DN ζ = 0.4 to 1.1
	max		0.15													0.15				0.26	
Butterfly valves PN 2.5-10 / PN 16-25	min	4				0.91	0.91	0.57	0.48	0.36	0.31	0.14	0.13	0.09	0.40	0.37	0.33	0.33	0.33	0.30	0.04	+ above PN 40
	max					1.20	1.01	0.80	0.70	0.42	0.54	0.51	0.42		0.45	0.40	0.76			0.43	0.28	
Globe valve, forged	min	5	10	→	4.0	→	→	4.0															ζ ≈ 2 to 3 can be achieved for optimised valve
	max		4.0		6.8			6.8															
Glebe valves, cast	min	6	0.3	0.4	0.6	0.6	1.0	→	→	→	→	→	→	→	3.0								
	max		0.3	0.9	1.3	1.9									6.0								
Compact valve	min	7	2.0	→	→	2.1	3.4	3.6	4.1	4.4	4.7	4.0	5.3	5.7	6.0	6.1	6.6						
	max		2.1																				
Angle valves	min	8	1.5	→	→	→	→	1.9	1.5	1.6	1.5		1.1										
	max		2.4							2.0		2.5											
Y-valve	min	9	0.6	→	→	→	→	→	→	→	→	→	0.8	0.8		0.6							
	max		1.6										1.7			1.6							
Straight-through valves	min	10	0.8	→	→	→	→	→	→	→	→	→	→	1.0									
	max		2.2												4.0								
Diaphragm valves	min	11	10	→	4.5	4.6	1.5	3.2	3.7	5.0	7.5	4.5	3.0		4.5	1.3							Axially expanded as from DN 125
	max		14	1.4			2.0	4.2	5.0	6.4	8.3	4.6				4.6							
Non-return valves, straight/seat	min	12	2.5	2.4	2.2	2.1		1.9	1.7	1.4	1.5		1.5		1.5								
	max		3.0												2.0								
Non-return valve, disc	min	15	0.5	0.5			0.5	0.4	0.5	0.5	0.7	0.6	0.5	0.4	0.4	0.4			7.0	6.5	4.5	4.0	‡) longer pipe
	max		1.0					2.0								1.0							
Swing check valve	min	16	0.5	→	→	→	0.5	0.4	→	→	→	→	→	→		0.4	→	+	→	0.3	0.3	Swing check valve without seat and weight.‡)
	max		1.0													1.0				3.0			
Hydrant types v = 4 m/s / v = 3 m/s / v = 2 m/s		17						0.9 / 1.8 / 5.0		3.0 / 4.0 / 6.0		2.0 / 4.5 / 8.0	2.5 / 4.0 / 7.5	2.5 / 4.0 / 6.5	1.2 / 1.8 / 6.0	2.2 / 3.4 / 7.0							
Filters		18	1.5			1.5										2.8							In clean condition
Strainers		19	1.9			1.0										1.0							

Table Loss coefficients ζ in elbows and bends

Round elbow	α	15°		30°		45°		60°		90°	
		Surface smooth	rough	Surface smooth	rough	Surface smooth	rough	Surface smooth	rough	Surface smooth	rough
	ζ for R = 0	0.07	0.10	0.14	0.20	0.25	0.15	0.50	0.70	1.15	1.30
	ζ for R = d	0.03	–	0.07	–	0.14	0.14	0.19	0.46	0.21	0.51
	ζ for R = 2 d	0.03	–	0.06	–	0.09	0.19	0.12	0.26	0.14	0.30
	ζ for R = 5 d	0.03	–	0.06	–	0.08	0.16	0.10	0.20	0.10	0.20
Welded bend	Number of circumferential welds	–	–	–	–	2	–	3	–	3	–
	ζ	–	–	–	–	0.15	–	0.20	–	0.25	–

Note: For the branch fittings in Table 7 and the adapters of Table 8, one must differentiate between the irreversible pressure loss (reduction in pressure)

$$p_L = \zeta \cdot \varrho \cdot v_2^2/2 \qquad (16)$$

where

p_L	Pressure loss in Pa
ζ	Loss coefficient
ϱ	Density in kg/m³
v	Flow velocity in m/s

and the reversible pressure change of the frictionless flow according to Bernoulli's equation (see 3.2.1.1):

$$p_2 - p_1 = \varrho \cdot (v_1^2 - v_2^2)/2 \qquad (17)$$

For accelerated flows (for example a reduction in the pipe diameter), $p_2 - p_1$ is always negative, for decelerated flows (e.g. pipe expansion) it is always positive. When calculating the net pressure change as the arithmetic sum of p_1 and $p_2 - p_1$, the pressure losses from Eq. 16 are always to be subtracted.

Often the so-called k_v value is used instead of the loss coef-

Table Loss coefficients ζ for fittings

Combinations of pipe bends (elbows):

The ζ value of the single 90° elbow should not be doubled, but only be multiplied by the factors indicated to obtain the pressure loss caused by the combination of elbows illustrated.

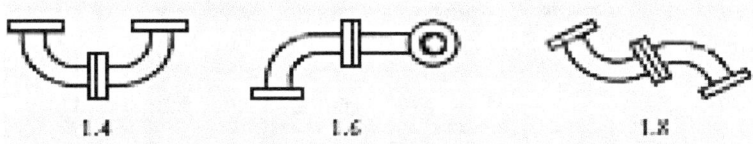

| 1.4 | 1.6 | 1.8 |

Expansion joints:

Bellows-type expansion joint with/without guide pipe	ζ = 0.3/2.0
Compensation tube bend	ζ = 0.6 to 0.8
Creased compensation tube bend	ζ = 1.3 to 1.6
Bellows-type compensation tube bend	ζ = 3.2 to 4

Inlet pipe fittings:

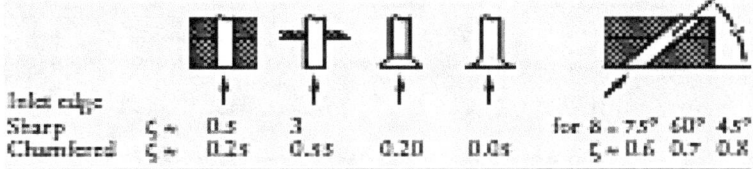

Inlet edge					for δ = 75° 60° 45°
Sharp	ζ = 0.5	3			
Chamfered	ζ = 0.25	0.55	0.20	0.05	ζ = 0.6 0.7 0.8

Discharge pipe fittings:

ζ = 1 downstream of an adequate length of straight pipe with an approximately uniform velocity distribution in the outlet cross-section

ζ = 2 for a very irregular velocity distribution, for example immediately downstream of a pipe fitting or a valve, etc.

Continued on next page

Flow rate and friction head loss for tubing and pipe sizes (Imperial) (based on 5 ft/s velocity)

Nom. dia. (in)	Inside dia. (in)	Flow rate (gpm)	Friction head loss (feet of head per feet of pipe)
1/4	0.311	1.2	0.58
1/2	0.527	3.4	0.29
3/4	0.745	6.8	0.187
1	0.995	12.1	0.13
1 1/2	1.6	31	0.071
2	2.067	52	0.051
2 1/2	2.469	75	0.041
3	3.068	115	0.031
4	4.026	198	0.022
6	6.065	450	0.014
8	8.125	808	0.009
10	10.25	1286	0.007
12	12.25	1837	0.0058
14	13.5	2230	0.052

Flow rate and friction head loss for tubing and pipe sizes (Imperial) (based on 15 ft/s velocity)

Nom. dia. (in)	Inside dia. (in)	Flow rate (gpm)	Friction head loss (feet of head per feet of pipe)
1/4	0.311	3.55	4.68
1/2	0.527	10.2	2.35
3/4	0.745	20.4	1.51
1	0.995	36.3	1.04
1 1/2	1.6	94	0.575
2	2.067	157	0.42
2 1/2	2.469	224	0.335
3	3.068	346	0.256
4	4.026	595	0.184
6	6.065	1351	0.111
8	8.125	2424	0.078
10	10.25	3858	0.059
12	12.25	5510	0.048
14	13.5	6692	0.042

Flow rate and friction head loss for tubing and pipe sizes (metric) (based on 1.5 m/s velocity)

Inside dia. (mm)	Flow rate (L/s)	Flow rate (m³/h) x 1000	Friction head loss (meter of head per meter of pipe)
15	1.1	3.98	0.24
20	1.97	7.1	0.17
25	3.1	11.1	0.13
32	5	18.1	0.09
40	7.87	28.3	0.07
50	12.3	44.2	0.052
65	20.8	74.8	0.038
80	31.5	113	0.029
100	49.2	177	0.022
125	76.8	276.7	0.017
150	110.7	398	0.014
200	196.8	708	0.0096
250	307	1107	0.0073
300	443	1594	0.0059

Flow rate and friction head loss for tubing and pipe sizes (metric) (based on 3 m/s velocity)

Inside dia. (mm)	Flow rate (L/s)	Flow rate (m³/h) x 1000	Friction head loss (meter of head per meter of pipe)
15	0.68	2.43	0.9
20	1.2	4.32	0.62
25	1.88	6.75	0.47
32	3.07	11.06	0.34
40	4.8	17.28	0.26
50	7.5	27.00	0.2
65	12.7	45.63	0.14
80	19.2	69.12	0.11
100	30	108.00	0.084
125	46.9	168.75	0.064
150	67.5	243.00	0.051
200	120	432.00	0.036
250	187	675.00	0.027
300	270	972.00	0.022

Flow rate and friction head loss for tubing and pipe sizes (metric) (based on 4.5 m/s velocity)

Inside dia. (mm)	Flow rate (L/s)	Flow rate (m³/h) x 1000	Friction head loss (meter of head per meter of pipe)
15	3.32	12	1.97
20	5.9	21.2	1.36
25	9.2	33.2	1.03
32	15.1	54.4	0.75
40	23.6	85	0.57
50	36.9	133	0.43
65	62.4	225	0.31
80	94.5	340	0.24
100	148	531	0.183
125	231	830	0.139
150	332	1196	0.112
200	590	2125	0.079
250	922	3321	0.06
300	1328	4782	0.048

Example calculation

Calculate the pipe friction loss of a 2 1/12" schedule 40 (2.469" internal pipe diameter) new steel pipe with a flow rate of 149 gpm for water at 60F and a pipe length of 50 feet. The roughness is 0.00015 ft and the viscosity is 1.13 cSt.

The average velocity v in the pipe is:

$$v(ft/s) = 0.4085 \times \frac{149}{2.469^2} = 9.98 \quad [1]$$

The Reynolds Re number is:

$$R_e = 7745.8 \times \frac{9.98 \times 2.469}{1.13} = 1.69 \times 10^4 \quad [2]$$

The friction parameter f is:

$$f = \frac{0.25}{\left(\log_{10}\left(\frac{0.00015 \times 12}{3.7 \times 2.469} + \frac{5.74}{(1.69 \times 10^4)^{0.9}}\right)\right)^2} = 0.02031 \quad [4]$$

The friction factor $\Delta H_{TM}/L$ is calculated with the Darcy-Weisback equation [5]

$$\frac{\Delta H_{TT}}{L}\left(\frac{ft\ fluid}{100 ft\ pipe}\right) = 1200 \times 0.02031 \times \frac{9.98^2}{2.469 \times 2 \times 32.17} = 15.34 \quad [5]$$

The pipe friction loss ΔH_{TP} is:

$$\Delta H_{TP}(ft, fluid) = 15.34 \times \frac{50}{100} = 7.67 \quad [6]$$

Table 12: Vapour pressure p_v, density ϱ and kinematic viscosity v of water at saturation conditions as a function of the temperature t

t °C	Pv bar	ϱ kg/m³	v mm²/s	t °C	pv bar	ϱ kg/m³	v mm²/s	t °C	pv bar	ϱ kg/m³	v mm²/s
0	0.00611	999.8	1.792	61	0.2086	982.6		145	4.155	921.7	
1	0.00656	999.9		62	0.2184	982.1		150	4.760	916.9	
2	0.00705	999.9		63	0.2285	981.6					
3	0.00757	1000.0		64	0.2391	981.1		155	5.433	912.2	
4	0.00812	1000.0		65	0.2501	980.5		160	6.180	907.4	0.1890
5	0.00872	1000.0		66	0.2614	980.0					
6	0.00935	999.9		67	0.2733	979.4		165	7.008	902.4	
7	0.01001	999.9		68	0.2856	978.8		170	7.920	897.3	
8	0.01072	999.8		69	0.2983	978.3					
9	0.01146	999.7		70	0.3116	977.7	0.413	175	8.925	892.1	
10	0.01227	999.6	1.307					180	10.027	886.9	0.1697
				71	0.3253	977.1					
11	0.01311	999.5		72	0.3396	976.6		185	11.234	881.4	
12	0.01401	999.4		73	0.3543	976.0		190	12.553	876.0	
13	0.01496	999.3		74	0.3696	975.4					
14	0.01597	999.2		75	0.3855	974.8		195	13.989	870.3	
15	0.01703	999.0		76	0.4019	974.3		200	15.550	864.7	0.1579
16	0.01816	998.8		77	0.4189	973.7					
17	0.01936	998.7		78	0.4365	973.0		205	17.245	858.7	
18	0.02062	998.5		79	0.4547	972.5		210	19.080	852.8	
19	0.02196	998.4		80	0.4736	971.8	0.365				
20	0.02337	998.2	1.004					215	21.042	846.6	
				81	0.4931	971.3		220	23.202	840.3	0.1488
21	0.02485	997.9		82	0.5133	970.6					
22	0.02642	997.7		83	0.5342	969.9		225	25.504	834.0	
23	0.02808	997.5		84	0.5557	969.4		230	27.979	827.3	
24	0.02982	997.2		85	0.5780	968.7					
25	0.03167	997.0		86	0.6010	968.1		235	30.635	820.6	
26	0.03360	996.7		87	0.6249	967.4		240	33.480	813.6	0.1420
27	0.03564	996.4		88	0.6495	966.7					
28	0.03779	996.1		89	0.6749	966.0		245	36.524	806.5	
29	0.04004	995.8		90	0.7011	965.3	0.326	250	39.776	799.2	
30	0.04241	995.6	0.801								

Fits and Tolerances–

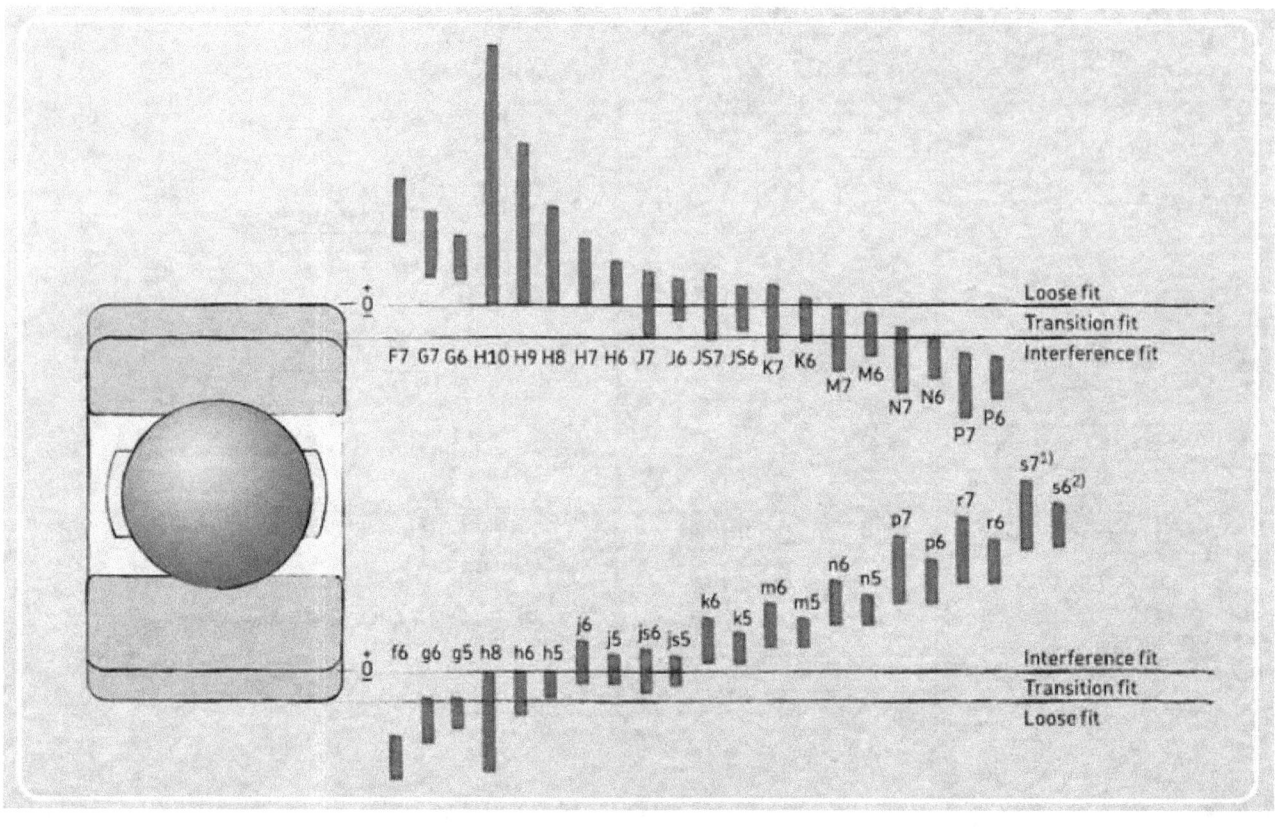

REFERENCES

1. Failure Cause Analysis Vol II, Stone and Webster – Electric power Institute Boston.

2. Energy Consultants & Corporation: Electric Power Research Institute Palo – CA.

3. Recommended Design Guide Lines for Feed Water Pumps – Energy Research Consultant corporation Palo CA.

4. Manual Centrifugal Pumps- Tata Steel Jamshed pur – India.

5. Flexible Couplings, Machineries, Scroll John.

6. Total Alignment: Dodd V R – Petrolium Publishing Co Tesla.

7. Basic Shaft Alignment Work BookProto waske John, Tuvac Inc – OH.

www.ingramcontent.com/pod-product-compliance
Lightning Source LLC
Chambersburg PA
CBHW081234180526
45171CB00005B/428